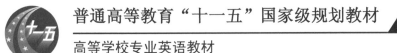

普通高等教育"十一五"国家级规划教材

高等学校专业英语教材

计算机专业英语教程
（第 7 版）

金志权　张幸儿　主编

张景祥　编

电子工业出版社

Publishing House of Electronics Industry

北京·BEIJING

内 容 简 介

本书以计算机学科的各分支为基础,选取分支中的典型素材、新技术的简介,以及最新的和常用的术语,使读者能通过学习,从英语角度巩固和扩大计算机专业知识,同时及时了解计算机领域的新技术和新术语。

本书配备的注释和习题旨在提高读者阅读与笔译专业英语文献资料的能力,以及提高"会看、会听、会说、会写"的英语"四会"能力。

本书素材取自近年来国内外计算机科学各领域的最新教材、专著、论文和计算机网络信息,内容新颖、与时俱进,覆盖面广,结构合理,系统性强。为了方便读者,本书提供辅助学习资料,主要内容包括参考译文、单词汇总、缩略语与术语索引,以及音频和视频素材等,扫描前言后的二维码即可查看,也可登录华信教育资源网(www.hxedu.com.cn)免费下载。为了方便教学,本书另配有教学资源,内容包括电子教案、中译英习题参考答案、视频素材的参考原文、若干课文参考译文、授课建议,向采用本书作为教材的教师免费提供,读者登录华信教育资源网即可下载。希望提高英语听力的读者可扫描书中二维码获取38篇英文朗读课文和12篇英文视频听力材料及其参考原文。

本书可以作为高等院校计算机、软件工程、信息系统、电子商务等相关专业的专业英语教材;对于高等院校其他学科希望了解计算机科学,特别是人工智能进展的师生,本书也值得一看;本书还可供计算机专业人员、IT专业人员或其他有兴趣的读者学习参考。

未经许可,不得以任何方式复制或抄袭本书之部分或全部内容。
版权所有,侵权必究。

图书在版编目(CIP)数据

计算机专业英语教程/金志权,张幸儿主编;张景祥编. —7版. —北京:电子工业出版社,2020.9
高等学校专业英语教材
ISBN 978-7-121-38589-6

Ⅰ. ①计⋯　Ⅱ. ①金⋯ ②张⋯ ③张⋯　Ⅲ. ①电子计算机-英语-高等学校-教材　Ⅳ. ①TP3

中国版本图书馆 CIP 数据核字(2020)第 032614 号

责任编辑:秦淑灵
印　　刷:涿州市般润文化传播有限公司
装　　订:涿州市般润文化传播有限公司
出版发行:电子工业出版社
　　　　　北京市海淀区万寿路 173 信箱　邮编 100036
开　　本:787×1092　1/16　印张:21.75　字数:718 千字
版　　次:2000 年 8 月第 1 版
　　　　　2020 年 9 月第 7 版
印　　次:2025 年 2 月第 8 次印刷
定　　价:59.00 元

凡所购买电子工业出版社图书有缺损问题,请向购买书店调换。若书店售缺,请与本社发行部联系,联系及邮购电话:(010)88254888,88258888。

质量投诉请发邮件至 zlts@phei.com.cn,盗版侵权举报请发邮件至 dbqq@phei.com.cn。

本书咨询联系方式:qinshl@phei.com.cn。

前　　言

　　这些年来人工智能的进步令人瞩目。AlphaGo 所向披靡，轻取围棋世界冠军；机器人 Sophia 外形酷似人类，与人交谈给人印象深刻；中科大的讯飞语言翻译已达到相当高的水平；中央电视台的机器人播音的声音与真人几乎相同；人脸识别已在日常生活中应用；Watson 人工智能医生诊断病人的正确率高于一般专家。近年来，人工智能学院在全国纷纷成立，众多的人工智能专业相继设置。为了适应这一新的发展趋势，与时俱进，提升对人工智能的理解和认识，本版对人工智能单元做了较大改动，新增 4 篇选文，分别是 Deep Learning、Robot Sophia、AlphaGo Zero 与 Big Data Analytics。此外，随着增强现实（AR）技术的成熟，AR 与行业的融合越来越深入。从设计到营销，从教育到医疗，从出行到文化，AR 正在重新定义各产业的思维方式和运行方式。新版教材增加了选文 Augmented Reality，并配有增强现实视频。量子计算（唯一真正的新型计算模式）发展至今，已引发人们极大的研究兴趣，也展现出一定的商业价值，但其将来的发展速度、方向和实际应用还有待观察。书中选文 Quantum Computing 能让读者对量子计算有一个大致了解。安卓编程已成功生成约 110 万个 App，不少人通过自学开发了 Android App。书中选文 Introduction to Android Programming 会对读者学习安卓编程有所帮助。

　　程序设计语言 Python 以其简洁性、易读性以及可扩展性，已经成为最受欢迎的程序设计语言之一；区块链在互联网中很热门，且正成为我国的发展重点，但许多人仍然不清楚区块链的概念。为此，本书以术语形式对它们进行简单介绍。第 6 版增加了已被收入 Wikipedia 的 WeChat（微信）术语，本版再加入已被 Wikipedia 收入的 Internet plus（互联网＋）术语。本版新增加的术语有 GPU (Unit 1，以下用数字简单表示)、OLED(2)、Python(3)、Android (operating system)、iOS、Network operating system (4)、Internet plus、5G(5)、NoSQL database、Sensor database system、Real-time database(7)、BCI (9)、MOOC、O2O(13)、Blockchain (16)、Supercomputer、DNA computing(17)。

　　为了节省篇幅，把第 6 版的 JavaScript Tutorial 等 3 篇课文放在网上，读者可以通过扫描下一页的二维码查看以前所有的已删除课文。为了便于读者查阅、提高听力，第 7 版把以前放在光盘上的内容也放在华信教育资源网上，读者可登录 www.hxedu.com.cn 免费下载或扫描下一页的二维码阅读，主要内容如下：

- 参考译文
- 单词汇总
- 缩略语与术语索引
- 音频(38 课 5 种不同朗读风格的英文朗读材料)
- 视频素材(12 个测试听力的视频材料)

　　本版保持了前 6 版的编排格式和基本风格。各校教师可根据自己的具体情况，因材施教，在每个单元里挑选课文进行教学。本书每个单元都给出了看、听、说、写四种练习题。针对英语口语练习，可结合所列题目，组织学生进行课堂小组讨论或讲述。为了方便教学，本书另配有教学资源,内容包括电子教案、中译英练习题参考答案、视频素材的参考原文、若干课文参考译文、授课建议,向采用本书作为教材的教师免费提供,读者登录华信教育资源网 www.hxedu.com.cn 即可下载。

全书共17单元,第1、2、4、5、6、8、10、16、17单元由金志权编写,第3、7、9、11、12、13、14、15单元由张幸儿编写,彼此进行了互审。本书的电子教案由济南大学张景祥制作。

在此十分感谢南京大学徐洁磐教授在我们准备第7版时参与的讨论以及给出的建议;也感谢对本书的编写给予帮助的南京大学陈珮珮、李存珠、陆钟楠、张福炎、李宣东、徐永森、蔡士杰、邵栋、黄皓、宋健建等老师,南京大学外国语学院王守成、杨治中、侯焕谬、张子清等老师,南京师范大学顾铁成老师;感谢张凌绮、Jimmy Jin、Amanda Fang、Steven Fang 在课文朗读、视听素材收集和整理上给予的支持;感谢天津大学计算机科学与技术学院戴维迪博士、苏州科技学院傅朝阳博士、广西科技大学阳树洪博士、海南大学冯思玲老师、南京理工大学鲁加军老师、河北唐山学院王永强老师、浙江财经大学东方学院张琼妮老师等的关心与支持;同时还要感谢丁正全、张万华、韩杰等提供的帮助。

限于作者水平,书中难免会有不妥和错误之处,敬请读者批评指正。反馈意见请发邮件至:ise_zhangjx@ujs.edu.cn。

编　者
于南京大学

辅助学习资料

CONTENTS

Unit 1　Hardware Ⅰ ·· (1)
   1.1　A Closer Look at the Processor and Primary Storage ····················· (1)
   1.2　Integrated Circuit—Moore's Law ·· (6)
   1.3　Multicore Processors ·· (10)
   1.4　Computer Architectures ·· (12)
   Exercises ·· (20)

Unit 2　Hardware Ⅱ ··· (21)
   2.1　Optical Storage Media: High-Density Storage ································ (21)
       2.1.1　Optical Laser Disks ·· (21)
       2.1.2　DVDs ·· (22)
   2.2　Display Devices ··· (26)
   2.3　3D Printing ··· (29)
       2.3.1　Manufacturing an Object with 3D Printer ························· (29)
       2.3.2　General Principles ··· (29)
       2.3.3　Applications ··· (30)
   2.4　The External Interface: USB ·· (33)
   Exercises ·· (40)

Unit 3　Programming and Programming Languages ······································· (41)
   3.1　Computer Programming ·· (41)
   3.2　C++ and Object-Oriented Programming ······································· (44)
   3.3　Introduction to Java ··· (49)
   3.4　Introduction to Android Programming ·· (52)
   3.5　Characteristics of Web Programming Languages ···························· (56)
   Exercises ·· (62)

Unit 4　Operating System ·· (63)
   4.1　Summary of OS ··· (63)
   4.2　Using the Windows Operating System ·· (65)
   4.3　Window Managers ·· (71)
   4.4　Myths of UNIX ··· (73)
   4.5　Using Linux in Embedded and Real-time Systems ·························· (75)
   Exercises ·· (83)

Unit 5　Computer Networks ··· (84)
   5.1　Internet ·· (84)
   5.2　Extending Your Markup: An XML Tutorial ·································· (91)

    5.3　Network Protocols ······ (96)
        5.3.1　Protocol Hierarchies ······ (96)
        5.3.2　WAP—The Wireless Application Protocol ······ (100)
    5.4　Mobile Internet, Mobile Web ······ (104)
    Exercises ······ (115)

Unit 6　Network Communication ······ (116)
    6.1　Two Approaches to Network Communication ······ (116)
    6.2　Carrier Frequencies and Multiplexing ······ (117)
    6.3　Internet of Things ······ (120)
    6.4　Wireless Network ······ (123)
    Exercises ······ (131)

Unit 7　Database ······ (133)
    7.1　An Overview of a Database System ······ (133)
    7.2　Introduction to SQL ······ (136)
    7.3　Object-relational Database ······ (138)
    7.4　Data Warehouse ······ (143)
        7.4.1　Data Warehouse ······ (143)
        7.4.2　What is Data Mining? ······ (146)
    7.5　Big Data ······ (149)
    Exercises ······ (152)

Unit 8　Multimedia ······ (154)
    8.1　Introduction ······ (154)
        8.1.1　Main Properties of a Multimedia System ······ (154)
        8.1.2　Multimedia ······ (155)
    8.2　Audio ······ (157)
        8.2.1　Computer Representation of Sound ······ (158)
        8.2.2　Audio Formats ······ (159)
        8.2.3　MP3 Compression ······ (159)
    8.3　Video ······ (162)
        8.3.1　Video Compression ······ (162)
        8.3.2　MP4 ······ (164)
    8.4　Synchronization ······ (168)
    Exercises ······ (171)

Unit 9　Artificial Intelligence ······ (172)
    9.1　Overview of Artificial Intelligence ······ (172)
    9.2　About Expert System ······ (175)
    9.3　Deep Learning ······ (178)
    9.4　Robot Sophia ······ (184)
    9.5　AlphaGo Zero: Learning from Scratch ······ (190)

9.6	Big Data Analytics	(195)
	Exercises	(201)

Unit 10 Data Structure and Algorithms (202)

10.1	Abstract Data Types and Algorithms	(202)
10.2	Spanning Trees	(205)
10.3	Block Sorting Algorithms: Parallel and Distributed Algorithm	(208)
10.4	Divide-and-Conquer	(211)
	Exercises	(213)

Unit 11 Fundamentals of the Computing Sciences (215)

11.1	Set Theory	(215)
11.2	Predicates	(219)
11.3	Languages and Grammars	(223)
11.4	Finite-State Machines	(226)
	Exercises	(230)

Unit 12 Computer Applications Ⅰ (232)

12.1	Computer Graphics	(232)
12.2	Computer-Aided Design	(235)
12.3	Graphical User Interface	(238)
12.4	The Virtual Reality Responsive Workbench	(241)
12.5	Augmented Reality	(246)
	Exercises	(252)

Unit 13 Computer Applications Ⅱ (253)

13.1	Distance Education Technological Models	(253)
13.2	Electronic Business	(256)
13.3	E-Government — Introduction	(260)
13.4	Office Automation	(264)
	Exercises	(265)

Unit 14 Computer Applications Ⅲ (267)

14.1	Geographic Information Systems(GIS): A New Way to Look at Business Data	(267)
14.2	Introduction to GPS	(269)
14.3	Management Information System (MIS)	(274)
14.4	Enterprise Resource Planning	(276)
	Exercises	(279)

Unit 15 Software Development (281)

15.1	Overview of Software Engineering	(281)
15.2	Unified Modeling Language	(284)
15.3	Integrated Computer Aided Software Engineering	(286)
15.4	Agile Software Development Methods	(290)
15.5	Middleware	(294)

  Exercises ··· (300)

Unit 16 Network Security ··· (301)
  16.1 What Do I Need to Know about Viruses? ··· (301)
  16.2 Modern Cryptography—Data Encryption ··· (303)
  16.3 Firewalls and Proxies ··· (306)
  Exercises ··· (311)

Unit 17 Computer Systems ··· (313)
  17.1 Embedded Systems ·· (313)
  17.2 Distributed Systems ··· (317)
  17.3 Cloud Computing and Cloud Storage ·· (321)
    17.3.1 Cloud Computing ·· (321)
    17.3.2 Cloud Storage ·· (322)
  17.4 Quantum Computing ·· (329)
  Exercises ··· (333)

参考资料 ··· (335)

Unit 1   Hardware I

1.1   A Closer Look at the Processor and *Primary Storage*

We have learned that all computers have similar capabilities and perform essentially the same functions, although some might be faster than others. We have also learned that a computer system has input, output, storage, and processing components; that the *processor* is the "intelligence" of a computer system; and that a single computer system may have several processors. We have discussed how data are represented inside a computer system in electronic states called *bits*. We are now ready to expose the inner workings of the nucleus of the computer system—the processor.

The internal operation of a computer is interesting, but there really is no mystery to it. The mystery is in the minds of those who listen to *hearsay* and believe science-fiction writer. The computer is a nonthinking electronic device that has to be plugged into an electrical power source, just like a toaster or a lamp.

Literally hundreds of different types of computers are marketed by *scores of* manufacturers.[1] The complexity of each type may vary considerably, but in the end each processor, sometimes called the **central processing unit or CPU**, has only two fundamental sections: the *control unit* and the *arithmetic and logic unit*. Primary storage also plays an *integral part* in the internal operation of a processor. These three — primary storage, the control unit, and the arithmetic and logic unit — work together. Let's look at their functions and the relationships between them.

Unlike magnetic secondary storage devices, such as *tape* and *disk*, primary storage has no moving parts. With no mechanical movement, data can be accessed from primary storage at electronic speeds, or close to the speed of light. Most of today's computers use DRAM (Dynamic Random-Access Memory) technology for primary storage. *A state of the art* DRAM *chip* about one eighth the size of a postage stamp[2] can store about 256,000,000 bits, or over 25,600,000 characters of data!

Primary storage, or main memory, provides the processor with temporary storage for programs and data. All programs and data must be transferred to primary storage from an input device (such as a VDT) or from *secondary storage* (such as a disk) before programs can be executed or data can be processed. Primary storage space is always *at a premium*; therefore, after a program has been executed, the storage space it occupied is *reallocated* to another program awaiting execution.

Figure 1-1 illustrates how all input/output (I/O) is "read to" or "written from" primary storage. In the figure, an inquiry (input) is made on a VDT. The inquiry, in the form of a message, is routed to primary storage over a channel (such as a *coaxial cable*). The message is interpreted, and the processor initiates action to retrieve the appropriate program and data from secondary storage.[3] The *program* and *data* are "loaded", or moved, to primary storage from secondary storage. This is a nondestructive read process. That is, the program and data that are read reside in both primary storage (temporarily) and secondary storage (permanently). The data are manipulated according to program *instructions*, and a report is written from primary storage to a printer.

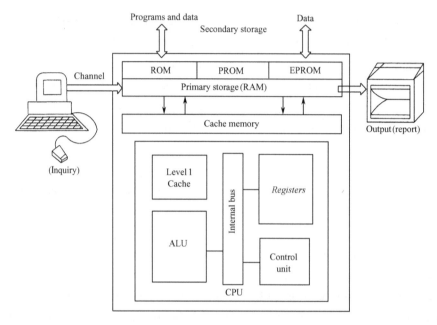

Figure 1-1   Interaction between primary storage and computer system components
(All programs and data must be transferred from an input device or from secondary storage before programs can be executed and data can be processed. During processing, instructions and data are passed between the various types of internal memories, the control unit, and the arithmetic and logic unit. Output is transferred to the printer from primary storage)

A program instruction or a piece of data is stored in a specific primary storage *location* called an **address**. Addresses permit program instructions and data to be located, accessed, and processed. The content of each address is constantly changing as different programs are executed and new data are processed.

Another name for primary storage is random-access memory, or RAM. A special type of primary storage, called **read-only memory (ROM)**, cannot be altered by the programmer. The contents of ROM are "*hard wired*" (designed into the logic of the memory chip) by the manufacturer and can be "read only". When you turn on a microcomputer system, a program in ROM automatically readies the computer system for use. Then the ROM program produces the initial display screen prompt.

A variation of ROM is **programmable read-only memory** (**PROM**). PROM is ROM into which you, the user, can load "read-only" programs and data. Once a program is loaded to PROM, it is seldom, if ever, changed. [4] However, if you need to be able to revise the contents of PROM, there is **EPROM**, erasable PROM. Before a write operation, all the storage cells must be erased to the same initial state.

A more attractive form of *read-mostly* memory is **electrically erasable programmable read-only memory** (**EEPROM**). It can be written into at any time without erasing prior contents; only the byte or bytes addressed are updated. [5]

The EEPROM combines the advantage of *nonvolatility* with the flexibility of being *updatable in place*, [6] using ordinary bus control, address, and data lines.

Another form of *semiconductor* memory is ***flash memory*** (so named because of the speed). Flash memory is intermediate between EPROM and EEPROM in both cost and *functionality*. Like EEPROM, flash memory uses an electrical erasing technology. An entire flash memory can be erased in one or a few seconds, which is much faster than EPROM. In addition, it is possible to erase just blocks of memory rather than an entire chip. However, flash memory does not provide *byte-level* erasure. [7] Like EPROM, flash memory uses only one transistor per bit, and so achieves the high density of EPROM.

Cache Memory

Program and data are loaded to RAM from secondary storage because the time required to access a program instruction or piece of data from RAM is significantly less than from secondary storage. Thousands of instructions or pieces of data can be accessed from RAM in the time it would take to access a single piece of data from disk storage. [8] RAM is essentially a high-speed holding area for data and programs. In fact, nothing really happens in a computer system until the program instructions and data are moved to the processor. This transfer of instructions and data to the processor can be time-consuming, even at microsecond speeds. To facilitate an even faster transfer of instructions and data to the processor, most computers are designed with **cache memory**. Cache memory is employed by computer designers to increase the computer system ***throughput*** (the rate at which work is performed).

Like RAM, cache is a high-speed holding area for program instructions and data. However, cache memory uses SRAM (Static RAM) technology that is about 10 times faster than RAM and about 100 times more expensive. With only a fraction of the capacity of RAM, cache memory holds only those instructions and data that are likely to be needed next by the processor. Two types of cache memory appear widely in computers. The first *is referred to as* internal cache and is built into the CPU chip. The second, external cache, is located on chips placed close to the CPU chip. A computer can have several different levels of cache memory. Level 1 cache is *virtually* always built into the chip. Level 2 cache used to be external cache but is now typically also built into the CPU like level 1 cache.

Figure 1-2 shows structure inside a typical PC system unit around 2000.

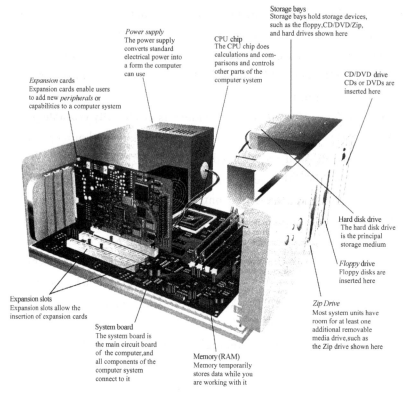

Figure 1-2 Structure inside a typical PC system unit around 2000 (The system unit *houses* the CPU, memory, and other important pieces of hardware)

Words and Expressions

processor [ˈprəusesə] *n.* 处理器
primary storage   主存储器
bit [bit] *n.* 位，二进制位，比特
hearsay [ˈhiəsei] *n.* 传闻，谣传
scores of   许多
control unit   控制部件，控制器
arithmetic and logic unit   算术逻辑部件
integral parts   不可缺的部分；组成部分
tape and disk   磁带和磁盘
a state of the art (the state of the art)   目前工艺水平，最新发展水平
chip [tʃip] *n.* 芯片
secondary storage   辅助存储器，二级存储器
at a premium   非常珍贵
reallocate [riːˈæləkeit] *vt.* 重新分配
capacity [kəˈpæsiti] *n.* 容量
coaxial cable   同轴电缆
program and data   程序和数据

instruction [in'strʌkʃən] n. 指令
register ['redʒistə] n. 寄存器；记录；登记簿；登记；注册
location [ləu'keiʃən] n. 单元；位置
hardwired ['hɑːd'waiəd] adj. 硬接线的，实线的；硬件(线路)实现的
read-mostly 以读为主的，大多数为读的
nonvolatility [nɔn'vɔlə'tiliti] n. 非易失性
updatable ['ʌpˌdeitəbl] adj. 可修改的
in place 在适当的地方；存在
semiconductor [semikən'dʌktə] n. 半导体
flash memory 闪存
functionality ['fʌŋkʃəneiliti] n. 功能，功能性，函数性
byte-level 字节级
cache [kæʃ] n. 高速缓存；隐藏
throughput [θruː(ː)'put] n. 吞吐量；生产量；生产能力
be referred to as 称作，叫作
virtually ['vəːtjuəli] adv. 事实上，实际上
house [haus] vt. 存放；安置；给……房子住；收藏 vi. 住 n. 住宅；房
expansion [iks'pænʃən] n. 扩充；开展
peripheral [pə'rifərəl] adj. 外围的 n. 外围设备，外设
slot [slɔt] n. 插槽，槽
power supply 电源，供电
system board 系统板 = mother board 主板
storage bay 存储机架
floppy ['flɔpi] n. 软盘；floppy drive 软盘驱动器
Zip drive Zip驱动器

Abbreviations

CPU (Central Processing Unit) 中央处理器
DRAM (Dynamic RAM) 动态随机存取存储器
SRAM (Static RAM) 静态随机存取存储器
VDT (Video Display Terminal) 视频显示终端
RAM (Random Access Memory) 随机存取存储器
ROM (Read Only Memory) 只读存储器
EPROM (Erasable PROM) 可擦可编程只读存储器
PROM (Programmable Read-Only Memory) 可编程ROM
EEPROM (Electrically Erasable Programmable ROM) 电可擦可编程ROM

Notes

1. 这里are marketed可译为"被销售"，literally译为"不加夸张地讲,确实地"。全句可

译为:不加夸张地讲,市场上有几百种不同类型的计算机在销售。

2. 本句中 about one eighth the size of a postage stamp 是介词短语,修饰前面的 DRAM chip,即约 1/8 邮票大小(的)。

3. 本句中 retrieve the appropriate … 可译为"取出所需的……",initiate 译为"启动,初始化"。本句可译为:消息被解释,处理器从辅助存储器取出所需的程序和数据。

   本句的上一句中 route 译为"发送,路由"。全句可译为:查询以消息的形式通过通道(如同轴电缆)发送到主存储器。

4. it is seldom, if ever, changed 中插入的 if ever 是常见用法,可译为"它简直从不改变"。

5. 本句中 only the byte or bytes addressed 中 addressed 修饰前面的 the byte or bytes。本句可译为:EEPROM 在任何时候都可写入,不需擦除原先内容,且只更新寻址到的字节或多个字节。

6. 本句中 being updatable in place 是 of 的介词短语。in place 是指需要更新的地方,因此短语的含义是"可更新、需要更新的字节"。本句可译为:EEPROM 把非易失性优点和可在需要更新处更新的灵活性结合起来,修改时使用普通的总线控制线、地址线和数据线。

7. 本句中 flash memory does not provide byte-level erasure(闪存不提供字节级的擦除)是针对 EEPROM 可对字节修改,即提供字节级的擦除;而 EPROM 若要修改字节,则必须先擦除整块 EPROM 的原先内容,所以三种存储的擦除单位分别是:

   EPROM　　　　　　整个存储器
   flash memory　　　 块(类似于硬盘)
   EEPROM　　　　　 字节(可能多个字节)

   目前的移动 U 盘,数码相机等的闪存卡[CF(Compact Flash)卡、Smart Media 卡、xD(eXtreme Digital)卡、记忆棒(Memory Stick)、SD(Secure Digital)卡]等都用闪存。

8. 本句中 it would take to access … 是定语从句,修饰前面的 the time,其前面省略了关系代词 that。it 是引导词,作形式主语,真实主语是动词不定式 to access … 。access 译为"访问,存取"。全句可译为:从磁盘存储器中存取单个数据所花的时间,可以从 RAM 中存取几千条指令或数据。

注:本节主要介绍计算机的 CPU、主存及闪存等内容。Fig. 1-2 中的 hard disk drive 现在已被 SSD(固态盘,见 Unit 2 的 Terms)代替。

1.2　Integrated Circuit — *Moore's Law*

The basic elements of a digital computer, as we know, must perform storage, movement, processing, and control functions. Only two fundamental types of components are required (see Figure 1-3): *gates* and memory *cells*. A gate is a device that implements a simple *Boolean* or logical function, such as IF A AND B ARE TRUE THEN C IS TRUE (AND gate). Such devices are called gates because they control data flow in much the same way that canal gates do.[1] The memory cell is a device that can store one bit of data; that is, the device can be in one of two stable states at any time. By interconnecting large numbers of these fundamental devices, we can construct a computer.

Thus, a computer consists of gates, memory cells, and interconnections among these elements.

The gates and memory cells are, in turn, constructed of simple digital *electronic components*.

The integrated circuit *exploits* the fact that such components as *transistors*, *resistors*, and *conductors* can be *fabricated* from a semiconductor such as *silicon*. It is merely an extension of the solid-state art to fabricate an entire circuit in a tiny piece of silicon rather than assemble *discrete* components made from separate pieces of silicon into the same circuit.[2] Many transistors can be produced at the same time on a single wafer of silicon. Equally important, these transistors can be connected with a process of *metallization* to form circuits.[3]

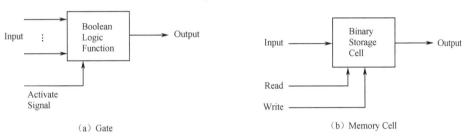

Figure 1-3 Computer elements

Figure 1-4 depicts the key concepts in an integrated circuit. A thin *wafer* of silicon is divided into *a matrix of* small areas, each a few millimeters square.[4] The identical circuit pattern is fabricated in each area, and the wafer is broken up into chips. Each chip consists of many gates and/or memory cells plus a number of input and output attachment points. This chip is then *packaged* in *housing* that protects it and provides *pins* for attachment to devices beyond the chip.[5] A number of these packages can then be interconnected on a *printed circuit board* to produce larger and more complex circuits.

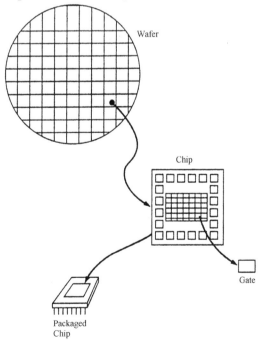

Figure 1-4 Relationship between Wafer, Chip, and Gate

7

Initially, only a few gates or memory cells could be reliably manufactured and packaged together. These early integrated circuits are referred to as small-scale integration (SSI). As time went on, it became possible to pack more and more components on the same chip. The growth in density is one of the most remarkable technological trends ever recorded.[6] It reflects the famous Moore's law, which was *propounded* by Gordon Moore, *cofounder* of Intel, in 1965. Moore observed that the number of transistors that could be put on a single chip was doubling every year and correctly predicted that this *pace* would continue into the near future[7]. To the surprise of many, including Moore, the pace continued year after year and decade after decade. The pace slowed to a doubling every 18 months in the 1970s, but has *sustained* that rate ever since. So long as this law holds, *chipmakers* can *unleash* a new *generation* of chips every three years with four times as many transistors[8]. In memory chip, this has *quadrupled* the capacity of dynamic random-access memory (DRAM), still the basic technology for computer main memory, every three years.[9]

The consequences of Moore's law are *profound*:

1. The cost of a chip has remained virtually unchanged during this period of rapid growth in density. This means that the cost of computer logic and memory *circuitry* has fallen at a dramatic rate[10].
2. Because logic and memory elements are placed closer together on more densely packed chips[11], the electrical path length is shortened, increasing operating speed.
3. The computer becomes smaller, make it more convenient to place in a variety of environments.
4. The interconnections on the integrated circuit are much more reliable than *solder* connections. With more circuitry on each chip, there are fewer *interchip* connections.

Words and Expressions

Moore's law   摩尔定律
gate [geit] *n.* [计算机]逻辑门；门电路
cell [sel] *n.* 位元；细胞；单元；晶格
Boolean ['buːliən] *n.* 布尔型；布尔；布尔运算
electronic component   *n.* 电子元件
exploit [ik'sploit] *vt.* 利用
transistor [træn'sistə(r)] *n.* 晶体管
resistor [ri'zistə(r)] *n.* 电阻(器)
conductor [kən'dʌktə] *n.* 导体；导线
fabricate ['fæbrikeit] *vt.* 制作
silicon ['silikən] *n.* 硅
discrete [di'skriːt] *adj.* 分立的；分离的；离散的
metallization ['metəlaizeiʃən] *n.* 金属化

wafer ['weifə(r)] *n.* 薄片，晶片，圆片　　wafer of silicon　硅晶片
a matrix of　……的阵列；一组
package ['pækidʒ] *n.* 封装；包装；把……打包
housing ['hauziŋ] *n.* 壳，套
pin [pin] *n.* 引脚
printed circuit board　印制电路板
propound [prə'paund] *vt.* 提出；建议
cofounder [kəu'faundə] *n.* 共同创办人，合伙创办人
pace [peis] *n.* 速度；进度；姿态
sustain [sə'stein] *vt.* 维持；继续；持续，保持；支撑，承受住；认可；遭受
chipmaker ['tʃipmeikə] *n.* 芯片制造商
unleash [ʌn'liːʃ] *vt.* 发布，释放，发动
generation [dʒenə'reiʃən] *n.* 代
quadruple ['kwɔdrupl] *n.* 四倍　　*adj.* 四倍的；四重的　　*vt.* 使……成四倍　　*vi.* 成为四倍
profound [prə'faund] *adj.* 深远的
circuitry ['səːkitri] *n.* 电路
solder ['sɔldə(r)] *v.* 焊接　　*vt.* 使连接在一起　　*n.* 焊料；接合物
interchip [ˌintə'tʃip] *n.* 芯片间

Abbreviations

IC (Integrated Circuit)　集成电路
SSI (Small-Scale Integrated circuit)　小规模集成电路
LSI (Large-Scale Integrated circuit)　大规模集成电路
VLSI (Very LSI)　超大规模集成电路

Notes

1. 本句可译为：这样一些元件（或设备）被称为逻辑门，因为它们控制数据流的方式与运河的闸门相同。
2. 本句中 it 代表动词不定式 to fabricate …，是形式主语。assemble … into … 译为"把……装配成……"。made from … 修饰前面的 discrete components。全句可译为：把整个电路制作在一块很小的硅片上，而不是用分立元件（它们由个别的硅片组成）组装成相同的电路，这只不过是固态技术的扩展。
3. 本句可译为：同样重要的是，这些晶体管能够通过金属化过程来连接，以形成电路。
4. 本句中 a matrix of small areas 可译为"小区域阵列"。each a few millimeters square 可译为"每个区域有几平方毫米"。
5. This chip is then packaged in housing 可译为"然后将这块芯片封装在外壳中"。attachment to devices beyond the chip 可译为"连接芯片外部的设备"。
6. ever recorded"曾经有过记录的"。全句可译为：这种密度的增长是（有记载的）最显著的技术发展趋势之一。

7. 本句中 this pace would continue into the near future 可译为"这种速度将持续到不远的将来"。下一句可译为：令许多人(包括摩尔在内)惊奇的是，这种态势年复一年地继续下来。

8. 本句中 with four times as many transistors 可译为"(新一代的)晶体管数为上一代的4倍"。

9. 本句中 this 指上一句所说的，晶体管数每3年提高到4倍；still ... 是插入的补充说明。全句可译为：对于内存芯片，这使得动态随机存取存储器(仍是计算机主存的基本技术)的容量每3年提高到4倍。

10. 本句中 has fallen at a dramatic rate 可译为"显著下降"。

11. 本句中 more densely packed chips 可译为"集成度更高的芯片"。

注：集成电路(IC)把整个电路制作在一块很小的硅片上。著名的摩尔定律指出，单块芯片上的晶体管数每年翻一番。Pentium(奔腾)微处理器上的晶体管数高达几百万、上千万个，因此可在很小的设备(如手机、数码相机)中放入一个处理机。本节内容是 IC 的简单介绍。

1.3 *Multicore Processors*

In 1965, when he first *set out* what we now call Moore's Law, Gordon Moore (who later co-founded Intel Corp.) said the number of components that could be packed onto an integrated circuit would double every year or so(later amended to 18 months)[1].

In 1971, Intel's 4004 CPU had 2,300 transistors. In 1982, the 80286 *debuted* with 134,000 transistors. Now, *run-of-the-mill* CPUs count *upward of* 200 million transistors, and Intel is *scheduled* to *release* a processor with 1.7 billion transistors for later this year.[2]

For years, such progress in CPUs was clearly predictable: successive generations[3] of semiconductor technology gave US bigger, more powerful processors on ever-thinner silicon *substrates* operating at increasing clock speeds. These smaller, faster transistors use less electricity, too.

But there's a catch[4]. It turns out that as operating voltages get lower, a significant amount of electricity simply leaks away and ends up generating excessive heat, requiring much more attention to processor cooling and limiting the potential speed advance—think of this as a *thermal barrier*.

To break through that barrier, processor makers are adopting a new strategy, packing two or more complete, independent processor cores, or CPUs, onto a single chip. This multicore processor plugs directly into a single *socket* on the motherboard, and the operating system sees each of the execution cores as a discrete logical processor that is independently *controllable*. Having two separate CPUs allows each one to run somewhat slower, and thus cooler, and still improve overall throughput for the machine in most cases.

From one *perspective*, this is merely an extension of the design thinking that has for several years given us *n*-way servers using two or more standard CPUs;[5] we are simply making the packaging smaller and the integration more complete. In practice, however, this multicore strategy represents a major *shift* in processor architecture that will quickly *pervade* the computing industry. Having two CPUs on the same chip rather than plugged into two separate sockets greatly speeds communication between them and cuts waiting time.

The first multicore CPU from Intel is already on the market. By the end of 2006, Intel expects multicore processors to make up 40% of new desktops, 70% of mobile CPUs and a *whopping* 85% of all server processors that it ships. Intel has said that all of its future CPU designs will be multicore. Intel's major competitors — including Advanced Micro Devices Inc. (AMD), Sun Microsystems Inc. and IBM — each appears to be betting the farm on multicore processors. [6]

Besides running cooler and faster, multicore processors are especially well suited to tasks that have operations that can be divided up into separate *threads* and run in parallel. [7] On a dual-core CPU, software that can use multiple threads, such as database *queries* and graphics *rendering*, can run almost 100% faster than it can on a single-CPU chip.

However, many applications that process in a linear fashion, including communications, backup and some types of numerical computation, won't benefit as much and might even run slower on a dual-core processor than on a faster single-core CPU. [8]

Words and Expressions

multicore processor　多核处理器
set out　陈述，阐明；宣布；出发
debut ['deibjuː] [法] vi. 初次登场，初次表演　n. 首次演出；首次露面，开张
run-of-the-mill　普通的，一般的，平凡的
upward of　……以上，多于，超过
schedule ['ʃedjuːl; 'skedʒul] v. 安排；预定；确定时间
release [ri'liːs] n. 发布；释放
substrate ['sʌbstreit] n. 衬底；底层；下层
thermal ['θəːməl] barrier ['bæriə]　热障
socket ['sɔkit] n. 插座，插槽；套接字
controllable [kən'trəuləbl] adj. 可管理的，可操纵的，可控制的
perspective [pə'spektiv] n. 观点,看法；视角；远景；透视；透视图；观察
shift [ʃift] vt. 移动，转移，转变　n. 移位
pervade [pə'veid] vt. 遍及；扩散；渗透
whopping [(h)wɔpiŋ] adj. 巨大的，庞大的　adv. 非常地，异常地
thread [θred] n. 线程；线
query ['kwiəri] n. 询问；查询　vt. 询问；质问　vi. 询问；表示怀疑
render ['rendə] vt. 渲染；浓淡处理；着色；提出；呈递
linear ['liniə(r)] adj. 线的，直线的，线性的

Notes

1. 本句中 component "成分，组成部分"，这里指 "元器件"。pack "包，打包"，这里指 "封装"。or so "左右"。全句可译为：1965 年，Gordon Moore 首次提出了今天我们所说的

摩尔定律，他（后来与人共同创建了 Intel 公司）说，能够封装进集成电路的元器件数大约每年翻一番（后来改为每 18 个月）。

2. 本句中 count "计数，数数"。全句可译为：现在，一般的 CPU 有 2 亿只以上的晶体管，Intel 公司预定在今年晚些时候推出有 17 亿只晶体管的处理器。

3. 本句中 successive generations of … "连续几代的"，可译为 "一代一代的"。

4. catch 在口语中用为 "隐情，圈套，疑难"。本句可译为：但是，存在潜在问题。下一句中 it turns out 的意思是 "结果是，最后证明是"。

5. 本句中 design thinking 有两个修饰成分：一个是后面紧跟的 that 引导的从句，that 代表 design thinking；另一个是 given us … 分词短语，即给我们 n 路服务器的设计思想。全句可译为：从某种观点看，这种多核处理器只是沿用多年的、采用两个或多个标准 CPU 的多路服务器设计思想的延伸。

6. 本句中 each appears to be betting … on … 其中 betting … on … 译为"对某事打赌"。the farm 这里是十分有把握的意思。本句意思是，每个公司看来都正把赌注（或宝）押在多核处理器上。

7. divide up 意为"分配，分担"。全句可译为：多核处理器除了运行温度低、速度快，还特别适合那些操作可以分成不同线程并且能并行运行的任务。

8. 本句中 numerical computation "数值计算"。slower … than … 是比较级，即在双核处理器上运行比在速度更快的单核 CPU 上运行更慢。全句可译为：但是，很多以线性方式处理的应用（程序），如通信、备份和某些类型的数值计算都不会获益很多，并且甚至可能在双核处理器上运行比在速度更快的单核 CPU 上运行更慢。

1.4 Computer Architectures

Michael J. Flynn's 1972 *taxonomy* of computer architectures is still the most generally used method of classifying parallel computers. Flynn divided computers according to whether they used single or multiple "*streams*" *of data*, and (*orthogonally*) single or multiple "streams" of instructions (see Table 1-1).

Table 1-1　Flynn's Taxonomy

	Single Instruction	Multiple Instruction
Single Data	SISD (*von Neumann*)	MISD
Multiple Data	SIMD (DAP)	MIMD (Sequent, iPSC/2)

　An SISD computer carries out one instruction on one datum at a time. This is the conventional von Neumann architecture.

　An MISD computer, on the other hand, would apply several instructions to each datum it *fetches* from memory. No computers conforming to this model have yet been constructed.

　The third of Flynn's four groupings is SIMD. In an SIMD computer, many processors simultaneously execute the same instructions, but on different data. For example, if the instruction is ADD A B, here A and B are matrix, each processor adds its own value of B to its own value of A.

The significant computers of this type produced to date are the Distributed Array Processor (or DAP), the Connection Machine (CM) etc. These machines are built of very simple processing elements (PEs) but compensate for this simplicity by using many of them together.[1] A single master processor broadcasts program instructions to the individual PEs, which carry out the instructions on their own data. PEs can be disabled temporarily so that operations are only carried out on part of the data; this provides a way of making computations *data-dependent*, like the IF statement in most languages.

Processing elements can also transfer data amongst themselves. In the DAP, PEs are connected in a *square array*. Each PE can simultaneously shift one bit in one direction and receive a bit from the opposite direction. Repeated shifts can move large volumes of data from any part of the *grid* to any other. One of the strengths of SIMD computers is that as more processing elements are added, so are more *interprocessor links*, so that the total communications bandwidth of the machine rises in proportion to its size.[2] Such *scaling properties* are an important consideration in MIMD computers as well.

Experience has shown that SIMD are very good at some things, but inefficient at others. For example, many of the algorithms used in image processing involve performing the same operations on each *pixel* of the image, such as taking a *weighted average* of its value and the values of its four nearest neighbors. If each pixel is *mapped to* a separate PE, an SIMD machine can carry out the calculation for each pixel simultaneously, producing the clean image in much less time than a serial machine would require. On the other hand, when the task is not well *load-balanced* the SIMD architecture can be inefficient. For example, in ray-tracing some light rays never intersect an object and leave the scene immediately, while others have a very complicated path involving many reflections/refractions.[3] While *schemes* exist for many applications which may reduce the imbalance, those PEs which are assigned straightforward tasks must still wait for the other processors to finish.

MIMD computers are an evolutionary step forward from SISD computers. An MIMD computer contains several independent (and usually equi-powerful) processors, each of which executes its individual program. For example, when p processors are available and n elements are to be *sorted* (n is much larger than p), an initial data distribution among the p ranked processors $P_1, P_2, \ldots, P_p$ is $X = \langle x_1, x_2, \ldots, x_i \rangle$, where x_i is a *block* of $M = \lceil n/p \rceil$ elements stored in processor P_i's local memory. To sort the n elements, each processor can first perform a local sort by using any fast serial sort algorithm independently. After all processors complete their local sort, they cooperate to carry out a global merge sort.

There are several different ways of building an MIMD computer with current technology. Some manufacturers *couple* many conventional microprocessors (typically the Intel 80x86) to create a machine with mainframe performance at minicomputer cost. The primary distinction of different ways is the relationship of processors to memory. Other characteristics, such as the way processors are connected, follow on from this initial distinction.

SM — MIMD

So long as the number of processors remains small, engineers can connect them all to a single memory store. This leads to a shared-memory computer (SM, see Figure 1-5), in which every processor can access any part of the whole machine's memory. (In less extreme examples, processors have some private memory, which may be compared to a private office, but also share some memory, like programmers share a library or a coffee lounge). Shared-memory computers are attractive because they are relatively simple to program.

However, computers with physically shared memory have one great flaw: they cannot be *scaled* up indefinitely. As the number of processors trying to access the memory increases, so do the odds that processors will be contending for such access.[4] Eventually, access to memory becomes a bottleneck limiting the speed of the computer. The use of local cache memory can alleviate this problem by permitting commonly used data to be stored locally on each processor. This approach, when taken to its logical limit, distributes all of the memory between the processors, so reducing the required memory bandwidth.[5] This leads to a shared-nothing computer (SN).

SN — MIMD

In basic shared-nothing design, each processor has a private memory. Processors communicate via a high-speed interconnection network (see Figure 1-6). However, accessing memory locations on remote nodes poses a new problem. In some SN computers, shared memory is *emulated* on top of the distributed memory architecture by having a global address space and interprocessor communication are hidden from the user.[6] However, it is more common in distributed memory machines, such as the Intel iPSC/2, for each processor to maintain its own memory and for the user program to specifically request information from another node.

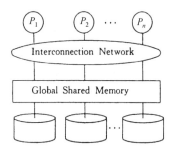
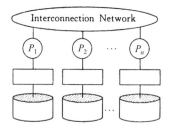

Figure 1-5 Shared memory multiprocessor   Figure 1-6 Shared nothing multiprocessor

How to link these processors in a SN computer together is another problem. Connecting them all to a single bus, or through a single switch, leads to the same sorts of bottlenecks as discussed above for large numbers of processors. Similarly, linking each processor to every other is not an acceptable solution, because the number of connections (and hence the cost) rises with the square of the number of processors. The only practical solution is to connect each processor to some small subset of its fellows.

Words and Expressions

taxonomy [tækˈsɔnəmi] *n.* 分类法；分类
orthogonally [ɔːˈθɔɡənəli] *adj.* 正交地
streams of data   数据流
von Neumann   冯·诺伊曼
fetch [fetʃ] *vt.* 取；取数
data-dependent   数据依赖的，数据相关的
square array   方阵
grid [ɡrid] *n.* 网格，栅格
interprocessor link   处理机间的连接，处理单元间的连接
scaling property   扩缩性，缩放性
pixel [ˈpiksəl] *n.* 像素，像元
weighted average   加权平均
map to   映射到
load-balanced   负载平衡的
scheme [skiːm] *n.* 方案；模式；图解   *v.* 计划；设计
sort [sɔːt] *v.* 排序；分类   *n.* 种类
block [blɔk] *n.* 块，代码块
couple [ˈkʌpl] *vt.* 连接；使耦合；结合   *vi.* 结合
scale [skeil] up (scaleup)   扩缩，伸缩（伸缩比），扩大
emulate [ˈemjuleit] *n.* 仿效，仿真

Abbreviations

SISD (Single-Instruction Single-Data stream)   单指令单数据流
MISD (Multiple-Instruction Single-Data stream)   多指令单数据流
DAP (Distributed Array Processor)   分布式阵列处理机
SIMD (Single-Instruction Multiple-Data stream)   单指令多数据流
MIMD (Multiple-Instruction Multiple-Data stream)   多指令多数据流
SPMD (Single Program Multiple Data)   单程序多数据
MIPS (Million Instructions Per Second)   每秒百万条指令
OEM (Original Equipment Manufacturer)   原始设备制造商
PnP (Plug and Play)   即插即用
SM (Shared Memory) 共享存储器
SN (Shared-Nothing) 无共享

Notes

1. compensate for "补偿"。全句可译为：这些机器由一些很简单的处理单元（PE）构成，

但一台机器可以用很多个 PE 来弥补由其简单带来的不足。

2. strength "力,力量,强度,趋势"。so are ... links 是 "so＋be＋主语"句型的倒装结构,意思是"也这样"。全句可译为:SIMD 计算机的一个特点是,随着处理单元的增加,处理单元之间的连接数也会增加,因此机器的总通信带宽与它的规模成比例增加。

3. ray-tracing 可译为"射线跟踪,光线跟踪"。本句可译为:例如,在射线跟踪中,某些光线从不与物体相交,稍纵即逝,而另一些光线具有很复杂的路径,包含了许多反射和折射路径。

   下一句中 which 引导的从句修饰 schemes。全句可译为:虽然对许多应用问题有一些方案可以减少这种不平衡,但是那些被分配完成简单任务的处理机,仍然必须等待其他处理机完成它们的任务。

4. 本句中 so do the odds 是 "so＋do＋主语"句型的倒装结构。后面的 that 引导的从句修饰 odds(可能性,机会,不等)。全句可译为:随着试图访问存储器的处理机数的增加,处理机争夺这类访问的可能性也增大。

5. 本句中 take to 可译为"走向"。本句可译为:这种方法,当走向极端时,把所有内存分配给各处理机,因此减少了所需的内存带宽。

说明:本节对 SM 结构的改变描述为:SM→SM＋每个处理机带有自己的 cache→把共享内存分给各处理机(变为 SN 机)。

6. 本句译为:在一些 SN 计算机中,通过一个全局地址空间在分布式存储器体系结构上模仿共享存储器,而用户不知道处理机之间的通信。

说明:SN 计算机无共享存储器,上述模仿的共享存储器实际上是用软件来实现的,即系统对各处理机的内存进行全局统一编址,用户只需用全局地址访问所需内容,而由系统软件确定该内存在哪台处理机上,若在他机上,再由系统通过通信把它们取来。

注:本节侧重介绍多处理机系统 SIMD 和 MIMD 的大致工作方式及简单应用例子,未学过这方面内容的读者可作为新内容学习。

Terms

SMP

Acronym for symmetric multiprocessing. A computer architecture in which multiple processors share the same memory, which contains one copy of the operating system, one copy of any applications that are in use, and one copy of the data. Because the operating system divides the workload into tasks and assigns those tasks to whatever processors are free, SMP speeds transaction time. See also architecture, processor.

MPP(Massively Parallel Processing)

A computer architecture in which each of a large number of processors has its own RAM, which contains a copy of the operating system, a copy of the application code, and its own part of the data, on which that processor works independently of the others. Compare symmetric multiprocessing.

Speedup(加速，Parallelism Goals and Metrics)

Given a fixed job run on a small system, and then run on a larger system, the speedup given by the larger system is measured as:

$$\text{Speedup} = \text{small system elapsed time} / \text{big system elapsed time}$$

Speedup is said to be linear, if an N-times large or more expensive system yields a speedup of N.

Scaleup(Parallelism Goals and Metrics)

Speedup holds the problem size constant, and grows the system. Scaleup measures the ability to grow both the system and the problem. Scaleup is defined as the ability of an N-times larger system to perform an N-times larger job in the same elapsed time as the original system. The scaleup metric is:

$$\text{Scaleup} = \text{small system elapsed time on small problem} / \text{big system elapsed time on big problem}$$

If this scaleup equation evaluates to 1, then the scaleup is said to be linear.

Word size

A computer word is the amount of data (measured in bits or bytes) that a CPU can manipulate at one time. Most newer CPU chips are designed for 64 bit words.

FSB (FrontSide Bus，前端总线)

Today's computers typically have a specific system bus to connect the CPU to RAM called frontside bus.

BSB (BackSide Bus)

The backside bus transfer data between the CPU and external cache.

Chipset

Most CPUs today use a chipset of one or more chips that help bridge or connect the various buses to the CPU.

DIMM (Dual In-line Memory Module，双列直插内存模块)

Like the CPU, RAM consists of circuits etched onto chips. These chips are arranged onto circuit board called Single In-line Memory Module (SIMM), Dual In-line Memory Module (DIMM), or Rambus In-line Memory Module (RIMM), depending on the type of memory and type of circuit board used.

DRAM

A dynamic RAM is made with cells that store data as charge on capacitors. The presence or absence of charge in a capacitor is interpreted as a binary 1 or 0.

SRAM
In a static RAM, binary values are stored using traditional flip-flop（触发器）logic-gate configurations. A SRAM will hold its data as long as power is supplied to it.

DDR (Double Data Rate)
DDR is an enhanced version of SDRAM (synchronous DRAM) which is a new type of DRAM. SDRAM is much faster than DRAM because it is synchronized to the system clock. DDR2 is twice faster than DDR and is used in laptop.

KB, MB, GB, TB, PB, EB, ZB, YB, BB, NB, DB
Kilo Byte (KB=2^{10} or 10^3 Bytes, 千字节) Mega Byte (MB=2^{20} or 10^6 Bytes, 兆字节)
Giga Byte (GB=2^{30} or 10^9 Bytes, 吉字节) Tera Byte (TB=2^{40} or 10^{12} Bytes, 太字节)
Peta Byte (PB=2^{50} or 10^{15} Bytes) Exa Byte (EB=2^{60} or 10^{18} Bytes)
Zetta Byte (ZB=2^{70} Bytes, 泽字节) Yotta Byte (YB=2^{80} Bytes, 尧字节)
Bronto Byte (BB=2^{90} Bytes) Nona Byte (NB=2^{100} Bytes)
Dogga Byte (DB=2^{110} Bytes)

NAND and NOR Flash
Though they are both considered leading non-volatile Flash memory technologies, NAND and NOR Flash meet completely different design needs based on their individual attributes. NOR offers faster read speed and random access capabilities, making it suitable for code storage in devices such as PDAs and cell phones. However, with NOR technology, write and erase functions are slow compared to NAND. NOR also has a larger memory cell size than NAND, limiting scaling capabilities and therefore achievable bit density compared to NAND. Since code storage tends to require lower density memory than file storage, NOR's larger cell size is not considered a concern when used in these applications.

Conversely, NAND offers fast write/erase capability and is slower than NOR in the area of read speed. NAND is, however, more than sufficient for a majority of consumer applications such as digital video, music or data storage. NAND's fast write/erase speed combined with its higher available densities and a lower cost-per-bit than NOR make it the favored technology for file storage in a host of consumer applications. Offering users the ability to rewrite data quickly and repeatedly, NAND is typically used for storing large quantities of information in devices such as Flash drives, MP3 players, multi-function cell phones, digital cameras and USB drives.

CISC and RISC（复杂指令集计算机和精简指令集计算机）
Acronym for Reduced Instruction Set Computer. A microprocessor design that focuses on rapid and efficient processing of a relatively small set of simple instructions that comprises most of the instructions a computer decodes and executes. **RISC** architecture optimizes each of

these instructions so that it can be carried out very rapidly—usually within a single clock cycle. RISC chips thus execute simple instructions more quickly than general-purpose **CISC** (Complex Instruction Set Computer) microprocessors, which are designed to handle a much wider array of instructions. RISC chips are, however, slower than CISC chips at executing complex instructions, which must be broken down into many machine instructions that RISC microprocessors can perform.

Tablet computer (平板计算机)

A tablet computer, or simply tablet, is a mobile computer with display, circuitry and battery in a single unit. Tablets come equipped with sensors, including cameras, a microphone, an accelerometer and a touchscreen, with finger or stylus gestures substituting for the use of computer mouse and keyboard. Tablets may include physical buttons (for example: to control basic features such as speaker volume and power) and ports (for network communications and to charge the battery). They usually feature on-screen, pop-up virtual keyboards for typing. Tablets are typically larger than smart phones at 7 inches (18 cm) or larger, measured diagonally.

Internet users owned a tablet, used mainly for viewing published content such as video and news. Among tablets available in 2012, the top-selling line of devices was Apple's iPad with 100 million sold between its release in April 2010 and mid-October 2012.

Ultrabook (超极本)

Ultrabook is a specification and trademarked brand by Intel for a class of high-end subnotebooks which are designed to feature reduced bulk without compromising battery life. They use low-power Intel Core processors, solid-state drives, and unibody chassis to help meet these criteria. Due to their limited size, they typically omit common laptop features such as optical disc drives and Ethernet ports.

UEFI

The **Extensible Firmware Interface (EFI)** is a specification that defines a software interface between an operating system and platform firmware. EFI is a replacement for the older BIOS firmware interface present in all IBM PC—compatible personal computers. The EFI specification was originally developed by Intel, and is now evolved to **Unified** Extensible Firmware Interface (**UEFI**). UEFI is not restricted to any specific processor architecture, and can run on top of or instead of traditional BIOS implementations.

GPU (Graphics Processing Unit)

GPU is a specialized electronic circuit designed to rapidly manipulate and alter memory to accelerate the creation of images in a frame buffer intended for output to a display device. Modern GPUs are very efficient at manipulating computer graphics and image processing. Their highly parallel structure makes them more efficient than general-purpose CPUs for algorithms that process large blocks of data in parallel. In a personal computer, a GPU can be present on a video card or embedded on the motherboard. In certain CPUs, they are embedded on the CPU die (a small block of semiconducting material on which a given functional circuit is fabricated).

The term GPU has been used from at least the 1980s; it was popularized by Nvidia in 1999, who marketed the GeForce 256 as "the world's first GPU". It was presented as a "single-chip processor with integrated transform, lighting, triangle setup/clipping, and rendering engines".

Exercises

1. Translate the text of Lesson 1.4 into Chinese.
2. Topics for oral workshop.
   - Talk about the main components of a computer and their functions.
   - Explain how the computer memory is organized into a hierarchy: register, cache, primary storage, and secondary storage.
   - Discuss with your partners the differences among ROM, EPROM, flash memory, and the applications of flash memory. (hint: mobile disk, digital camera, digital video camera recorder, ...)
3. Translate the following into English.

计算机系统由硬件系统和软件系统组成。计算机的硬件系统通常分为三个主要部分或三个主要子系统：CPU、存储器子系统和输入/输出子系统。

CPU 执行许多操作并控制计算机。存储器子系统用来存储正在被 CPU 执行的程序，连同程序的数据。输入/输出子系统允许 CPU 与诸如个人计算机的键盘和监视器那样的输入和输出设备交互。计算机的这些组成部分都连接到总线。

计算机执行大量数据处理操作的部件叫作中央处理器，也可称作(be referred to as)CPU。在微机中，它经常被称为微处理器。CPU 由三个主要部件组成：控制器、ALU 和寄存器组。

存储器也被认为是(be known as)内存或主存。它是指计算机中的电路，这些电路保持的无论是(whatever)程序还是数据，都是 CPU 立即可使用的。

输入/输出子系统包括输入/输出设备和接口。有各种各样的输入/输出设备，如鼠标、打印机、传感器、磁盘等。输入/输出接口提供一种在内存储器和外部输入/输出设备之间传输信息的方法。为了与 CPU 接口，连接到计算机的外部设备需要专用的通信连接。该通信连接的目的是解决中央计算机和每个外部设备之间存在的差别。

4. Listen to the video "Computer tour" and write down the first two paragraphs.

Unit 2　Hardware II

2.1　Optical Storage Media: High-Density Storage

2.1.1　Optical Laser Disks

Optical laser disk technology eventually may make magnetic-disk and magnetic-tape storage obsolete. With this technology, the read/write head used in magnetic storage is replaced by two lasers. One *laser beam* writes to the recording surface by *scoring microscopic pits* in the disk, and another laser reads the data from the *light-sensitive* recording surface. [1] A light beam is easily *deflected* to the desired place on the optical disk, so an *access arm* is not needed.

Optical laser disks are becoming a very *inviting* option for users. They are less sensitive to environmental *fluctuations*, and they provide more direct-access storage at a cost that is much less per megabyte of storage than the magnetic-disk alternative. [2] Optical laser disk technology is still *emerging* and has yet to *stabilize*; however, at present there are three main categories of optical laser disks. They are CD-ROM, recordable CD-R and rewritable CD-RW, and *magneto-optical disk*.

CD-ROM

Introduced in 1980, the extraordinarily successful CD, or compact disk, is an optical laser disk designed to enhance the recorded reproduction of music. [3] To make a CD recording, the analog sounds of music are translated into their digital equivalents and stored on a 4.72-inch optical laser disk. Seventy-four minutes of music can be recorded on each disk in digital format by 2 billion digital bits. With its tremendous storage capacity, computer-industry *entrepreneurs* immediately recognized the potential of optical laser disk technology. In effect, anything that can be digitized can be stored on optical laser disk: data, text, voice, still pictures, music, graphics, and *video*.

CD-ROM (pronounced cee-dee-ROM) is a *spin-off* of *audio* CD technology. CD-ROM stands for compact disk-read only memory. The name implies its application. CD-ROM disks, like long-playing *record albums*, are "pressed" at the factory and distributed with their prerecorded contents (for example, *the complete works of Shakespeare* or the first 30 minutes of *Gone With the Wind*). [4] Once inserted into the *disk drive*, the text, video images, and so on can be read into primary storage for processing or display; however, the data on the disk are fixed—they cannot be altered. This is in contrast, of course, to the read/write capability of magnetic disks.

The tremendous amount of low-cost direct-access storage made possible by optical laser disks has opened the door to many new applications.

CD-R and CD-RW

CD-R and CD-RW both allow user to store data on compact discs, but only data on *rewritable* CDs can be erased and *overwritten*. **CD-R** discs can only be written to once. Both types of discs look very similar to CD-ROM discs. Recordable CDs are commonly used to store music files, allowing home users to make high-quality *personalized* music CDs. **CD-RW** discs can be written to and erased similar to a floppy disk. Consequently, they are a good *alternative* for large file storage, as well as for creating a "master" disc before *burning* a CD-R disc (most CD-RW drives can write to both CD-RW and CD-R discs, but some CD and CD-R drives cannot read CD-RW discs). CD-R and CD-RW drives also read CD-ROM discs. However, CDs will be *eclipsed* by DVDs, once that technology becomes improved and standardized.

Magneto-Optical Disk

Magneto-optical disk offers promise that optical laser disks will become commercially viable as a read-and-write storage technology. [5] The 5¼-inch disks can store up to 1,000 Mb. At present, magneto-optical disks are too expensive and do not offer anywhere near the kind of reliability that users have come to expect of magnetic media. [6] In addition, the access times are relatively slow, about the same as a *low-end Winchester disk*.

As optical laser disk technology matures to reliable, *cost-effective*, read/write operation, it eventually may dominate secondary storage in the future as magnetic disks and tape do today.

2.1.2 DVDs

The *acronym* **DVD** (for *digital versatile* disc, or *digital video disc*) refer to a relatively new high-capacity optical storage format that can hold from 4.7 GB to 17 GB, depending on the number of recording layers and disc sides being used. A standard *single-layered*, *single-sided* DVD can store 4.7 GB of data; a two-layered standard enhances the single-sided layer to 8.5 GB. DVD can be double-sided with a maximum storage of 17 GB per disc. The DVD was initially developed to store the full contents of a standard two-hour movie, but is now also used to store computer data and software. DVD-ROM technology is seen by many as the *successor* to music CDs, computer CD-ROMs, and *prerecorded* VHS *videotapes* that people buy and rent for home viewing — in other words, read-only products. DVD-R is DVD recordable (similar to CD-R). The user can write to the disk only once. DVD-RW is DVD rewritable (similar to CD-RW). The user can erase and rewrite to the disk multiple times. Only one-sided disks can be used for both DVD-R and DVD-RW. Most DVD drives can play both computer and audio CDs, but you can't play DVDs in a CD drive.

DVD Forum Receives Top Information Technology Industry Award for Creation of Unified Specification for Next Digital Multimedia Era

The DVD Forum today announced that it has received the 1997 PC Magazine Award for Technical Excellence in the category of "Standards", in recognition of the Forum's successful development of the DVD-ROM *specification*.[7]

"New standards are particularly important, since they promise to bring higher levels of technology innovation and market compatibility to today's technology users," said Michael J. Miller, editor-in-chief of PC Magazine. "DVD-ROM is a *compelling* technology that was chosen because it's a familiar format that brings *a wealth of* new computing, educational, gaming and entertainment possibilities to the user."

In the Award citation to DVD-ROM technology, PC Magazine referred to DVD as the format that will "replace the CD-ROM as the primary means of PC content distribution".[8] Representatives of three companies involved in development of the specification, Hitachi Ltd., Matsushita Electric Industrial Company (Panasonic) and Toshiba Corporation, accepted the award on behalf of the DVD Forum in a ceremony held on November 17 at COMDEX'97 in Las Vegas.

"Products based on specifications defined by the DVD Forum are now shipping in volume to the worldwide computer and *consumer electronics* markets, and the Award for Technical Excellence adds to the market's validation of the success of the standards process",[9] said Koji Hase, General Manager of the DVD Products Division at Toshiba Corporation and a founding member of the DVD Forum. "We are extremely pleased to see the work of the DVD Forum recognized as one of the key technical achievements in the personal computer industry, particularly as the Forum expands the scope of its work with a larger, global membership in 1998."

"The members of the DVD Forum developed the DVD-ROM specification as the best technical approach and also the best approach for customers in the marketplace," said Sakon Nagasaki, director of the DVD Business Development Office of Matsushita Electric Industrial Co. Ltd. (Panasonic). "Acceptance of the format illustrates how standards-making efforts advance the goals of the entire electronics industry."

In addition to its role in development of the DVD-ROM and DVD-Video standards, the Forum has proposed the format for recordable DVD, known as DVD-R, and rewritable DVD, known as DVD-RAM, to international standards bodies. Work is also continuing on definition of a DVD Audio specification.

"The mission of the DVD Forum is to define a smooth migration path from CD to DVD technology by working with the widest possible representative group of manufacturers and technology end-users in the converging industries of computers and consumer electronics",[10] said Dr. Yoshita Tsunoda, a member of the Executive Staff of Hitachi, Ltd. and Chairman of the DVD Forum's DVD-RAM Working Group. "The different working groups have already completed definition of three separate DVD technology standards, and we have begun work on developing next generation specifications that will provide compatible products well into the next century."

Recipients of the PC Magazine Awards for Technical Excellence are named by a team of editors, senior contributors and PC Labs personnel after months of evaluation and discussion. PC Magazine, the sponsor of the Technical Excellence Awards, is a 1.175 million circulation magazine published by Ziff-Davis Inc. PC Magazine is published 22 times a year in print, quarterly on CD, and continuously on the World Wide Web.

Words and Expressions

optical laser disk　光盘，激光盘
laser beam　激光束
score [skɔː] *n.* 刻痕；划线
microscopic pit　微小的凹点
light-sensitive　光敏感的
deflect [diˈflekt] *v.* （使）偏转；（使）偏斜
access arm　存取臂
inviting [inˈvaitiŋ] *adj.* 令人动心的；有魅力的
fluctuation [ˌflʌktjuˈeiʃən] *n.* 波动；起伏；动摇不定
emerge [iˈməːdʒ] *vi.* 显现；浮现；暴露；形成；（事实）显现出来
stabilize [ˈsteibilaiz] *v.* 稳定
gigabyte (GB)　千兆字节，吉字节（10^9 字节）
magneto-optical disk　磁光盘
entrepreneur [ɔntrəprəˈnəː(r)] *n.*（法文）企业家
video [ˈvidiəu] *adj.* 视频的，电视的　*n.* 电视，视频
spin-off [spin-ɔf]　有用的副产品
audio CD　激光唱盘，镭射唱盘
audio [ˈɔːdiəu] *n.* 音频，声频　*adj.* 声音的，听觉的
record album　唱片，唱片集
disk drive　光盘驱动器，磁盘驱动器
rewritable [riːˈraitəbl] *adj.* 可重写的，可改写的
overwrite [ˈəuvəˈrait] *vt.* 盖写，写在……上面
personalize [ˈpəːsənəlaiz] *vt.* 个人化，使成为个人所有，个性化
alternative [ɔːlˈtəːnətiv] *adj.* 备选的；两者择一的　*n.* 可供选择的方法，两者择一
burn [bəːn] *vt.* 烧；烧刻；烧录
eclipse [iˈklips] *vt.* 超越；使黯然失色　*n.* 食，日食，月食；蒙蔽；衰落
low-end Winchester disk　低档温彻斯特盘
cost-effective　成本效益好的，效能价格合算的
acronym [ˈækrənim] *n.* 只取首字母的缩写词
single-layered, single-sided　单层，单面
digital versatile disc　数字通用盘
digital video disc　数字视频光盘
successor [səkˈsesə] *n.* 继承者，接任者

prerecorded [ˈpriːriˈkɔːdid] *adj.* 预录的
videotape [ˈvidiəuteip] *n.* 录像带
specification [spesifiˈkeiʃən] *n.* 规范；规格说明；说明书；详细计划书
compelling [kəmˈpeliŋ] *adj.* 激发兴趣的；使人非相信不可的
a wealth of   大量，丰富
consumer electronics   消费电子产品

Abbreviations

CD（Compact Disc）光盘，紧密盘
WORM（Write-Only-Read-Many）一次写，多次读
VHS（Video Home System） 家用录像系统
WWW（World Wide Web） 环球网，万维网

Notes

1. 本句中 by scoring ... 译为"以刻制……为手段"。全句可译为：一束激光通过在光盘上刻制微小的凹点，在记录表面写数据；而另一束激光用来从光敏感的记录表面读取数据。

2. 本句中 that is much ... 是定语从句，that（代表 a cost）为主语，其中 alternative 是"可选择的，选择物"，不必译出。全句可译为：它们（光盘）对环境变化不太敏感，并且它们以比磁盘低得多的每兆字节存储器价格，提供更多的直接存取存储（空间）。

3. 本句中 designed to enhance ... 是过去分词短语，修饰前面的 disk。reproduction "再现，再生，复制"。短语含义是：设计来提高音乐的录音重放质量。

4. *the complete works of Shakespeare*，即《莎士比亚全集》，*Gone with the Wind*，即电影《飘》。本句可译为：只读光盘与（能长期播放的）唱片一样，在工厂里"压制"并带着预先录好的内容（如《莎士比亚全集》或电影《飘》的前 30 分钟部分）发行出去。

5. 本句中 offer promise that ... 直译是"呈现出希望，这种希望是激光盘将成为……"。全句可译为：磁光盘有希望使激光盘成为商业上可行的读/写存储技术。

6. 本句中 do not offer anywhere ... 译为"根本没有提供……"。句中 that users ... 修饰 reliability。全句可译为：目前磁光盘太贵且根本未达到用户对磁介质所期望的那种可靠性。

7. 本句中 forum "论坛"。in the category of "Standards" 意为"制定标准的那类（奖）"。PC 杂志奖分为几类，其中之一是制定各类标准。in recognition of ... 译为"因……"。全句可译为：DVD 论坛今天宣布，因该论坛成功制定了 DVD-ROM 规范而获得 1997 年 PC 杂志的（制定标准的）卓越技术奖。

8. 本句可译为：在对 DVD-ROM 技术的颁奖词中，PC 杂志把 DVD 称为"将代替 CD-ROM 作为 PC 内容传播主要手段的"载体形式。

9. 本句很长，其中 shipping in volume to 译为"大量运往"。the Award ... adds to the market's validation ... 译为"卓越技术奖增加了市场对（制定）标准过程成功的认可度"。后半句中的 recognized as ... 是分词短语，作为 see 的宾语补足语。

10. 本句中 by ... "通过……"，converge "会聚，使集中于一点"。全句可译为：DVD 论坛的任务是，在计算机工业和消费电子产品工业中，通过与具有最广泛代表性的制造商

与技术最终用户小组一起工作，确定从 CD 技术向 DVD 技术平滑过渡的途径。

注：本节包含两部分：一是介绍光盘种类，二是 DVD 的 1997 年获奖报道（作为报道类英文学习）。现在 DVD 已普及，光盘技术也得到进一步发展，但在发展中出现过 DVD 论坛（Forum）和 DVD 联盟（Alliance）两大阵营，它们分别推出 DVD-R，DVD-RW 和 DVD＋R，DVD＋RW 两类标准。现在的 DVD 驱动器对这两类光盘都能读/写。之后又出现 HD-DVD 和蓝光盘两种下一代数字光盘，结果蓝光盘胜出。

2.2　Display Devices

Resolution

A key characteristic of any display device is its *resolution*, or *sharpness* of the *screen image*. Most *monitors* form images by *lighting up* tiny dots on their screens called pixels (from the phrase "picture elements"). Because pixels are so finely packed, when viewed from a distance they appear to *blend* together to form continuous images.[1] The density of the pixels, called the dot *pitch*, is measured by the distance between the pixels in millimeters. Many displays in use today have a dot pitch of .26 or .28. A smaller dot pitch results in a better, *sharper* image.

The term resolution is also frequently used when referring to the number of pixels used to display images on the screen.[2] A display resolution of 640 by 480 — usually referred to as 640×480 — means that the screen consists of 640 columns by 480 rows of pixels. The higher the number of pixels displayed, the more data that can be displayed on the screen,[3] but everything displayed on the screen is displayed smaller. Higher resolutions also usually result in sharper images.

To *accommodate* the variety of screen sizes and users' personal preferences, most monitors allow the user to select from a few different resolutions, almost always 640×480 and 800×600; possibly also 1,024×768, 1,280×1,024, or more. At 800×600 resolution, the screen contains about half a million pixels. Very high-resolution monitors, such as those used for viewing digital X-rays and other specialized applications, can use more than 200 pixels per inch for a total of 9 million pixels or more on the screen.

Graphics Standards

Computer graphics standards specify such characteristics as the possible resolutions and number of colors that can be used. For example, the older VGA display (for video graphics array) uses a resolution of 640 by 480 pixels and allows a screen to display at most 256 colors. Most monitors today use the SVGA standard (for super VGA), which took over when Pentium computers and 17-inch monitors became popular. It allows for higher resolutions and true color (16.7 million colors).

The system's video card, sometimes called a graphics card, connects the display to the system unit and will support a particular graphics standard. In addition, the card usually contains memory (frequently called video RAM or VRAM), though some systems use some of the PC's regular RAM instead for video RAM. The amount of memory located on the card must be enough to support the number of colors and resolution desired—usually about 32 MB today.

On Windows computers, the number of colors and screen resolution can be changed using the Display option in the Control Panel.

CRT vs. *Flat-Panel* Displays

Monitors for most desktop systems *project* images on large *picture tube* displays similar to standard TV sets. This type of monitor is commonly called a CRT (cathode-ray tube) display. CRT technology uses an *electron* gun that is fired at the screen, so CRT monitors have to be very deep. Over the years, CRTs have become very inexpensive and have gained capabilities for excellent color output. However, they also are *bulky*, *fragile*, and consume a great deal of power.

An alternative type of display device forms images by manipulating charged chemicals or gases *sandwiched* between thin *panes* of glass instead of firing a bulky electron gun.[4] These much *slimmer* alternatives to CRTs are called flat-panel displays. Flat-panel displays take up little space, are lightweight, and require less power than CRT monitors. Because of these features, they are commonly found on portable computers and in a variety of consumer products, such as handheld video games, handheld television sets, and *electronic photo frames*.

Most flat-panel displays on ordinary computer systems use liquid *crystal* display (LCD) technology. LCD displays light up charged liquid crystals located between two sheets of material, and special color *filters* manipulate this light to draw the appropriate images on the screen. Until recently, LCD screens for desktop computers have been too costly for many consumers compared to CRTs. The price gap is getting smaller, however, and it is largely expected that LCDs will *overtake* CRTs on the desktop in the not-too-distant future. In addition to the size and portability advantages already named, desktop LCDs provide a sharper picture than CRTs and emit less *radiation*. Large flat-panel displays use a similar technology called gas *plasma*, which uses a layer of gas instead of liquid crystals.

Many flat-panel displays use either *active-matrix* or *passive*-matrix color-display technology. Active-matrix displays provide much sharper screen images than passive-matrix screens—especially when viewed from an angle other than directly in front of the display—but they provide this benefit at a higher cost.

Words and Expressions

resolution [ˌrezəˈluːʃən] *n.* 分辨率；清晰度；解决；解答
sharpness [ˈʃɑːpnis] *n.* 清晰度；锐度；锐利
screen [skriːn] *n.* 屏幕，银幕，显示屏
light up   使发光；点燃
monitor [ˈmɔnitə] *n.* 监视器，监控器

blend [blend] *vt.* 混合
pitch [pitʃ] *n.* 行距；节距；程度；斜度
sharp [ʃɑ:p] *adj.* 轮廓清晰的；锐利的；锋利的；线条分明的
accommodate [əˈkɔmədeit] *vt.* 使适应；提供；供应，供给
flat-panel display 平板显示器
project [ˈprɔdʒekt] *vt.* 投映，投射；放映；计划 *n.* 项目；方案；课题；工程
picture tube （电视机）显像管
electron [iˈlektrɔn] *n.* 电子
bulky [ˈbʌlki] *adj.* 庞大的；笨重的；体积大的
fragile [ˈfrædʒail] *adj.* 易碎的，脆的
sandwich [ˈsænwidʒ,-tʃ] *n.* 夹心面包，三明治 *vt.* 夹入中间
pane [pein] *n.* 长方块；显示窗格；窗格玻璃
slim [ˈslim] *adj.* 苗条的，纤细的；薄的
photo frame 相框 electronic photo frame 电子相框
crystal [ˈkristl] *adj.* 结晶状的 *n.* 晶体，水晶
filter [ˈfiltə] *n.* 滤波器，过滤器，滤光器
overtake [ˌəuvəˈteik] *vt.* 赶上，追上；压倒
radiation [ˌreidiˈeiʃən] *n.* 辐射，放射；放射线，放射物
plasma [ˈplæzmə] *n.* 等离子体；[解]血浆
active-matrix 主动矩阵
passive-matrix 被动矩阵

Abbreviation

CRT(Cathode-Ray Tube) 阴极射线管

Notes

1. 本句中 finely "很好，（时间或空间方面）几乎不留余地，细小地"。pack 意为"包装，打包，使挤在一起，挤满"。When viewed ... 是 "when＋过去分词" 结构，省略无人称代词和 be，即 when (they are) viewed ...。全句可译为：因为（这些）像素排得很紧，因此从远处看它们好像连在一起形成连续图像。

2. refer to "谈到，提到，涉及，参考，查阅"。when referring to ...，即 when (it is) referring to ...。全句可译为：术语分辨率也经常用来指屏幕上显示图像的像素数。下一句中的 refer to as 意为"称作，叫作"。

3. 本句是 "the＋比较级 ＋ the＋比较级" 结构，表示"越……就越……"。句子可译为：所显示的像素数越高，则能显示在屏幕上的数据就越多。

4. charge "充电，加电，导电，收费"。chemical "化学的，化学制品"。全句可译为：另一类显示设备通过操作夹在薄的玻璃显示窗格之间的带电化学物质或气体来形成图像，而不是通过激发庞大的电子枪。

2.3　3D Printing

3D printing or additive manufacturing (AM) refers to any of the various processes for printing a three-dimensional *object*. Primarily additive processes are used, in which successive layers of material are laid down under computer control.[1] These objects can be of almost any shape or geometry, and are produced from a 3D model or other electronic data source. A 3D printer is a type of industrial robot, but has recently been popularized by the low-cost open source RepRap 3D printer project.[2]

2.3.1　Manufacturing an Object with 3D Printer

The object to be manufactured is built up a layer at a time. A layer of powder is automatically deposited in the *model tray*. The print head then applies *resin* in the shape of the object.[3] The layer dries solid almost immediately. The model tray then move down the distance of a layer and another layer of powder is deposited in the position, in the model tray. The print head again applies resin in the shape of the object, *binding* it to the first layer. This sequence occurs one layer at a time until the object is complete, see Figure 2-1.

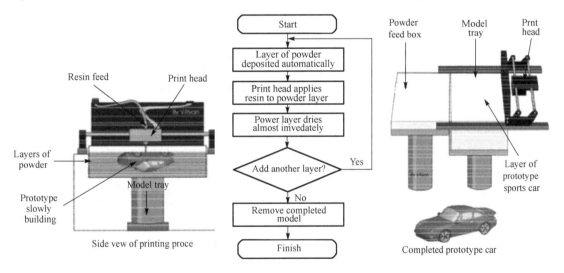

Figure 2-1　3D printing

2.3.2　General Principles

Modeling [4]

3D printable models may be created with a computer aided design package or via 3D scanner. The manual modeling process of preparing geometric data for 3D computer graphics is similar to plastic arts such as sculpting. 3D scanning is a process of analyzing and collecting digital data on the shape and appearance of a real object. Based on this data, three-dimensional models of the scanned object can then be produced.

Both manual and automatic creation of 3D printable models is difficult for average consumers. This is why several 3D printing marketplaces have emerged over the last years. Among the most popular are Shapeways, Thingiverse and MyMiniFactory.

Printing

Before printing a 3D model from an STL file, it must first be processed by a piece of software called a *"slicer"* which converts the model into a series of thin layers and produces a G-code file containing instructions tailored to a specific printer.[5] Several open source slicer programs exist, including Skeinforge, Slic3r, KISSlicer, and Cura.

The 3D printer follows the G-code instructions to lay down successive layers of liquid, powder, paper or sheet material to build the model from a series of *cross sections*[6]. These layers, which correspond to the virtual cross sections from the CAD model, are joined or automatically *fused* to create the final shape. The primary advantage of this technique is its ability to create almost any shape or geometric feature.

Printer resolution describes layer thickness and X-Y resolution in dots per inch (dpi) or micrometers (μm). Typical layer thickness is around 100 μm (250 dpi), although some machines such as the Objet Connex series and 3D Systems' ProJet series can print layers as thin as 16 μm (1,600 dpi). X-Y resolution is comparable to that of laser printers. The particles (3D dots) are around 50 to 100 μm (510 to 250 dpi) in diameter.

Construction of a model with *contemporary* methods can take anywhere from several hours to several days, depending on the method used and the size and complexity of the model. Additive systems can typically reduce this time to a few hours, although it varies widely depending on the type of machine used and the size and number of models being produced simultaneously.

Traditional techniques like *injection moulding* can be less expensive for manufacturing *polymer* products in high quantities, but additive manufacturing can be faster, more flexible and less expensive when producing relatively small quantities of parts.[7] 3D printers give designers and concept development teams the ability to produce parts and concept models using a desktop size printer.

Finishing

Though the printer-produced resolution is sufficient for many applications, printing a slightly *oversized* version of the desired object in standard resolution and then removing material with a higher-resolution *subtractive* process can achieve greater precision.[8]

Some additive manufacturing techniques are capable of using multiple materials in the course of constructing parts. Some are able to print in multiple colors and color combinations simultaneously. Some also utilize *supports* when building. Supports are removable or *dissolvable* upon completion of the print, and are used to support *overhanging* features during construction.

2.3.3 Applications

Industrial production roles within the *metalworking* industries achieved significant scale for the first time in the early 2010s.[9] Since the start of the 21st century there has been a large growth in the sales of AM machines, and their price has dropped substantially. According to Wohlers Associates, a *consultancy*, the market for 3D printers and services was worth ＄2.2 billion worldwide in 2012, up 29％ from 2011. There are many applications for AM technologies, including architecture, *construction*, industrial design, automotive, aerospace, military, engineering, dental and medical industries, *biotech* (human tissue replacement), fashion, *footwear*, jewelry, *eyewear*, education, geographic information systems, food, and many other fields.

In 2005, a rapidly expanding *hobbyist* and home-use market was established with the *inauguration* of the open-source RepRap and Fab@Home projects. Virtually all home-use 3D printers released to-date have their technical roots in the *on-going* RepRap Project and associated open-source software initiatives.[10] In distributed manufacturing, one study has found that 3D printing could become a mass market product enabling consumers to save money associated with purchasing common household objects. For example, instead of going to a store to buy an object made in a factory by injection molding (such as a measuring cup or a funnel), a person might instead print it at home from a downloaded 3D model.

Words and Expressions

object ['ɔbdʒikt] *n.* 实物；物体；对象；目标
model tray   模型盘
resin ['rezin] *n.* 树脂
bind [baind] *vt.* 使黏合，使结合
modeling ['mɔdliŋ] *vt.* 建模
model ['mɔdl] *vt.* 建模，制作模型；将……做成模型 *n.* 模型
slicer ['slaisə] *n.* 切片程序；切片机
cross section   横截面
fuse [fju:z] *vt.* 熔合，熔(化)；融合 *n.* 保险丝，熔丝
contemporary [kən'tempərəri] *adj.* 现代的，当代的
injection moulding   注射成型；注塑
polymer ['pɔlimə] *n.* 聚合物
oversized ['əuvə'saizd] *adj.* 过大的；超大型的
subtractive [səb'træktiv] *adj.* 去除的，减去的
support [sə'pɔ:t] *n.* 支持；支撑物，支柱，支座 *vt.* 支持；支撑
dissolvable [di'zɔlvəbl] *adj.* 可溶解的；可熔化的；可消散的
overhang ['əuvə'hæŋ] *vt.* 悬垂；突出；伸出
metalwork ['metəl,wə:k] *n.* 金属加工
consultancy [kən'sʌltənsi] *n.* 咨询公司；咨询机构；顾问工作
construction [kən'strʌkʃən] *n.* 建造业；建造；建筑物；结构
biotech ['baiəutek] *n.* 生物技术；生物工艺学
footwear ['futwɛə] *n.* [总称]鞋类
eyewear ['aiwɛə] *n.* [总称]眼镜；眼镜防护；护目镜

hobbyist [ˈhɔbiist] *n.* 癖好者，业余爱好者
inauguration [inˌɔgjəˈreiʃən] *n.* 开始；开展；就职
on-going [ˈɔnˌgəuiŋ] *adj.* 正在进行的；继续的，持续的

Abbreviation

AM(Additive Manufacture)　增材制造

Notes

1. process 有很多解释，如"过程，进程，处理，（制作）方法，步骤"等；successive layers 是指"连续（或相继）的各层"。全句可译为：主要使用增材制作，这种技术在计算机的控制下将材料一层一层地放置下去。
2. RepRap(replicating rapid prototyper)项目指英国倡议开发一种 3D 打印机，这种打印机能打印它自身大多数部件。由 RepRap 项目产生的所有设计都在自由软件许可条款下发布。
3. 本句中 apply 意思是"应用，涂，施"；in the shape of the object 中的 in 表示"方式"。全句可译为：然后打印头以实物的形状喷涂树脂。
4. 这里 modeling 是指"3D 建模"。在 3D 计算机图形学中，3D 建模是借助专用软件生成一个物体的 3D 表面的数学表示过程。这是 3D 打印的第一步。
5. STL(Stereo Lithography，立体平板印刷术)是一种文件格式，出自立体平板印刷术 CAD 软件，由 3D 系统创建，许多软件包都支持这种格式。它被广泛地用于快速成型和计算机辅助制造。STL 文件只描述 3D 物体的表面几何形状。G-code 是用得最广的数控程序设计语言名字，它主要用在计算机辅助制造领域，用来控制自动机床。全句可译为：在根据 STL 文件打印 3D 模型之前，首先必须由一个称为切片程序的软件来处理它，切片程序把该模型转换成一系列薄层，并产生一个 G-code 文件，该文件包含了特定打印机的指令。
6. 这里 model 的"模型"意思是"原型，样机"。3D 打印使用的材料可以是液体、粉末、纸或者薄板等，例如，打印战斗机的部件所用的材料是钛合金，打印机通过激光将钛熔化并一层层喷出所需的部件。句中 a series of cross sections（一系列横截面）就是上面讲的切片程序"切"出的一系列薄层（切片）。
7. 全句可译为：像注射成型等一些传统技术对制造大量聚合物产品可能花费较少，但是当生产相对小量部件时，增材制造可能更快、更灵活、更便宜。注：3D 打印也称为快速成型（原型）。
8. 本句后半句中 can achieve 的前面都是主语，后半句可译为：用标准分辨率打印出比实物大一点的同样物体，然后用更高分辨率的减成法去除（多余的）材料可以达到更高的精度。
9. scale"大小，规模"。全句可译为：金属加工行业的工业生产（角色）在 21 世纪 10 年代初首次达到显著规模。
10. 本句中 released to-date 修饰前面的 3D printer。全句可译为：实际上，至今发布的所有家用 3D 打印机的技术根源在不断发展的 RepRap 项目和相关的开源软件倡议。

注：3D打印已应用于建筑、工业设计、航天航空、军工、汽车业、牙科、医疗工业和生物技术等领域。例如，3D打印使用钛合金材料打印出飞机部件、人体骨骼；在医疗领域，3D打印机使用塑料材料打印出心脏模型；在美国，家用3D打印机使用融化的树脂、金属或者陶瓷等材料可打印出家用物品。本文简单介绍3D打印及其应用。

2.4　The External Interface: USB

The interface to a peripheral from an I/O *module* must be *tailored* to the nature and operation of the peripheral.[1] One major characteristic of the interface is whether it is *serial* or *parallel* (see Figure 2-2). In a parallel interface, there are multiple lines connecting the I/O module and the peripheral, and multiple bits are transferred *simultaneously*, just as all of the bits of a word are transferred simultaneously over the data bus. In *serial interface*, there is only one line used to transmit data, and bits must be transmitted one at a time. A parallel interface is commonly used for higher-speed peripherals, such as disk and tape. The serial interface is more common for printers and *terminals*.

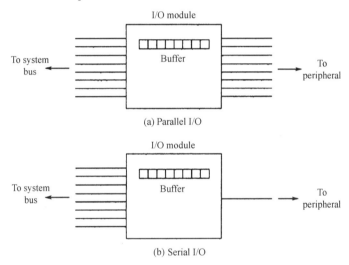

Figure 2-2　Parallel and serial I/O

What is USB?

Anyone who has been around computers for more than two or three years knows the problem that the Universal Serial Bus is trying to solve—in the past, connecting devices to computers has been a real headache!

- Printers connected to parallel printer *ports*, and most computers only came with one.
- *Modems* used the serial port, but so did some printers and a variety of *odd* things like *Palm Pilots*[2] and digital cameras. Most computers have at most two serial ports, and they are very slow in most cases.
- Devices that needed faster connections came with their own *cards*, which had to fit in a card slot inside the computer's *case*.[3] Unfortunately, the number of card slots is limited and you needed a Ph.D. to install the software for some of the cards.

The goal of USB is to end all of these headaches. The Universal Serial Bus gives you a single, standardized, easy-to-use way to connect up to **127 devices** to a computer.

Connecting a USB device to a computer is simple — you find the USB *connector* on the back or front of your machine and plug the USB connector into it. [4]

If it is a new device, the **operating system** *auto-detects* it and asks for the driver disk. If the device has already been installed, the computer *activates* it and starts talking to it. USB devices can be connected and disconnected at any time.

A USB cable has two wires for power (+5 volts and ground) and a *twisted pair of* wires to carry the data. *Low-power* devices (such as mice) can *draw* their power directly from the bus. High-power devices (such as printers) have their own power supplies and draw minimal power from the bus. Individual USB cables can run as long as 5 meters; with hubs, devices can be up to 30 meters (six cables' worth) away from the host.

Many USB devices come with their own *built-in* cable, and the *cable* has an "A" connection on it. If not, then the device has a socket on it that accepts a USB "B" connector. The USB standard uses **"A" and "B" connectors** to avoid confusion:

- "A" connectors head *"upstream"* toward the computer. [5]
- "B" connectors head *"downstream"* and connect to individual devices.

By using different connectors on the upstream and downstream end, it is impossible to ever get confused—if you connect any USB cable's "B" connector into a device, you know that it will work. Similarly, you can plug any "A" connector into any "A" socket and know that it will work.

USB 2.0

The standard for USB version 2.0 was released in April 2000 and serves as an *upgrade* for USB 1.1. USB 2.0 (**High-speed USB**) provides additional bandwidth for multimedia and storage applications and has a data transmission speed 40 times faster than USB 1.1. To allow a smooth *transition* for both consumers and manufacturers, USB 2.0 has full forward and backward *compatibility*[6] with original USB devices and works with cables and connectors made for original USB, too.

Supporting three speed modes (1.5, 12 and 480 megabits per second), USB 2.0 supports low-bandwidth devices such as **keyboards** and **mice**, as well as high-bandwidth ones like high-resolution **Webcams**[7], **scanners**, **printers** and high-capacity **storage systems**. The *deployment* of USB 2.0 has allowed PC industry leaders to *forge ahead* with the *development* of next-generation PC peripherals to *complement* existing high-performance PCs. The transmission speed of USB 2.0 also facilitates the development of next-generation PCs and applications. In addition to improving functionality and encouraging *innovation*, USB 2.0 increases the productivity of user applications and allows the user to run multiple PC applications at once[8] or several high-performance peripherals simultaneously.

Data Transfer

When the host *powers* up, it queries all of the devices connected to the bus and assigns each one an address. This process is called ***enumeration*** — devices are also enumerated when they connect to the bus. The host also finds out from each device what type of data transfer it wishes to perform:

- **Interrupt** — A device like a mouse or a keyboard, which will be sending very little data, would choose the *interrupt* mode.
- **Bulk** — A device like a printer, which receives data in one big packet, uses the *bulk* transfer mode. A block of data is sent to the printer (in 64-byte chunks) and verified to make sure it is correct.
- **Isochronous** — A streaming device (such as speakers) uses the *isochronous* mode. Data streams between the device and the host in real-time, and there is no error correction. The host can also send commands or query *parameters* with **control packets.**

Words and Expressions

module ['mɔdju:l] *n.* 模块；模；模数
tailor ['teilə] *vt.* 定制，专门制作；使适应，使合适；修改；剪裁   *n.* 裁缝
serial ['siəriəl] *adj.* 串行的
parallel ['pærəlel] *adj.* 并行的
simultaneously [ˌsiməl'teiniəsli] *adv.* 同时；同时发生
terminal ['tə:minl] *n.* 终端；[计] 终结符   *adj.* 终端的
port [pɔ:t] *n.* 端口；港口   *vt.* [计] 移植
modem ['məudem] *n.* [计] 调制解调器
odd [ɔd] *adj.* 奇数的；单只的；额外的；零散的
Palm Pilot   掌上计算机，掌上通；面向手持设备（Palm 公司的产品名）
card [kɑ:d] *n.* 卡；插件
case [keis] *n.* 机箱；外壳；案例
connector [kə'nektə] *n.* 连接器；插头；插座
auto-detect   自动检测
activate ['æktiveit] *vt.* 激活
twisted pair   双扭线，双绞线
low-power   低功率，小功率
draw [drɔ:] *vt.* 获得；提取；拉；拖；画
built-in [ˌbilt'in]   内置的，嵌入的
cable ['keibl] *n.* 电缆；有线电视
upstream [ʌp'stri:m] *adv.* 向上游，上行   *adj.* 向上游的；逆流而上的
downstream ['daunstri:m] *adv.* 下游地，下行；顺流而下   *adj.* 下游的；顺流的
upgrade [ʌp'greid] *n.* 升级；上升   *vt.* 使升级；提升   *adj.* 向上的   *adv.* 往上
transition [træn'ziʃən] *n.* 过渡，转移；转换
compatibility [kəmˌpætə'biləti] *n.* 兼容性
Webcam   网络摄像头，网络摄像机（Web camera 的缩写）

deployment [di'plɔimənt] *n.* 使用，采用；部署
forge [fɔ:dʒ] *vi.* 稳步前进　*v.* 锻造　forge ahead　突然加速前进
development [di'veləpmənt] *n.* 开发
complement ['kɔmplimənt] *n.* 补充，补足
innovation [ˌinə'veiʃən] *n.* 革新
power up　开机，加电
enumeration [iˌnju:mə'reiʃən] *n.* 计数；列举，枚举
interrupt [intə'rʌpt] *n.* 中断　*vt.* 中断；打断　*vi.* 打断；打扰
bulk [bʌlk] *n.* 大块；大批，成批
isochronous [ai'sɔkrənəs] *adj.* 等时的；等步的
parameter [pə'ræmitə(r)] *n.* 参数，参量

Abbreviation

USB (Universal Serial Bus)　通用串行总线

Notes

1. tailor 用于软件，常指把一种软件进行必要的修改，使之更好地适用于某一特定应用，或为某一应用"定制"软件。对硬件也类似。本句可译为：从 I/O 模块到外设的接口都必须适应外设的性质和操作。
2. Palm Pilot 是 Palm 公司的 PDA（Personal Digital Assistant，个人数字助理）产品。PDA 以前指记事本等设备，后来其范围和类型日益扩大，有些人把 Palm PC（掌上PC）、Handheld PC（手持式PC）、Palm-Size PC 等设备统称为 PDA。
3. 本句中 fit in 指"符合，使适合"，这里可译为"插入……相符合的……"。全句可译为：需要较快连接的设备有它们自己的卡，这些卡必须插入计算机机箱内相符合的卡槽口。
4. 本句中前一个 connector 译为"插座"，后一个译为"插头"，或均译为"插口"。
5. 本句中 head 指"对着"。本句可译为："A"插口上行连接计算机。下一句可译为："B"插口下行连接各个设备。
6. forward and backward compatibility 译为"向前和向后兼容性"。
7. high-resolution Webcams "高分辨率网络摄像头（机）"。
8. at once "同时，立刻"。
注：本节介绍串行接口 USB，它是外部接口技术。USB 接口适用范围宽，可接低速、高速设备。USB 使用网络 Hub 连接技术。
接口技术近期发展很快，特别是串行技术，它为用户带来的好处有，串行接口体积小，所用频率可比并行接口高得多。

Terms

Blu-ray Disc（蓝光盘）
　　Blu-ray Disc (BD) is a digital optical disc data storage format designed to supersede the

DVD format, in that it is capable of storing high-definition video resolution (1080 p). The plastic disc is 120 mm in diameter and 1.2 mm thick, the same size as DVDs and CDs. Conventional Blu-ray Discs contain 25 GB per layer, with dual layer discs (50 GB) being the industry standard for feature-length video discs. Triple layer discs (100 GB) and quadruple layers (128 GB) are available for re-writer drives. The name Blu-ray Disc refers to the blue laser used to read the disc, which allows information to be stored at a greater density than is possible with the longer-wavelength red laser used for DVDs. The main application of Blu-ray Discs is as a medium for video material such as feature films and physical distribution of video games.

Sony unveiled the first Blu-ray Disc prototypes in October 2000. During the high definition optical disc format war, Blu-ray Disc competed with the HD DVD format. Toshiba, the main company that supported HD DVD, conceded in February 2008, releasing its own Blu-ray Disc player in late 2009.

TFT-LCD（薄膜晶体管液晶显示器）

TFT-LCD is a variant of LCD which is now the dominant technology used for computer monitors. TFT-LCD is an abbreviation of Thin Film Transistor liquid crystal display.

Smart card（智能卡）

A smart card is a credit-card-size piece of plastic that contains some type of computer circuitry, typically including a processor, memory, and storage. This circuitry can store electronic data and programs, but the storage capacity varies widely — usually from a few kilobytes to a few megabytes. Essentially, a smart card is a wafer-thin computer. Many smart cards used today store a prepaid amount of money for retail purchases via vending machines, gas stations, fast-food restaurants, toll booths, and online purchases using a smart cell phone or smart card reader. Every time the card is used, the available amount of money is reduced. To refill a reusable card, a kiosk that accepts cash, credit, or debit cards is usually used. Smart cards can also be loaded with identification data for accessing facilities or computer networks, as well as for housing an individual's medical history and insurance information for fast treatment and hospital admission in an emergency.

CES（消费者电子展）

International CES, more commonly known as the Consumer Electronics Show (CES), is an internationally renowned electronics and technology trade show, attracting major companies and industry professionals worldwide. The annual show is held each January at the Las Vegas Convention Center in Las Vegas, Nevada, United States. Not open to the public, the Consumer Electronics Association-sponsored show typically hosts previews of products and new product announcements. CES rose to prominence after a rival show, COMDEX, was canceled.

Sector

Formatting divides the disk surface into pie-shaped **sectors**, and thereby prepares it for use with a particular operating system.

Cluster

On many PC systems, the part of a track that crosses a fixed number of contiguous sectors—anywhere from two to eight sectors is typical—forms a unit called a **cluster.**

Cylinder

A disk **cylinder** is the collection of one particular track on each disk surface, such as the first track or the tenth track on each disk surface.

Disk cache

When disk caching is being used, during any disk access the computer system also fetches program or data contents located in neighboring disk areas (such as the entire track) and transports them to a dedicated part of RAM known as a **disk cache.**

Partitioning

Partitioning a hard drive enable you to logically divide the physical capacity of a single drive into separate areas called *partitions*. You can then treat each of the partition as an independent disk drive, such as a C drive and a D drive.

FAT

The computer system automatically maintains a disk's *file directory or file allocation table* (**FAT**), which keeps track of the files stored on the disk. This directory shows the name of each file, its size, and the cluster at which it begins, so the computer can retrieve the file when requested.

FAT32

Windows computers using the **FAT32** file system are much more efficient than those using the original FAT system, since FAT32 system allow cluster sizes to be as small as 4 KB each.

SSD

A solid-state disk or drive (SSD), also called a flash drive, is the next generation hard disk. Though the architecture of an SSD does not employ disks at all, the name is carried over from standard hard disks. In reality an SSD utilizes a special kind of memory chip with erasable, writeable cells that can hold data even when powered off. It might help to think of an SSD as the larger cousin of the memory stick.

Like standard disks, an SSD utilizes a special area for cache memory. Cache memory serves the function of increasing processing speeds by holding data that is needed repeatedly.

With the data close at hand in the cache, it does not need to be fetched from the main storage area each time.

IDE

Acronym for Integrated Device Electronics. A type of disk-drive interface in which the controller electronics reside on the drive itself, eliminating the need for a separate adapter card.

SCSI

Acronym for small computer system interface, a standard high-speed parallel interface defined by the X3T9.2 committee of the American National Standards Institute (ANSI). A SCSI interface is used to connect microcomputers to SCSI peripheral devices, such as many hard disks and printers, and to other computers and local area networks. Compare enhanced small device interface, IDE.

SAN and NAS

Both Storage Area Networks (SANs) and Network Attached Storage (NAS) provide networked storage solutions. A NAS is a single storage device that operates on data files, while a SAN is a local network of multiple devices that operate on disk blocks. A SAN commonly utilizes Fiber Channel interconnects. A NAS typically makes Ethernet and TCP/IP connections.

SATA

SATA or Serial ATA (Advanced Technology Attachment) is the next generation drive interface, following the traditional Parallel ATA which is known also as IDE.

LED (Light Emitting Diode, 发光二极管)

An LED is a small semiconductor device which emits light when an electric current passes through it. LEDs are energy saving and have a long service life. They are good for displaying numerical images because they can be relatively small, and they do not burn out.

All-In-One(一体机)

Multifunctional, All-In-One (AIO), or Multifunction Printer/Product (MFP), is an office machine which incorporates the functionality of multiple devices in one. A typical MFP may act as a combination of some or all of the following devices: Printer, Scanner, Photocopier, Fax, E-mail.

OLED (Organic Light-Emitting Diode)

OLED is a light-emitting diode in which the emissive electroluminescent layer is a film of organic compound that emits light in response to an electric current. This organic layer is situated between two electrodes; typically, at least one of these electrodes is transparent. OLEDs are used to create digital displays in devices such as television screens, computer monitors, smartphones and PDAs. A major area of research is the development of white OLED devices for use in solid-state lighting applications.

There are two main families of OLED: those based on small molecules and those employing polymers. Adding mobile ions to an OLED creates a light-emitting electrochemical cell (LEC) which has a slightly different mode of operation. An OLED display can be driven with a passive-matrix (PMOLED) or active-matrix (AMOLED) control scheme. In the PMOLED scheme, each row (and line) in the display is controlled sequentially, one by one, whereas AMOLED control uses a thin-film transistor backplane to directly access and switch each individual pixel on or off, allowing for higher resolution and larger display sizes.

An OLED display works without a backlight because it emits visible light. Thus, it can display deep black levels and can be thinner and lighter than a liquid crystal display (LCD). In low ambient light conditions (such as a dark room), an OLED screen can achieve a higher contrast ratio than an LCD, regardless of whether the LCD uses cold cathode fluorescent lamps or an LED backlight.

Exercises

1. Translate the text of Lesson 2.1.2 into Chinese.
2. Topics for oral workshop.
   - Discuss the categories of secondary storage: tape, disk, optical laser disk, mobile storage; and their different applications.
   - Describe the difference between primary storage and secondary storage in terms of characteristic, function, speed, size.
   - Talk about external interfaces, especially USB; compare the serial interface and parallel interface.
3. Translate the following into English.

存储硬盘提供由计算机检索的信息和程序的永久存储。硬盘驱动器把信息存储在嵌(embed)在磁盘上的磁微粒(particle)中。通常，计算机的永久部分——硬盘驱动器能够存储大量的信息，并且能很快地取回那些信息。

虽然固定硬盘驱动器系统比光盘提供更快的存取并且有更高的存储容量，但是光盘系统使用的是可移动介质——一个明显的好处。光盘——主要是CD和DVD——比可移动硬盘驱动器系统用得更广泛，它们是现今软件交付的标准，同样一般用来存放高容量音乐和视频文件。也存在可用于家庭音响和家庭影院使用的既是CD也是DVD驱动器的版本。光盘常被称作压缩盘。

CD和DVD由CD和DVD驱动器读取。CD或DVD驱动器的速度被定速为24倍速、32倍速、36倍速等。这些标记描述了该驱动器比起第一个版本驱动器快多少。例如，一个36倍速驱动器的速度是最初被制造的基线驱动器的36倍。大多数光盘仅在一面上印刷标题和其他文字，并且将印刷面向上(facing up)插入驱动器中。插入这样的CD或DVD时，注意不要将其弄脏、留有手印、擦刮，或附有其他可能阻碍盘的表面光反射性的任何东西。

4. Listen to the video "Computer tour" and write down the third and fourth paragraphs.

Unit 3 Programming and Programming Languages

3.1 Computer Programming

Computer *programming* (often shortened to **programming** or *coding*) is the process of designing, writing, and *debugging* the source code of computer programs. This source code is written in a programming language. The purpose of programming is to create a program that exhibits a certain desired behavior.[1] The process of writing source code often requires expertise in many different subjects, including knowledge of the application *domain*, specialized algorithms and formal logic.

Within *software engineering*, programming (the implementation) is regarded as one phase in a software development process. Whatever the approach to software development may be, the final program must satisfy some fundamental properties.[2] The following properties are among the most relevant:

• Efficiency/performance: the amount of system resources a program consumes (processor time, memory space, slow devices such as disks, network *bandwidth* and to some extent even user interaction): the less, the better. This also includes correct *disposal* of some resources, such as cleaning up temporary files and lack of *memory leaks*.

• Reliability: how often the results of a program are correct. This depends on conceptual correctness of *algorithms*, and minimization of programming mistakes, such as mistakes in resource management (e.g., *buffer overflows* and *race conditions*) and logic errors (such as division by zero or off-by-one error[3]).

• *Robustness*: how well a program *anticipates* problems not due to programmer error. This includes situations such as incorrect, inappropriate or *corrupt* data, unavailability of needed resources such as memory, operating system services and network connections, and user error.

• Usability: the *ergonomics*[4] of a program: the ease with which a person can use the program for its intended purpose, or in some cases even unanticipated purposes. Such issues can make or break its success even regardless of other issues. This involves a wide range of textual, graphical and sometimes hardware elements that improve the clarity, intuitiveness, *cohesiveness* and completeness of a program's user interface.

• *Portability*: the range of computer hardware and operating system platforms on which the source code of a program can be *compiled*/interpreted and run. This depends on differences in the programming facilities provided by the different platforms, including hardware and operating system resources, expected behavior of the hardware and operating system, and availability of platform specific compilers (and sometimes *libraries*) for the language of the source code.[5]

- Maintainability: the ease with which a program can be modified by its present or future developers in order to make improvements or customizations, fix *bugs* and security holes, or adapt it to new environments. Good practices during initial development make the difference in this regard. This quality may not be directly apparent to the end user but it can significantly affect the *fate* of a program over the long term.

In computer programming, readability of the source code refers to the ease with which a human reader can comprehend the purpose, control flow, and operation of source code. It affects the aspects of quality above, including portability, usability and most importantly maintainability. Readability is important because programmers spend the majority of their time reading, trying to understand and modifying existing source code, rather than writing new source code. Unreadable code often leads to bugs, inefficiencies, and duplicated code. A study found that a few simple readability transformations made code shorter and *drastically* reduced the time to understand it.

Following a consistent programming style often helps readability. However, readability is more than just programming style. Many factors contribute to readability. Some of these factors include:

- Different *indentation* styles (whitespace)
- Comments
- Decomposition
- *Naming conventions* for objects (such as *variables*, classes, procedures, etc)

The academic field and the engineering practice of computer programming are both largely concerned with discovering and implementing the most efficient algorithms for a given class of problem. For this purpose, algorithms are classified into orders using so-called *Big O notation*, $O(n)$, which expresses resource use, such as execution time or memory consumption, in terms of the size of an input. Expert programmers are familiar with a variety of well-established algorithms and their respective complexities and use this knowledge to choose algorithms that are best suited to the circumstances.

Different programming languages support different styles of programming (called programming *paradigms*). The choice of language used [6] is subject to many considerations. Ideally, the programming language best suited for the task at hand will be selected.

It is very difficult to determine what are the most popular of modern programming languages. Some languages are very popular for particular kinds of applications (e.g., COBOL is still strong in the corporate data center, often on large mainframes, FORTRAN in engineering applications, *scripting languages* in web development, and C in *embedded applications*), while some languages are regularly used to write many different kinds of applications. Also many applications use a mix of several languages in their construction and use.

Many computer languages provide a mechanism to *call functions* provided by libraries such as in .dlls. Provided the functions in a library follow the appropriate run time conventions (e.g., method of passing arguments), then these functions may be written in any other language. [7]

Debugging is a very important task in the software development process, because an incorrect program can have significant consequences for its users. Some languages are more prone to some kinds of faults because their *specification* does not require compilers to perform as much checking as other languages. Use of a static analysis tool can help detect some possible problems.

Words and Expressions

programming ['prəugræmiŋ] *n.* 程序设计；规划；设计
code [kəud] *vt.* 编码；将……译成电码 *n.* 代码；编码
debug [ˌdiː'bʌg] *vt.* 调试，排错；排除故障
domain [dəu'mein] *n.* （活动、学问等的）范围，领域；[数]域，定义域；领土；领地
software engineering 软件工程
bandwidth ['bændwidθ] *n.* 带宽
disposal [dis'pəuzəl] *n.* 处理，处置
leak [liːk] *vt.* 使泄漏；使渗漏 *vi.* 漏；渗；泄漏出去 *n.* 泄漏；漏洞
memory leak 内存泄漏
algorithm ['ælgəriðəm] *n.* 算法
buffer overflow 缓冲器溢出
race condition 竞争条件
robustness [rəu'bʌstnis] *n.* 健壮性；坚固性；鲁棒性
anticipate [æn'tisipeit] *vt.* 预料；预感；期望
corrupt [kə'rʌpt] *adj.* （语言、版本等）讹用的；多讹误的；腐烂的；腐败的 *vt.* 讹用（语、词等）
ergonomics [ˌəːgə'nɔmiks] *n.* 人类工程学，人体工程学；工效学
cohesiveness [kəu'hiːsivnis] *n.* 凝聚性；内聚性；黏结性
portability [ˌpɔːtə'biliti] *n.* 可移植性；可携带性；轻便
compile [kəm'pail] *vt.* 编译；编辑；编制
library ['laibrəri] *n.* （程序）库；图书馆
bug [bʌg] *n.* 错误；（机器等）故障
fate [feit] *n.* 命运
drastically ['dræstikəli] *adv.* 大大地；彻底地；激烈地
indentation [ˌinden'teiʃən] *n.* 凹进，缩格；凹口
naming convention 命名约定
variable ['vɛəriəbl] *n.* 变量；可变因素 *adj.* 可变的，变化的
Big O notation 大 O 表示法
paradigm ['pærədaim] *n.* 范式；范例；示例
scripting language 脚本语言
embedded application 嵌入式应用
call [kɔːl] *vt.* 调用
function ['fʌŋkʃən] *n.* 函数；功能
specification [ˌspesifi'keiʃən] *n.* 规格说明；详细的计划书；说明书

Notes

1. 本句中，不定式 to create 引导不定式短语，作表语。关系代词 that 引导限制性定语从句，修饰其前的 program。过去分词 desired 作定语，修饰其后的 behavior。全句可译为：程序设计的目的是创建一个展示某种所期望行为的程序。
2. whatever 是连接形容词，这里含意为"不管什么样的"。全句可译为：不论软件开发的方法是什么，最终的程序必须满足某些基本性质。
3. off-by-one error(OBOE)可译为："差一次离开(循环)错误"，是一种逻辑错误，往往在计算机程序运行时，由于对循环控制变量的初值与终限未设置好，而使重复次数比预期的多一次或少一次结束循环。
4. ergonomics 意为"人类工程学"。人类工程学又称人体工程学，是第二次世界大战后发展起来的一门新学科。它以人机关系为研究对象，以实测、统计、分析为基本研究方法。从室内设计的角度来说，人体工程学的主要功用在于通过对生理和心理的正确认识，使室内环境因素适应人类生活活动的需要，进而达到提高室内环境质量的目标。程序的人类工程学可以仿此来理解。
5. 本句中的 provided 是过去分词，引导分词短语，修饰前面的 facilities；其后的 including 引导分词短语，作 facilities 的同位语，说明 facilities 是哪些。注意本句中的过去分词和现在分词的用法。
6. 此 used 是过去分词，作定语，意为"所使用的"。下一句中的分词短语 best suited for the task at hand 作定语，修饰前面的 language，意为"最适合于手头上任务的"。
7. 本句中 provided 是连接词，意为"以……为条件，如果"。passing 引导动名词短语 passing arguments，作介词 of 的宾语，含意为"传递诸变元(的)"。全句可译为：如果在(程序)库中的诸函数都遵循一些适当的运行时刻约定(如传递变元方法)，那么，这些函数可以用其他任何一种语言来写。

3.2 C++ and Object-Oriented Programming

Object-oriented programming concepts cross language boundaries.[1] Microsoft Quick Pascal, for example, was one of the first languages to allow the use of objects. What does C++ have that makes it a suitable language for *developing* object-oriented programs?[2] The answer is, as previously mentioned, the *class* data type. It is C++'s class type, built upon C's struct type, that gives the language the ability to build objects. Also, C++ brings several additional features to object-oriented programming not included in other languages that simply make use of objects.[3] C++'s advantages include *strong typing*, operator *overloading*, and less emphasis on the *preprocessor*. It is true that you can do object-oriented programming with other products and in other languages, but with C++ the benefits are outstanding. This is a language that was designed, not *retrofitted*, for object-oriented programming.

Object-oriented programming is a programming technique that allows you to view concepts as a variety of objects. By using objects, you can represent the tasks that are to be performed, their interaction, and any given conditions that must be observed. A data structure often forms the basis of an object; thus, in C or C++, the struct type can form an elementary object. Communicating with objects can be done through the use of *messages*, as mentioned earlier. Using messages is similar to *calling* a *function* in a *procedure*-oriented program. When an object receives a message, methods contained within the object respond. Methods are similar to the functions of procedure-oriented programming. However, methods are part of an object.

The C++ class is an extension of the C and C++ struct type and forms the required *abstract data type* for object-oriented programming. The class can contain closely related *items* that share *attributes*. Stated more formally[4], an object is simply an *instance* of a class.

Ultimately, there should emerge class libraries containing many object types. You could use instances of those object types to piece together[5] program *code*.

Before you examine these terms in closer detail, it is a good idea to become familiar with several additional concepts that relate to C++ and object-oriented programming, as described in the next few sections.

Encapsulation

Encapsulation refers to the way each object combines its member data and member functions (methods) into a single structure. Figure 3-1 illustrates how you can combine data *fields* and methods to build an object.

Data fields	Methods
Data	Member function
	Member function
Data	Member function
Data	Member function
	Member function
Data	Member function

Figure 3-1 Data fields and methods combined to build an object

Typically, an object's description is part of a C++ class and includes a description of the object's internal structure, how the object relates with other objects, and some form of protection that isolates the functional details of the object from outside the class. The C++ class structure does all of this.

In a C++ class, you control functional details of the object by using private, public, and/or protected *descriptors*. In object-oriented programming, the public section is typically used for the *interface information* (methods) that makes the class reusable across applications. If data or methods are contained in the public section, they are available outside the class. The private section of a class limits the availability of data or methods to the class itself. A protected section containing data or methods is limited to the class and any derived *subclasses*.

Class Hierarchy

The C++ class actually serves as a *template* or pattern for creating objects. The objects formed from the class description are instances of the class. It is possible to develop a class *hierarchy where there is a parent class and several child classes*. In C++, the basis for doing this revolves around *derived classes*. Parent classes represent more generalized tasks, while derived child classes are given specific tasks to perform. For example, the Lincoln class discussed earlier might contain data and methods common to the entire Lincoln line, such as engines, *instrumentation*, batteries, braking ability, and handling.[6] Child classes derived from the parent, such as Town Car, Mark VIII, and Continental, could contain items specific to the class. For example, the 1995 Continental was the only car in the line with an active suspension system.

Inheritance

Inheritance in object-oriented programming allows a class to inherit properties from a class of objects. The parent class serves as a pattern for the derived class and can be altered in several ways. (In the next Lesson you will learn that member functions can be overloaded, new member functions can be added, and member access *privileges* can be changed.) If an object inherits its attributes from a single parent, it is called single inheritance. If an object inherits its attributes from multiple parents, it is called multiple inheritance. Inheritance is an important concept since it allows reuse of a class definition without requiring major code changes.[7] Inheritance encourages the reuse of code since child classes are extensions of parent classes.

Polymorphism

Another important object-oriented concept that relates to the class hierarchy is that common messages can be sent to the parent class objects and all derived subclass objects. In formal terms, this is called *polymorphism*.

Polymorphism allows each subclass object to respond to the message *format* in a manner appropriate to its definition. Imagine a class hierarchy for gathering data. The parent class might be responsible for gathering the name, social security number, occupation, and number of years of employment for an individual. You could then use child classes to decide what additional information would be added based on occupation. In one case a supervisory position might include yearly salary, while in another case a sales position might include an *hourly rate* and *commission* information. Thus, the parent class gathers general information common to all child classes while the child classes gather additional information relating to specific job descriptions. Polymorphism allows a common data-gathering message to be sent to each class. Both the parent and child classes respond in an appropriate manner to the message. Polymorphism encourages *extendability* of existing code.

Virtual Functions

Polymorphism gives objects the ability to respond to messages from *routines* when the object's exact type is not known. In C++ this ability is a result of *late binding*. With late binding, the addresses are determined dynamically at run time, rather than statically at *compile* time, as in traditional compiled[8] languages. This static (fixed) method is often called *early binding*. Function names are replaced with memory addresses. You accomplish late binding by using *virtual functions*. Virtual functions are defined in the parent class when subsequent derived classes will overload the function by redefining the function's implementation. When you use virtual functions, messages are passed as a *pointer* that points to the object instead of directly to the object.[9]

Virtual functions utilize a table for address information. The table is initialized at run time by using a *constructor*. A constructor is invoked whenever an object of its class is created. The job of the constructor here is to link the virtual function with the table of address information. During the compile operation, the address of the virtual function is not known; rather, it is given the position in the table (determined at run time) of addresses that will contain the address for the function.[10]

Words and Expressions

object-oriented *adj.* 面向对象的
develop [di'veləp] *vt.* 开发；发展
class [klɑːs] *n.* 类
strong typing 强类型
overload ['əuvə'ləud] *vt.* 重载；使超载 *n.* 重载；超载
preprocessor [priː'prɔsesə] *n.* 预处理器，预处理程序
retrofit ['retrəˌfit] *vt.* （对……）翻新改进 *n.* 式样翻新，花样翻新
message ['mesidʒ] *n.* 消息，信息；文电，报文
call [kɔːl] *vt.* 调用；呼叫
function ['fʌŋkʃən] *n.* 函数；功能
procedure [prə'siːdʒə] *n.* 过程；手续
abstract data type 抽象数据类型
item ['aitem, 'aitəm] *n.* 项，条；项目；条款；(新闻等的)一条，一则
attribute [ə'tribjuː(ː)t] *n.* 属性；品质；特征，表征
instance ['instəns] *n.* 实例
code [kəud] *n.* 代码；编码 *vt.* 编码
encapsulation [inˌkæpsjuː'leiʃən] *n.* 封装；用胶囊包
field [fiːld] *n.* 域，领域；字段
descriptor [dis'kriptə] *n.* 描述符
interface ['intə(ː)ˌfeis] *n.* 界面，接口
information [ˌinfə'meiʃən] *n.* 信息，情报
subclass ['sʌbklɑːs] *n.* 子类
template ['templit] *n.* (=templet) 模板
derived class 派生类
instrumentation [ˌinstrumen'teiʃən] *n.* 检测仪表，测试设备
inheritance [in'heritəns] *n.* 继承；遗传；遗产
privilege ['privilidʒ] *n.* 特权 *vt.* 给予……特权
polymorphism [ˌpɔli'mɔːfizəm] *n.* 多态性，多态现象

format ['fɔːmæt] *n.* 格式　*vt.* 安排……的格局(或规格)；格式化(磁盘)
hourly rate　小时工资
commission [kə'miʃən] *n.* 佣金；委任；委托
extendability [ikˌstendə'biliti] *n.* 可扩充性
virtual function　虚函数
routine [ruː'tiːn] *n.* 例行程序　*adj.* 日常的，例行的，常规的
late binding　迟绑定；后期约束；迟联编
compile [kəm'pail] *vt.* 编译；编辑；编制
early binding　早绑定；早期约束
pointer ['pɔintə] *n.* 指针，指引元，指示器
constructor [kən'strʌktə] *n.* 构造符，构造器

Notes

1. cross 含意为"穿过,跨越",此处意指无边界限制。全句可译为：一些面向对象的程序设计概念在语言间渗透。
2. that 引导的限制性定语从句修饰 what,这里 a suitable language for … 又是 it 的补足语。全句可译为：C++ 有什么(特性)使得它成为一种适合于开发面向对象程序的语言？
3. included 引导的分词短语作 features 的定语,其后的 that 引导的限制性定语从句修饰 other languages。
4. stated more formally 是独立成分,相当于句子状语。
5. piece together 意为"拼合"。
6. 句中 common 引导的形容词短语作定语,修饰前面的 data and methods。Lincoln line 译为"林肯系列"。全句可译为：例如,早先讨论的林肯类也许包含整个林肯系列公共的数据和方法,诸如引擎、检测仪表、电池、制动能力和操纵。
7. 全句可译为：继承是一个重要概念,因为它使得无须对(程序)代码做大的改变就能重用类定义。
8. compiled 是过去分词作定语,意为"编译型的"。
9. 全句可译为：当使用虚函数时,消息不是直接传给对象,而是作为指向对象的指针传递。
10. 全句可译为：在编译期间虚函数的地址是未知的；更确切地说,给出的是(在运行时刻确定的)地址表中的位置,该位置将包含该函数的(入口)地址。

Terms

3GL

　　Short for third-generation language. A programming language one step above assembly language, characterized by being readable by humans. Some examples are C, Pascal, and BASIC. Also called high-level language (HLL). Compare 4GL, assembly language.

4GL

　　Short for fourth-generation language. This name is given to a group of languages that allow users to specify what the output should be without describing all the details of how the data should be manipulated to produce that result.

Procedural language

A programming language in which the basic programming element is the procedure (a named sequence of statements, such as a routine, subroutine, or function). The most widely used high-level languages (C, Pascal, BASIC, FORTRAN, COBOL, Ada) are all procedural languages. Compare nonprocedural language.

Logic programming

A style of programming, best exemplified by Prolog, in which a program consists of facts and relationships from which the programming language is expected to draw conclusions.

Functional programming

A style of programming in which all facilities are provided as functions (subroutines), usually without side effects. Pure functional programming languages lack a traditional assignment statement; assignment is usually implemented by copy and modify operations. Functional programming is thought to offer advantages for parallel processing computers.

Python

Python is a widely used general-purpose, very-high-level dynamic object-oriented programming language. Python is sometimes called a scripting language or a glue language. This means Python is used often to run a series of commands as a script or used to create links between two other technologies as a glue. Python is interpreted, meaning it is executed one line at a time. This makes the program flow easy to understand. Also it is semantically dynamic which allows the syntax to be less restrictive and less formal when using declarations and variables types. These characteristics add to Python's ease-of-use for basic programming tasks. Python is increasingly used for science, data analysis, and engineering. The scientific community and ecosystem have been growing fast into the ultimate environment for students, scientists/engineers, and companies that build technological software.

3.3 Introduction to Java

Java is designed to meet the challenges of application development in the context of *heterogeneous*, network-wide distributed environments. *Paramount* among these challenges is secure *delivery* of applications that consume the minimum of system resources, can run on any hardware and software *platform*, and can be extended dynamically.

Java originated as part of a research *project* to develop advanced software for a wide variety of network devices and *embedded systems*[1]. The goal was to develop a small, reliable, *portable*, *distributed*, *real-time* operating platform. When the project started, C++

was the language of choice[2]. But over time the difficulties encountered with C++ grew to the point where the problems could best be addressed by creating an entirely new language platform. Design and architecture decisions drew from a variety of languages such as Eiffel, SmallTalk, Objective C, and Cedar/Mesa. The result is a language platform that has proven ideal for developing secure, distributed, network based *end-user* applications in environments ranging from network-embedded devices to the World-Wide Web and the *desktop*.

The design requirements of Java are driven by the nature of the computing environments in which software must be *deployed*.

The massive growth of the Internet and the World-Wide Web leads us to a completely new way of looking at development and distribution of software. To live in the world of electronic commerce and distribution, Java must enable the development of secure, high performance, and highly robust applications on multiple platforms in heterogeneous, distributed networks.

Operating on multiple platforms in heterogeneous networks invalidates the traditional schemes of binary distribution, release, upgrade, patch, and so on. To *survive* in this jungle, Java must be architecture neutral, portable, and dynamically adaptable. [3]

The Java system that emerged to meet these needs is *simple*, so it can be easily programmed by most developers; familiar, so that current developers can easily learn Java; object oriented, to take advantage of modern software development *methodologies* and to fit into distributed *client-server* applications; *multithreaded*, for high performance in applications that need to perform multiple *concurrent* activities, such as *multimedia*; and interpreted, for maximum portability and dynamic capabilities. [4]

Together, the above requirements comprise quite a collection of *buzzwords*, so let's examine some of them and their respective benefits before going on.

What's completely new is the manner in which Java and its run-time system have combined them to produce a flexible and powerful programming system.

Developing your applications using Java results in software that is portable across multiple machine architectures, operating systems, and graphical user interfaces[5], secure, and high performance. With Java, your job as a software developer is much easier—you focus your full attention on the end goal of shipping *innovative* products on time, based on the solid foundation of Java. The *better* way to develop software is here, now, brought to you by the Java language platform.

Very dynamic languages like Lisp, TCL, and SmallTalk are often used for *prototyping*[6]. One of the reasons for their success at this is that they are very robust—you don't have to worry *about* freeing or corrupting memory.

Similarly, programmers can be relatively fearless about dealing with memory when programming in Java. The *garbage* collection system makes the programmer's job vastly easier; with the burden of memory management taken off the programmer's shoulders, storage allocation errors go away. [7]

Another reason commonly given that languages like Lisp, TCL, and SmallTalk are good for prototyping is that they don't require you to *pin down* decisions early on—these languages are semantically rich.[8]

Java has exactly the opposite property: it forces you to make explicit choices. Along with these choices come a lot of assistance — you can write method invocations and, if you get something wrong, you get told about it at compile time. You don't have to worry about method invocation error.

Words and Expressions

heterogeneous [ˌhetərəʊˈdʒiːnjəs] *adj.* 异构的;异质的,由不同成分组成的;异类的
paramount [ˈpærəmaʊnt] *adj.* 首要的;最高的,至上的 *n.* 最高掌权者;元首
delivery [diˈlivəri] *n.* 交付,交货,投递,传送;转让
platform [ˈplætfɔːm] *n.* 平台,(车站)月台
project [ˈprɒdʒekt] *n.* 项目;课题;工程;方案 *vt.* 计划;投映,投射;放映
embedded system   嵌入式系统
portable [ˈpɔːtəbl] *adj.* 可移植的,便携式的
distributed [disˈtribjuːtid] *adj.* 分布式的
real-time *adj.* 实时的 *n.* 实时
end-user [ˈendˌjuːzə] *n.* 终端用户,最终用户,直接用户
desktop [ˈdesktɒp] *n.* 台式计算机,台式电脑;桌面,(计算机屏幕上的)桌面
deploy [diˈplɔi] *v.* 展开;调度;部署
performance [pəˈfɔːməns] *n.* 性能
survive [səˈvaiv] *vi.* 活下来;幸存 *vt.* 从……中逃生,幸免于;比……活得长
adaptable [əˈdæptəbl] *adj.* 能适应的,适应性强的;可改编的
methodology [meθəˈdɒlədʒi] *n.* 方法学,方法论
client-server   客户机-服务器;客户-服务器;客户端-服务器
multithreaded [ˈmʌltiˈθredid] *adj.* 多线程的
concurrent [kənˈkʌrənt] *adj.* 并发的;同时发生的 *n.* 同时发生的事件
multimedia [ˈmʌltiˈmiːdjə] *n.* 多媒体
buzzword [ˈbʌzwɜːd] *n.* 时髦的术语
innovative [ˈinəʊveitiv] *adj.* 创新的,革新的
prototype [ˈprəʊtətaip] *vt.* 制作原型,制造(某产品的)样品 *n.* 原型;样板;典型
garbage [ˈɡɑːbidʒ] *n.* 无用存储单元;垃圾,废物
pin down   使受约束,阻止,牵制

Abbreviation

GUI (Graphical User Interface)   图形用户接口,图形用户界面

Notes

1. embedded 是 embed 的过去分词,用作定语来修饰 systems,意为"嵌入式的"。类似的

有，structured 意为"结构式的"，distributed 意为"分布式的"。
2. of choice 含意为"精选的"。the language of choice 可译为"精选的语言"。
3. jungle 含意为"为生存而残酷斗争的地方"。全句可译为：为在激烈竞争中生存，Java 必须是总体结构不带偏向性的、可移植的，且有动态适应能力的。
4. simple, familiar, object oriented, multithreaded 与 interpreted 皆是并列的，作 is 的表语。
5. across 引导的介词短语作状语，修饰 portable，可译为"可在多种机器总体结构、操作系统和图形用户接口间移植的"。
6. prototyping 是 prototype 的动名词，译为"原型构造"，也可译为"原型构造技术"。
7. taken off 引导的分词短语作定语，修饰 the burden of memory management。全句可译为：无用存储单元收集系统使程序员的工作容易得多，卸下了程序员肩上的存储管理负担，存储分配错误不再发生。
8. given 引导的分词短语作定语，修饰 another reason。这里 that 又引导一个从句作 given 的宾语。全句可译为：一般认为 LISP、TCL 和 SmallTalk 这样一些语言十分适合于原型（构造）法的另一理由，是它们不要求你受早期决定的约束——这些语言的语义是很丰富的。

3.4　Introduction to Android Programming

Hi guys, I am back with my 5 *instructable* on *Android programming*. I know very well that android is easy to learn and about 65% of people know how to program for android. But this instructable is for those 35%. If you have any ideas, opinions or anything related to this ible[1], please comment it in the comment section.

In my dad's college there is *gathering* in which a two day *workshop* of android programming taken place. It is only for the college students but I requested and I got a chance to learn the android programming! I enjoyed very much. They've teached me four programs but here I am only teaching you one. Other three are advanced and I am going to describe that three programs in my next ibles. Okay, enough talk, let's make it.[2]

Step 1: Downloading (see Figure 3-2)

Figure 3-2　Step 1: Downloading

Requirements:

(1) You need the Java JDK, because we do 90% of coding in Java, 5% in XML and 5% in C.

(2) You need ADT (Android Development Toolkit) in which we code, it is made up of Eclipse[3].

After downloading, extract it and install it.

Step 2: Preparing Eclipse(ADT) (see Figure 3-3)

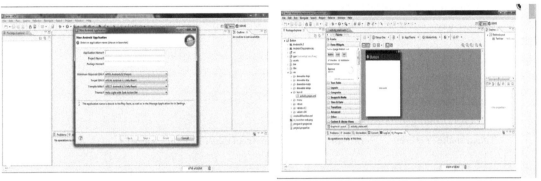

Figure 3-3　Preparing Eclipse (ADT)

Now after installing open Eclipse(ADT). It will ask you to make a workspace. Give it a desired name then you will see the Eclipse window. Workspace is a project *bundle* in which our projects are saved. Now go to **NEW-Android Application Project**. As you see in the first image it will ask you to enter Application name. Enter the name of project as **"button"** because our first project is Button. After typing, click finish and you will get the window like in the second image.

Step 3: Coding Part

Code:

package com. example. button;

import android. os. Bundle;
import android. app. Activity;
import android. view. Menu;
import android. view. View;
import android. view. View. OnClickListener;
import android. *widget*. Button;
import android. widget. Toast;

public class MainActivity extends Activity implements OnClickListener {
　Button a;
　@Override

```
protected void onCreate(Bundle savedInstanceState) {
    super. onCreate(savedInstanceState);
    setContentView(R. layout. activity_main);
    a=(Button)findViewById(R. id. button1);
    a. setOnClickListener(this);
}
@Override
public boolean onCreateOptionsMenu(Menu menu) {
    // Inflate the menu; this adds items to the action bar if it is present.
    getMenuInflater(). inflate(R. menu. activity_main, menu);
    return true;
}

@Override
public void onClick(View arg0) {
    // TODO Auto-generated method stub
    Toast. makeText(this,"Here I am writing on instructables. ",5000). show();
}
}
```

You can copy-paste this code or write it yourself. But remember, writing it helps you to learn the language. If you have any questions, you can ask me in the comment section.

So here's the explanation:

As you can see, in the **first line**, it shows package which we've entered earlier in the second step. Everything with **import** is the widgets we want to import in the project.[4] The **public class** shows in which class we are working with. **Button a** is the variable we have set on the button we have created earlier.[5]

The starting three lines of **First Override** is by default in our project and it is for making the *layout* of our project. The fourth line of **First Override** shows that we are registering our **variable a** to the **widget Button 1**. The fifth line shows that we want to do something when **Click** on our **Button.**

The **Center Override** is not really worth and even I don't know what it is.

The **last Override** is the main, because it shows which message we want to show on our screen. The **Toast**[6] is the method for showing our message. You can think it like **print** in C language. So, what it is happening over there is, the first word of bracket **this** is we are telling Android to work on the **Button.**[7] At the center, we are writing our message which we want to show. You can write anything in the double quotes in the Toast section. Like, I've written "Here I am writing on instructables. ". The **5000** is the milliseconds of exact time we want to show this message. They are milliseconds so if we convert it to seconds it will be 5 seconds. You can vary the time and can experiment with it. At last, you can see the **show()** which is very important to show our message on the screen.

Don't hesitate to ask me anything in the comment section or you can also PM me. Also, don't forget the **semicolon** at every end of line because it shows our language that which line ends where.

Step 4: AVD(Emulator)

Ok to emulate anything in Eclipse we have to make an AVD. AVD means Android Virtual Device. It is a virtual android device on which we can emulate our apps. Go to Window and then in AVD Manager, click new write a name and Target then click create AVD. In a few minutes our AVD will be created. Go to run and click run. Now your application is emulating … .

When the application opens click on the button and it will show your desired message! Congratulations you have created your very first ANDROID APPLICATION.

If you have any questions you can ask me in the comment box and below are some links from which you can learn the advanced knowledge of Android.

Words and Expressions

instructable [in'strʌkteibl] *n.* 指导 *adv.* 可教育的
Android programming 安卓程序设计
gathering ['gæðəriŋ] *n.* 集会；聚集(或搜集、采集等)
workshop ['wəːkʃɔp] *n.* 专题讨论会，研讨会；实/讲习班；创作室；车间；工场
bundle ['bʌndl] *n.* 包，捆，束；包袱 *vt.* 包，捆，扎
package ['pækidʒ] *n.* (程序)包；包；捆；包装用物 *vt.* 包装；把……打包 *adj.* 一揽子的
import [im'pɔːt] *n.* 移入；输入，进口；进口商品 *vt.* 移入；输入，进口；引入
widget ['widʒit] *n.* 控件，(窗口)小部件；桌面小程序
override [ˌəuvə'raid] *n.* 覆盖；代理佣金 *vt.* 覆盖；使无效；压倒；奔越过
inflate [in'fleit] *vt.* 使膨胀；使充气；抬高(物价)；使得意 *vi.* 膨胀
stub [stʌb] *n.* 存根，票根；桩；树桩；残端 *vt.* 连根挖(或拔)
layout ['leiaut] *n.* 布局；安排；设计；(报纸等的)版面编排

Abbreviations

JDK (Java Development Kit) Java 开发工具包
ADT (Android Development Toolkit) 安卓开发工具包
AVD (Android Virtual Device) 安卓虚拟设备
PM (Private Message) 私人信息，个人信息，主要指网络 BBS 中发送的个人信息；
  发送私人信息

Notes

1. ible 一般作为形容词结束标志，可作"能……的"解，例如，possible 意为"可能的"。本句中的 this ible 作"本指导"或"本文"解。
2. 本句中，make it 含意为"办成功，做到；赶到；规定时间"。全句可译为：好了，说得够

多了，让我们开始吧。

3. Eclipse 是一个开放源代码的、基于 Java 的可扩展开发平台，就其本身而言，它只是一个框架和一组服务，用于通过插件组件构建开发环境。不过，Eclipse 附带了一个标准的插件集，包括 Java 开发工具包(Java Development Kit，JDK)。尽管 Eclipse 是使用 Java 语言开发的，但它的用途并不限于 Java 语言；例如，支持诸如 C/C++、COBOL、PHP、Android 等编程语言的插件已经可用或将会推出。Eclipse 框架还可用来作为与软件开发无关的其他应用程序类型的基础，比如内容管理系统。

4. 本句中，we 之前省略了 that，它引导限制性定语从句，修饰其前的 widgets。全句可译为：import 后跟的都是我们要在项目中移入的控件。

5. 本句中，两处 we 之前都省略了 that，它们都引导限制性定语从句，分别修饰其前的 variable 和 button。全句可译为：Button a 是我们设置在早先创建的 button 上的变量。

6. Toast 是 Android 中的一种简易消息提示框。Toast 类帮助创建和显示消息。当视图显示时，在应用程序中显示为浮框。和 Dialog 不一样的是，它无法被点击。Toast 显示的时间有限，会在用户设置的显示时间后自动消失。先如下移入 Toast 类：

   import android.widget.Toast;

   为某点击事件加入下列代码：

   Toast.makeText (getApplicationContext(),
         "你想提示的信息",Toast.LENGTH_SHORT).show();

   则点击时会出现浮框提示。

7. 本句中，第3个 is 之后省略了 that，它引导名词性从句作表语。全句可译为：那么，那里正发生的是，括号中的第一个词 this 指明我们正指示 Android 对 Button 进行操作。

3.5　Characteristics of Web Programming Languages

Just as there is a diversity of programming languages available and suitable for conventional programming tasks, there is a diversity of languages available and suitable for Web programming. There is no reason to believe that any one language will completely *monopolize* the Web programming scene, although the varying availability and suitability of the current *offerings* is likely to favor some over others. Java is both available and generally suitable, but not all application developers are likely to prefer it over languages more similar to what they currently use, or, in the case of non-programmers, over higher level languages and tools.[1] This is OK because there is no real reason why we must *converge* on a single programming language for the Web any more than we must converge on a single programming language in any other *domain*.[2]

The Web does, however, place some specific constraints on our choices: the ability to deal with a variety of protocols and formats (e.g. graphics) and programming tasks; performance (both speed and size); safety; platform independence; protection of *intellectual property*; and the basic ability to deal with other Web tools and languages. These issues are not independent of one another. A choice which seemingly is optimal in one *dimension* may be sub-optimal or worse in another.[3]

The wide variety of computing, display, and software platforms found among clients necessitates a strategy in which the client plays a major role in the decision about how to process and/or display retrieved information, or in which servers must be capable of driving these activities on all potential clients.[4] Since the latter is not practical, a suite of Web *protocols* covering addressing conventions, presentation formats, and handling of foreign formats has been created to allow *interoperability*.[5]

HTML (Hyper Text Markup Language) is the basic language understood by all WWW (World Wide Web) clients. HTML is simple enough that nearly anyone can write an HTML document, and it seems almost everyone is doing so.[6]

HTML is a markup language rather than a complete programming language. An HTML document (program) is ASCII text with embedded instructions (markups) which affect the way the text is displayed.[7] The basic model for HTML execution is to fetch a document by its name (e.g. URL), interpret the HTML and display the document, possibly fetching additional HTML documents in the process, and possibly leaving *hot areas* in the displayed document that, if selected by the user, can accept user input and/or cause additional HTML documents to be fetched by URL.[8]

HTML applications, or what we might consider the HTML equivalent of an application, consist of a collection of related web pages managed by a single HTTP (HTTP is the TCP/IP protocol that defines the interaction of WWW clients and servers) server. This is an oversimplification, but the model is simple, and the language is simple, and that is one of its strengths.

Performance. Because of an HTML program's limited functionality, and the resulting shift of computational *load* to the server, certain types of applications perform poorly, especially in the context of clients connected to the Internet with rather low bandwidth dialup communications ($\leqslant 28.8$ Kbps).[9] The performance problems arise from two sources: (a) an application which is highly interactive requires frequently hitting the server across this low bandwidth line which can dramatically and, at times, unacceptably slow observed performance; and (b) requiring all computation to be done on the server increases the load on the server thereby reducing the observed performance of its clients.[10]

Today, most users have pretty *competent* client machines which are capable of accepting a larger *share* of the computational load than HTML allows[11]. For example, an *Internet-based* interactive game or simulation can be a frustrating experience for users with low speed connections, and can *overwhelm* the server that hosts it. If you were the developer of such a game, you'd be *inclined* to push more of the functionality to the client,[12] but, since HTML limits the possibilities, another *route* to supporting computation on the client must be found. The developer might make an executable client program available to users, which would be invoked via the HTML browser, but users might only be willing to accept such programs if they trust the source (e.g. a major vendor), as such programs are a potential safety concern. Also, users don't want to be continuously downloading client programs to be able to access web pages, so this solution has real practical limitations considering the size and *dynamism* of the Web. [13] If safe powerful high performance programs could be automatically downloaded to client platforms, in much the same way as HTML pages, the problem would be solved.

When code is to be executed on a client, there are two main considerations: what gets shipped and what gets executed. There are three main alternatives for each of these: source code, a partially compiled intermediate format (e. g. byte code), and binary code. Because compilation can take place on the client, what is shipped is not necessarily what is executed. [14]

Byte code, according to measurements presented at the JavaOne conference can be 2-3x smaller than comparable binary code. Execution performance clearly favors binary code over byte code, and byte code over source code. In general, binary code executes 10-100 times faster than byte code. Most Java VM developers are developing JIT (Just In Time) compilers to get the benefits of *bytecode* size and binary speed. Java bytecodes are downloaded over the net and compiled to native binary on the local platform. The binary is then executed, and, possibly, cached for later executions. [15]

Platform independence can be achieved by shipping either byte code or source code. One advantage of shipping byte code over source code is that a *plethora* of source languages would require the client machines to maintain many compilers and/or interpreters for the source languages, while fewer byte code formats would require fewer virtual machines.

HTML proved that downloading source code in a safe language and executing it with a trusted interpreter was safe. Note that safety can only be provided by the interpreter or virtual machine, not by the language or the language's compiler.

Conclusions. HTML is proving insufficient by itself to develop the *myriad* Web-based applications envisioned. [16] As extended by server and client programs, the task is feasible, yet awkward and sub-optimal in terms of performance and safety. The ability to easily develop sophisticated Web-based applications optimally segmented between client and server in the context of the heterogeneous and dynamic environment of the Web while not compromising safety, performance, nor intellectual property, is the goal of current efforts. [17] The first significant result of those efforts is Java, a C++-derived language with capabilities specialized for Web-based application development. Java is compiled by the developer to a platform-independent bytecode format, with bytecodes downloadable via HTML browsers to the client, and interpreted by a virtual machine which can guarantee its safety. Sun is working to improve the safety, performance, comprehensiveness, and *ubiquity* of Java, and the industry appears to be accepting their approach. Others, especially other language developers, vendors and users, are taking similar approaches to developing Web-based applications.

Words and Expressions

monopolize [məˈnɔpəlaiz] *vt.* 独占，垄断

offering [ˈɔfəriŋ] *n.* 提供；奉献物，献礼；祭品

converge [kən'və:dʒ] vi. 集中,会聚;收敛 vt. 使集中于一点;使会聚
domain [dəu'mein] n. (活动、学问等的)范围,领域;[数]域,定义域;领土;领地
intellectual property 知识产权
dimension [di'menʃən] n. 方面;维(数);度(数);尺寸;尺度;元
protocol ['prəutəkɔl] n. 协议;草案
interoperability ['intərɔpərə'biliti] n. 可互操作性;互用性;协同工作的能力
hot area 热区;高度放射性区域
load [ləud] n. 负担;重载;负荷,负载;工作量 vt. 装载;装满;使负担
competent ['kɔmpitənt] adj. 有能力的,胜任的
share [ʃɛə] n. 份额;一份;共享,分享 vt. 分享,共同具有;均分;分配
Internet-based 基于因特网的,基于互联网的
overwhelm ['əuvə'welm] vt. 压倒,制服;覆盖;淹没;倾覆
inclined [in'klaind] adj. 倾向……的
route [ru:t] n. 路线;路程;通道 vt. 按规定路线发送
dynamism ['dainəmizəm] n. 劲头;精力,推动力;物力论,力本论
bytecode ['bait'kəud] n. 字节码
plethora ['pleθərə] n. 过多,过剩;多血(症)
myriad ['miriəd] adj. 无数的;一万的;种种的 n. 无数;无数的人或物
ubiquity [ju:'bikwiti] n. (同时的)普遍存在,无处不在

Abbreviations

VM (Virtual Machine) 虚拟机
JIT (Just In Time) compiler 即时编译程序

Notes

1. 本句包含两个并列子句,两者由连接词 but 连接。在后一子句中,两处出现的介词 over 表示程度,各引导一个短语作状语修饰 prefer。全句可译为:Java 既是可用的,通常也是适合于 Web 程序设计的,但是很可能不是所有应用程序开发人员喜爱它的程度都超过与他们当前使用的语言类似的那些语言,或者,非程序设计人员喜爱它的程度超过一些更高级的语言和工具。
2. 本句中的 OK 可翻译为"没问题"。全句可译为:这没问题,因为对于 Web 我们必须集中于一个程序设计语言的理由,不比在其他任何领域我们必须集中于一个程序设计语言的理由更多。
3. 本句中的 which 是关系代词,引导一个限制性定语从句修饰其前的 choice。全句可译为:从某个方面看来是最佳的选择,从另一个方面看可能是次佳的,或许是更糟的。
4. 本句中的 found 是过去分词,引导一个短语作定语,修饰前面的 computing, display, and software platforms。retrieved 是过去分词作定语修饰 information,翻译为"所检索的"。本句中两个 which 引导的限制性定语从句用连接词 or 连接,并列地修饰 strategy。全句可译为:在诸客户机中找到了各种各样的计算、显示和软件平台,必须制定一种策略,按这种策略,客户机在决定如何处理和/或显示所检索的信息中起着主要的

作用，或者，按这种策略，服务器必须具有在所有潜在的客户机上驱动这些活动的能力。

5. covering 是现在分词，引导一个短语作定语修饰 Web protocols。addressing convention 译为"寻址约定"，presentation format 译为"展示格式"。全句可译为：因为后者不实际，所以已经创建了一组包括寻址约定、展示格式和处理外来格式的 Web 协议以允许可互操作性。

6. 本句中 seems 之后省略了 that，先行代词 it 所替代的就是它所引导的从句。全句可译为：HTML 足够简单，几乎每个人都能写 HTML 文档，而且看来几乎每个人都正在这样做。

7. 本句中的 which 引导的定语从句修饰其前的 instructions。the text is displayed 是定语从句，修饰其前的 the way。

8. 本句可译为：HTML 执行的基本模型是：根据其名（如 URL）取一个文档，解释该 HTML 文档，并且显示该文档，在此过程中可能取另一些 HTML 文档，还可能在所显示的文档中留一些热区，这些热区如果被用户选择的话，能接收用户的输入，并且/或者把另一些 HTML 文档（根据 URL）取过来。

9. 本句中的 connected 是过去分词，引导的短语作定语修饰其前的 clients。本句可译为：由于 HTML 程序的功能有限，以及因此而把计算负荷转移给服务器，所以某些类型的应用程序执行效率低下，特别是在用带宽相当低的拨号通信（≤28.8 Kbps）连接到因特网的客户机环境下更是如此。

10. 本句中(a)后的第一个 which 引导的定语从句修饰其前的 application。hitting 是动名词，作 requires 的宾语；第二个 which 引导的定语从句修饰其前的 line。(b)后的 requiring 是动名词，引导一个动名词短语作主语，谓语动词是 increases；reducing 引导一个现在分词短语作状语。全句可译为：性能问题起因于两个根源：(a)高度交互的应用程序需要频繁地通过该低带宽线访问服务器，这会显著地，有时会不可接受地降低了可觉察到的性能，并且(b)要求所有的计算在服务器上完成，就增加了服务器的负荷，因而降低了其各客户机可觉察到的性能。

11. 本句中的 which 引导的定语从句修饰其前的 machines；accepting 引导一个动名词短语，作介词 of 的宾语，可译为"承受比 HTML 允许的更大份额的计算负荷"。

12. were 是 be 的过去分词形式，这是虚拟条件句，表示假设与可能性不大的情况。

13. 本句中 be continuously downloading 是现在进行时态，表示"连续不断地下载"；to be able to 引导一个不定式短语修饰其前的 programs。considering 是现在分词，引导一个分词短语作状语。

14. 本句中的两个 what 都是关系代词，在引导的从句中作主语，两个从句分别作主语与表语。全句可译为：由于编译能发生在客户机上，所以被传输的未必是被执行的。

15. 本句中的 cached 是过去分词，与前面的 executed 并列，一起与之前的 is 构成被动语态。全句可译为：然后二进制（代码）被执行，并且可能为以后的多次执行而被高速缓存。

16. envisioned 是过去分词，作形容词，意为"所设想的"，修饰其前的 applications；by itself 意为"独立地，单独地"。

17. segmented 是过去分词，引导的短语修饰其前的 applications。下一句中的 specialized 情况与此类似。全句可译为：当前各种努力的目标是不费力地开发基于 Web 的复杂应用程序的能力，这种应用程序，当不危及安全性和性能，也不违背知识产权时，在异构的和动态的 Web 环境下把客户机与服务器最佳地分隔开。

Terms

Structured programming

Structured programming (sometimes known as *modular programming*) is a subset of procedural programming that enforces a logical structure on the program being written to make it more efficient and easier to understand and modify. Certain languages such as Ada, Pascal, and dBASE are designed with features that encourage or enforce a logical program structure.

Structured programming frequently employs a top-down design model, in which developers map out the overall program structure into separate subsections. A defined function or set of similar functions is coded in a separate module or submodule, which means that code can be loaded into memory more efficiently and that modules can be reused in other programs. After a module has been tested individually, it is then integrated with other modules into the overall program structure.

Program flow follows a simple hierarchical model that employs looping constructs such as "for", "repeat", and "while". Use of the "Go To" statement is discouraged.

Modeling language

A modeling language is any artificial language that can be used to express information or knowledge or systems in a structure that is defined by a consistent set of rules. The rules are used for interpretation of the meaning of components in the structure. A modeling language can be graphical or textual.

Graphical modeling languages use a diagram techniques with named symbols that represent concepts and lines that connect the symbols and that represent relationships and various other graphical annotation to represent constraints.

Textual modeling languages typically use standardised keywords accompanied by parameters to make computer-interpretable expressions.

An example of a graphical modeling language and a corresponding textual modeling language is EXPRESS-G and EXPRESS.

A large number of modeling languages appear in the literature.

Compiler(编译程序)

A compiler is a computer program (or set of programs) that transforms source code written in a programming language (the source language) into another computer language (the target language, often having a binary form known as object code). The most common reason for wanting to transform source code is to create an executable program.

The name "compiler" is primarily used for programs that translate source code from a high-level programming language to a lower level language (e.g. assembly language or machine code). If the compiled program can run on a computer whose CPU or operating system is different from the one on which the compiler runs, the compiler is known as a cross-compiler. A program that translates between high-level languages is usually called a language translator, source to source translator, or language converter.

A compiler is likely to perform many or all of the following operations: lexical analysis, preprocessing, parsing, semantic analysis (syntax-directed translation), code generation, and code optimization.

The term compiler-compiler is sometimes used to refer to a parser generator, a tool often used to help create the lexer and parser.

Ajax (Asynchronous JavaScript and XML)

Ajax is a web development technique for creating interactive web applications. The intent is to make web pages feel more responsive by exchanging small amounts of data with the server behind the scenes, so that the entire web page does not have to be reloaded each time the user makes a change. This is meant to increase the web page's interactivity, speed, and usability.

Exercises

1. Translate the text of Lesson 3.1 into Chinese.
2. Topics for oral workshop.
   - What advanced features do the object-oriented programming languages have?
   - Why need compilers for high-level programming languages? Talk about the reasons by using instances.
   - What are the major differences between C and C++?
3. Translate the following into English.

(a) 程序是指挥(direct)计算机执行所需的数据处理任务的一列指令或语句。程序设计是创建那列指令的多步骤过程。

(b) 重要的是理解类和类的对象之间的区别。类不过(simply)是创建一些对象的详细说明(specification)，因此，单个类可能创建多个对象。

(c) Java是面向对象的、网络友好的高级程序设计语言，它允许程序员创建在几乎任意的操作系统上都能运行的应用程序(application)。

(d) ActiveX是一组控件(control)或可重用的组件(component)，它使得一些程序或几乎任意种类的内容能被嵌入在Web页内。鉴于(whereas)每次你访问某个Web站点时必须下载一个Java，因此需要使用ActiveX，该组件仅下载一次，然后为方便以后的重复使用，被存储在你的硬盘中。

(e) 程序设计涉及大量的创造力。设计是对各个组件的功能或目的的指南(guide)，但是程序员在把设计实现成代码时有极大的灵活性。不管使用什么语言，每个程序组件涉及至少三个主要方面(aspect)：控制结构、算法和数据结构。

4. Listen to the video "Computer tour" and write down the Last three paragraphs.

Unit 4 Operating System

4.1 Summary of OS

An operating system is the software which acts as an interface between a user of a computer and the computer hardware. The purpose of an operating system is to provide an environment in which a user may execute programs. The primary goal of an operating system is thus to make the computer system convenient to use. A secondary goal is to use the computer hardware in an efficient way.

We can view an operating system as a *resource allocator*. A computer system has many resources which may be required to solve a problem: CPU time, memory space, file storage, input/output (I/O) devices, and so on. The operating system acts as the manager of these resources and allocates them to specific programs and users as necessary for their tasks. Since there may be many, possibly conflicting, requests for resources, the operating system must decide which requests are allocated resources to operate the computer system fairly and efficiently.

Early computers were (physically) very large machines run from a *console*. The programmer would write a program and then operate the program directly from the operator's console. Software such as *assemblers*, *loaders*, and compilers improved on the convenience of programming the system, but also required substantial *set-up time*. To reduce the setup time, operators were hired and similar jobs were batched together[1].

Batch systems allowed automatic job sequencing by a *resident monitor* and improved the overall utilization of the computer greatly. The computer no longer had to wait for human operation. CPU utilization was still low, however, because of the slow speed of the I/O devices relative to the CPU. *Off-line* operation of slow devices was tried.

Buffering was another approach to improving system performance by overlapping the input, output, and computation of a single job. Finally, *spooling* allowed the CPU to overlap the input of one job with the computation and output of other jobs.

Spooling also provides a pool of jobs which have been read and are waiting to be run. This job pool supports the concept of *multiprogramming*. With multiprogramming, several jobs are kept in memory at one time; the CPU is *switched* back and forth between them in order to increase CPU utilization and to decrease the total real time needed to execute a job.

Multiprogramming, which was developed to improve performance, also allows *time sharing*.

Time-shared operating systems allow many users (from one to several hundred) to use a computer system interactively at the same time. As the system switches rapidly from one user to the next, each user is given the impression that he has his own computer. Other operating systems types include real-time systems and multiprocessor systems.

A real-time system is often used as control device in a dedicated application[2]. Sensors bring data to the computer. The computer must analyze the data and possibly adjust controls to modify the sensor inputs. Systems which control scientific experiments, medical computer systems, industrial control systems, and some display systems are real-time systems. A real-time operating system has well-defined fixed time constraints. Processing must be done within the defined constraints, or the system will fail.[3]

A multiprocessor system has more than one CPU. The obvious advantages would appear to be greater computing power and reliability. There are various types of operating systems for multiprocessors and multicomputers. It is more-or-less possible to distinguish two kinds of operating systems for multiple CPU systems: *loosely coupled*, such as network operating system and *distributed operating* system, and *tightly coupled*, such as parallel operating system. As we shall see, loosely and tightly-coupled software is roughly analogous to loosely and tightly-coupled hardware.

The operating system must ensure correct operation of the computer system. To prevent user programs from interfering with the proper operation of the system, the hardware was modified to create two modes: user mode and monitor mode.[4] Various instructions (such as I/O instructions and halt instructions) are privileged and can only be executed in monitor mode. The memory in which the monitor resides must also be protected from modification by the user. A timer prevents infinite loops. Once these changes (dual mode, privileged instructions, memory protection, timer interrupt) have been made to the basic computer architecture, it is possible to write a correct operating system.[5]

Words and Expressions

resource allocator　资源分配程序(器)
console ['kɔnsəul] *n*. 控制台
assembler [ə'semblə] *n*. 汇编程序(器)
loader ['ləudə] *n*. 装入程序
set-up time　准备时间；建立时间
batch system　批处理系统
resident monitor　常驻(内存的)监控程序
off-line operation　脱机操作
buffering ['bʌfəriŋ] *n*. 缓冲技术，缓冲
spooling [spu:liŋ] *vt*. 假脱机　spool 的现在分词
multiprogramming [ˌmʌlti'prəugræmiŋ] *n*. 多道程序设计
switch [switʃ] *vt*. 切换；交换　*n*. 开关；交换机
time sharing (time-shared)　分时
loosely coupled　松耦合的，松散耦合的
tightly coupled　紧耦合的
distributed [di'stribju:tid] OS　分布式操作系统

Notes

1. 本句中 similar jobs were batched together 译为"相似的一些作业成批地排在一起"。
2. 本句中 dedicated application"专用应用系统"。
3. 本句中 defined constraints 意为"规定的时间限制"。本句可译为:处理必须在规定的时间内完成,否则系统将失效。
4. to prevent … from "为了阻止(或防止)……"。全句可译为:为阻止用户程序干扰系统的正常操作,对硬件做了修改,建立了两种方式:用户方式和监控方式。后面句子中的 protected from … 结构类似。
5. privileged"特权的"。全句可译为:一旦对基本的计算机体系结构做了这些改变(两种方式,特权指令,内存保护,时钟中断),就有可能写出正确的操作系统。

注:本篇是 OS 概述。

4.2　Using the Windows Operating System

After a PC is *booted* up, the computer waits to receive input from the user regarding what activity he or she would like to perform first. The manner in which an operating system or any other type of program *interacts* with its users is known as its *user interface*. Older software programs used a *text-based* user interface, which required the user to type precise instructions indicating exactly what the computers should do. Most programs today use a **graphical user interface or GUI** (pronounced "goo-ey"), which uses graphical objects (such as icons and buttons) and the mouse to much more easily tell the computer what the user wants it to do.

GUIs became the standard once hardware became sufficient to support it. Older systems could not *adequately* deliver the necessary WSYIWYG (what you see is what you get, pronounced "wizzy-wig") displays in which screen images resemble printed *documents*, since the output wasn't of acceptable quality. And later, when screen outputs started to look acceptable, it just took too long to display the graphics on the computer to be worthwhile.[1] These days, of course, most computers can rapidly display WYSIWYG displays, as well as other types of text and graphics.

The Windows Interface

One of the advantages of using Windows is that application software written for any version of the Windows operating system has a similar appearance and works essentially the same way as Windows does. Thus, if you are comfortable using the Windows interface and some Windows software, all other Windows software should seem familiar. Other graphical operating systems, such as Mac OS (used on Macintosh computers) and graphical versions of Linux (used on some PC computers) look and act similarly to Windows.

The Desktop

The Windows **desktop** appears on the screen after a computer using the Windows operating system has completed the boot process. The desktop is where documents, *folders*, programs, and other objects are displayed when they are being used, similar to the way documents and file folders are laid on a desk when they are being used[2]. Though the appearance of the Windows desktop can be *customized*, all desktops contain common elements, such as desktop icons, the *taskbar*, the Start button, windows, and task buttons (see Figure 4-1).

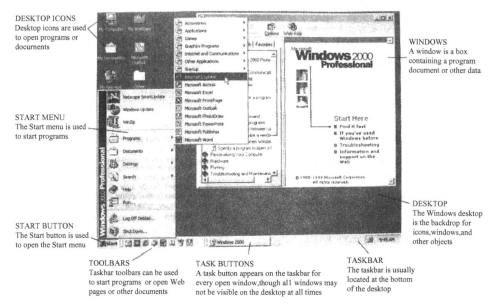

Figure 4-1 The Windows desktop

- The *taskbar* is located along the bottom of the desktop. It houses the Start button at the left edge, toolbars and task buttons in the center, and a clock and other indicators at the far right edge.
- The *Start button* is the main menu for Windows that is used to start programs.
- *Toolbars* (groups of icons that can be used to quickly *invoke* commands) may be displayed on the Windows taskbar.
- A *task button* appears on the taskbar for each window that is currently open, though the window may not be visible on the desktop at all times. As discussed later, these task buttons can be used to select which windows are visible on the screen.

Windows

The principle component of the GUI is the window. A **window** is a rectangular area of information that is displayed on the screen. These windows can contain programs and documents, as well as menus, *dialog boxes*, icons, and a variety of other types of data, as discussed next.

Menus

A **menu** is a set of *options* — usually text based — from which the user can choose to initiate a desired action in a program. At the top of many windows is a menu bar showing the main menu categories (see Figure 4-2). *Pull-down menus* (also called drop-down menus) display on the screen when the user selects an item on the menu bar. As shown on the right-most screen in Figure 4-2, in some Microsoft programs a feature called personalized menus can be used. With personalized menus, only the options that were most recently used are initially displayed on the menu. Waiting a moment or clicking on the down arrow symbol at the bottom of the menu displays all items that belong on that menu.

Options on pull-down menus either display another, more specific, menu; open a dialog box to prompt the user for more information; or execute a command. These options are discussed in the following sections along with other conventions that pull-down menus typically follow.

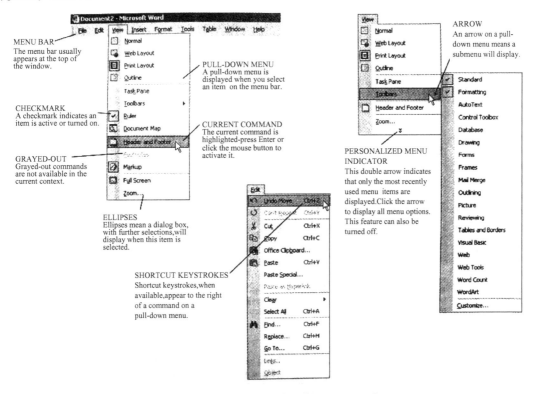

Figure 4-2 The menu bar and pulldown menu elements

Checkmarks

An item on a menu with a *checkmark* to the left of it means that the associated option is turned on. For instance, the checkmark by the Ruler option on the top left screen of Figure 4-2 means that the ruler for this program is displayed on the screen. If this menu item was not checkmarked, the ruler would not appear on the screen. You can *toggle* between *checking* and unchecking a checkmark item by clicking the item with the mouse.[3] Sometimes on a menu only one item in a group of items can be selected. When this is the case, another symbol—a dot or circle called a *radio button*—is used instead to indicate the selected item. Radio buttons will be discussed in more detail shortly.

Grayed-out Type

On the View pull-down menu at the top left of Figure 4-2, the Footnotes command is shown in *faded* or *grayed-out* type. This means that this particular choice is unavailable in the context of what you're currently doing. For example, if you haven't created any *footnotes* in a document, viewing them is impossible at this time.

Ellipses

On menus, *ellipses* (...) are often displayed to the right of a command—such as next to the Zoom option on the View menu shown in Figure 4-2. Ellipses mean that selecting the menu item will display a dialog box. As discussed shortly, a dialog box is displayed when additional information needs to be supplied; once the dialog box has been completed, the appropriate command will be carried out.

Shortcut Keystrokes

Some menu items display *shortcut keystrokes* next to them, such as Ctrl+Z (performed by holding down the Ctrl key while *tapping* the Z key[4]) for *Undo* and Ctrl+C for Copy on the Edit menu shown at the bottom of Figure 4-2. Shortcut keystrokes are used when the menu is not currently open to execute that command.

Icons and Other Navigational Objects

In addition to menus, other *navigational* objects can typically be used to allow the user to select the desired option or command. Figure 4-3 shows a variety of navigational objects you may find on the desktop or in Windows programs. These objects are described in more detail next.

Icons

Icons are small pictures that represent such items as a computer program or document. When you select an icon with the mouse, the software takes the corresponding action, often starting or opening the appropriate program or document. Programs are commonly represented as program icons displayed on the Windows desktop. To start a program or anything else represented by a desktop icon, the mouse is used. Depending on how Windows is set up, either a single click or a double-click of the left mouse button opens the program or document. Icons can also be located on toolbars, as discussed next.

Toolbars

A **toolbar** consists of a set of icons or buttons called toolbar buttons and usually *stretches horizontally* across the screen. Each toolbar button has a name, which is displayed if you point to the button (see Figure 4-3). In a software *suite*, such as Microsoft Office, the toolbar buttons that can be used to save a document, print a document, or perform other common tasks have the same appearance in all programs within the suite. As shown in Figure 4-1, there can also be toolbars displayed on the Windows taskbar.

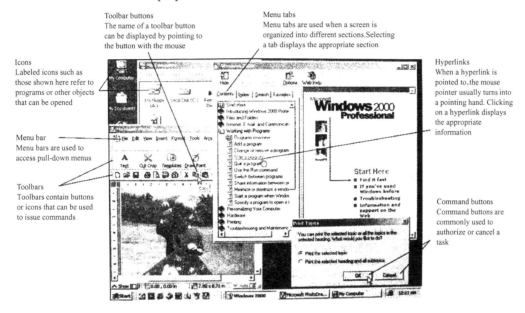

Figure 4-3　Icons and other navigational objects

Menu *Tabs*

Some programs or program elements contain menu tabs, such as on the Help screen in Figure 4-3. Menu tabs, such as Contents, Index, Search, and Favorites in the figure, organize a screen into file-folder-type tabs. By clicking any of the tabs, the information corresponding to that tab is displayed. For example, in Figure 4-3, the Contents tab on the Help screen is currently selected.

Hyperlinks

Both image-based *hyperlinks* and text-based hyperlinks (referred to as hypertext) are an increasingly common control option. Though initially found just on Web pages, hyperlinks now more frequently appear on other types of Windows applications, such as on the Help screen in Figure 4-3. Icons on the Windows desktop that are *underlined*, as well as underlined and/or different colored text on a Web page or Help screen, generally indicate hyperlinks.[5] When you move the mouse pointer to a hyperlink, the pointer's shape usually changes to a pointing hand. Clicking the hyperlink opens the appropriate program or document or displays some other type of new information on your screen. Hyperlinks found on Web pages are discussed in more detail later in this chapter.

Dialog Boxes

After selecting a menu item with an ellipsis (...) next to it, a dialog box appears on the

69

screen. Dialog boxes are windows that prompt the user to provide further information. Such information can be supplied by a variety of means, several of which are covered in the following sections.

Words and Expressions

boot [bu:t] *vt.* 自举，引导；自展　*n.* 引导程序，自展程序(=bootstrap)
cold boot　冷启动　warm boot　热启动
interact [ˌintər'ækt] *vi.* 交互，互相作用；互相影响
text-based　基于文本的
adequately ['ædikwitli] *adv.* 充分地；能满足需要地；足够地
document ['dɔkjumənt] *n.* 文档，文件；公文；档案
folder ['fəuldə] *n.* 文件夹；折叠器
customize [kʌstəmaiz] *vt.* [计]定制，用户化
taskbar [tɑːskbɑː(r)] *n.* [计]任务栏，任务条
invoke [in'vəuk] *vt.* 调用
dialog box　对话框
option ['ɔpʃən] *n.* 选项；选择权
pull-down menu　下拉(式)菜单
checkmark ['tʃekmɑːk] *n.* 记号 √
check [tʃek] *vt.* 检查，核对，在……上打经过查对无误的记号 √
toggle ['tɔgl] *vt.* 切换；轮转
radio button　单选钮，小圆按钮
grayed-out　变灰显示；打灰
footnote ['futnəut] *n.* 脚注
fade [feid] *vi.* 变淡，变暗；衰减
ellipse [i'lips] *n.* 省略号；[数]椭圆
shortcut keystroke　快捷键，快速键
undo [ʌn'duː] *vt.* 取消，撤销
tap [tæp] *vt.* 轻打，轻敲　*n.* 轻拍，轻叩
navigational　*adj.* 导航的；航行的
stretch [stretʃ] *v.* 展开；伸展
horizontally [ˌhɔri'zɔntli] *adv.* 水平地，地平地
suite [swiːt] *n.* 套件，组，套
tab [tæb] *n.* 标记，标志，标号；Tab 键，制表
hyperlink　*n.* [计]超链接，超链
underline [ʌndə'lain] *vt.* 在……下面画线　*n.* 下画线

Notes

1. worthwhile "值得花时间的，有价值的"。本句中 it just took too long ... to be

worthwhile 是"too+形容词+动词不定式"结构,其中动词不定式在翻译时要加上一个否定词。其意思是,显示图形花太长时间以至于没有价值。句中的 it 是先行代词,代表 to display ... 。全句可译为:之后,当屏幕输出看上去可以接受时,若在计算机上显示图形花时间太长,仍是没有价值的。
2. 本句中,the way documents ... 在 way 的后面省略了 in which,是定语从句修饰 way。
3. 全句可译为:你可以通过用鼠标单击记号项在选中和取消选中记号项之间切换。
4. 全句可译为:通过按住 Ctrl 键的同时单击 Z 键来完成。
5. 本句中,and/or 是指两种可能:一是用下画线且不同颜色的文本来表示超链;二是用不同颜色且无下画线的文本来表示超链。全句可译为:在 Windows 桌面上有下画线的图标,以及在网页或帮助界面有下画线并且/或者不同颜色的文本通常表示超链。
注:本节主要目的是给出 Windows 桌面上常用的英文单词和术语。

4.3　Window Managers

Window managers manage the devices used to exchange information between applications and users. Output devices include video displays and sound *synthesizers*. Input devices include keyboards and pointing devices such as mouse, *joystick*, *track ball*, or light pen. The window manger interacts with the device drivers of output devices to present information to the user, and with the device drivers for input devices to obtain messages which represent information being entered by the user.[1] Applications and *script execution engines* pass images expressed as bit maps or *PostScript* notation to the window manager which presents those images to the user. The window manager returns messages to the applications and script execution engine entered by the user via input devices.[2]

Window managers have become popular because they support many features which are useful to both end users and *application developers*.

Users Interact with Multiple Processes. Window managers allocate windows to each of several processes. Multiple processes can share a video screen if each process is associated with a window displayed within the screen. The user views the progress and controls each process by looking at the contents of the window allocated to the process and issuing commands to the process when its window is in focus.[3]

Users easily Move Information Between Applications. Windows provide a convenient way for users to transport information between applications. Figure 4-4 illustrates a screen with three windows. One window is allocated to a text editor, another is allocated to a *spreadsheet* program, and the third window is allocated to a database program. In this example, the user first retrieves some data from the database using the database application. The user then selects some of the retrieved data and moves that data to the spreadsheet application in the second window. The spreadsheet application calculates *aggregate* and summary information, which the user then moves to the window allocated to the text editing program. The user uses the text editor to integrate the summary information in a report being prepared using the text editor.

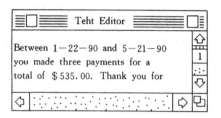

Figure 4-4 Three windows for three applications

Users Access Remote Applications. Some window managers can allocate a window to a process operating on a *remote* computer via a communication system. For example, a user at a workstation wishes to access a database on a mainframe. The user asks the window manager to create a window into which the user can enter a request to the database management system on the mainframe. After the request is processed, the window manager displays the results from the database management system in the window.

Users Have Multiple Views to a Process. Some window managers can allocate several windows to an application; each window displays some aspect of the application to the user. Multiple windows can be useful for providing various views of the application to the user. For example, in Figure 4-5, two windows have been allocated to a planning system. One window displays a graph showing the *precedence* among the tasks of a project. The other window displays the completion percentage for each task.

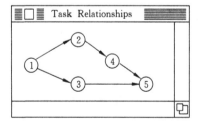 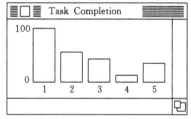

Figure 4-5 Two windows representing two views of an application

Users Receive Events. When a process detects some unusual event of which the user should be aware[4], the process notifies the window manager to display a message describing the event. A user will notice the appearance of the message and take appropriate action.

Sophisticated User Interfaces Can Be Constructed Using Window Managers. Most window managers have libraries containing a wide variety of *reusable* interaction objects which can be used to build sophisticated user interfaces.

Terminal Independence. Some window managers work on a large variety of terminals and workstations, and hide the differences between those terminals and workstations from the application.[5] Terminal independence increases the *transportability* of applications to a wider variety of terminals and workstations.

Words and Expressions

synthesizer ['sinθəsaizə] *n.* 合成器(程序)，综合器
joystick ['dʒɔistik] *n.* 控制杆，操纵杆
track ball   跟踪球，控制球
script execution engine   脚本执行引擎(器)
PostScript   软件名
application developer   应用程序开发者
spreadsheet *n.* [计]电子制表软件；电子数据表；电子表格
aggregate ['æɡriɡət] *n.* 聚集(体)；合计   *v.* (使)聚集，总计   *adj.* 聚集的；合计的
remote [ri'məut] *adj.* 远程的；遥远的
precedence ['presidəns] *n.* 先后次序；优先
reusable [riː'juːzəbl] *adj.* 可重用的
transportability [trænsˌpɔːtə'biliti] *n.* 可移植性；可移动性

Notes

1. 本句中 interacts with … , and with … 其中两个 with 是并列成分，即"与输出设备的设备驱动程序交互……和与……交互"。此外 being entered … 是现在分词被动语态，作定语修饰 information，即"输入的……"。
2. 本句中 entered by … 修饰前面的 engine。全句可译为：窗口管理程序把用户通过输入设备输入的消息返回给应用程序和脚本执行引擎。
3. 本句中 by looking … and issuing … ，这里 looking 和 issuing 是对等结构。全句可译为：用户通过观看分配给进程的窗口的内容和向其窗口处于激活状态的进程发命令来观察进展（即执行情况）和控制各进程。句中 when its window is in focus 就是指当该进程的窗口处于激活状态。
4. 本句中 of which … 是定语从句，which 代表 some unusual event，即 user should be aware of some unusual event。
5. hide sth. … from sb. 意为"把某事瞒着某人"。

注：本节所讲的终端独立性是指，应用程序不必知道不同终端设备之间的差别。本节介绍 Windows 的一些特性。

4.4 Myths of UNIX

UNIX *evokes* a wide range of emotions: loved by some for its power and flexibility, despised by others for its complex and *arcane* commands. UNIX has established a *checkered* reputation[1] in the world of computing.

Ease of Use. UNIX is *infamous* for its *glut* of arcane, *non-mnemonic*, and *cryptic* keyboard commands, each with many command-line switches, which can be incredibly confusing to remember[2]. Its SVR 4 implementation contains more than 2,000 commands. Many of these functions can be combined, using *pipes* and redirection. This illustrates one of UNIX's design fundamentals: the creation of *a large assortment of* very specialized and modular commands that can be combined to accomplish complex tasks.

While UNIX was essentially limited to use by software professionals at universities and in applications development houses, its complex command-line syntax and resulting flexibility were considered an advantage rather than a problem. But this same flexibility also creates a major drawback for using UNIX in a business-oriented market—the more flexible a system is, the more difficult it becomes to learn and operate.

UNIX's script languages provide one form of help. Using scripts, a system administrator can tailor the system to a set of users' needs quickly.

Another factor *mitigating* the difficulties of UNIX's arcane command language are the Graphic User Interfaces (GUIs), such as Motif, SunView, or OpenLook. GUIs, however, place another level of incompatibility problems on what is already a complex system. [3]

Motif has been *ported to* the most different architectures and (because it follows the Presentation Manager style) is perhaps closest in look and feel to a PC interface such as Microsoft Windows. SunView is also dominant because of its large installed base and the numbers of applications programmers who have become familiar with writing software to its specifications.

Binary Compatibility. UNIX marketers have looked with some envy at the huge-base of *shrink-wrapped* applications available in the DOS world and have promised that binary-compatible applications for systems is just *around the corner*. [4] These promises have yet to be met in any significant way. While binary compatibility is not yet available, it is getting easier to share data and applications across different machines.

Portability. Compared with most operating system sources, UNIX code is quite portable. It's written in C as opposed to assembly language[5], making it possible to move UNIX to different architectures. But a UNIX port to a new system is not a trivial matter, often taking several man-years of work, and can result in *glitches* and *malfunctions*, which may create very subtle inconsistencies in performance. These *bugs* are often difficult to identify and correct.

Having the source code available for your computer's operating system is beneficial and *detrimental*: if the OS lacks certain desirable features, having the sources in-house[6] greatly enhances a company's ability to make necessary changes.

On the other hand, the customized version of the operating system, with its new or modified features may later present compatibility problems with newer releases or purchased applications.

Words and Expressions

evoke [i'vəuk] *vt.* 引起，唤起
arcane [ɑː'kein] *adj.* 神秘的，秘密的
checkered (chequered) ['tʃekəd] *adj.* 多波折的，盛衰无常的
infamous ['infəməs] *adj.* 名声不好的，声名狼藉的
glut [glʌt] *vt.* 使充斥　*n.* 供过于求　a glut of 过剩的
non-mnemonic [nɔn-ni'mɔnik] *adj.* 不易记忆的，非记忆的
cryptic ['kriptik] *adj.* 含义模糊的；神秘的
pipe [paip] *n.* 管道，管
assortment [ə'sɔːtmənt] *n.* 花色品种；分类　a large assortment of　一批花色齐全的
mitigate ['mitigeit] *vt.* 使缓和，减轻；平息
port to　移植到
shrink-wrapped　简装的
around the corner　即将来临，即将面世
portability [ˌpɔːtə'biliti] *n.* 可移植性；可携带性；轻便
glitch [glitʃ] *n.* 小失灵，小故障
malfunction [mæl'fʌŋkʃən] *n.* 失灵，发生故障，运转不正常
bug [bʌg] *n.* 错误；故障
detrimental [ˌdetri'mentl] *adj.* 有害的；不利的

Notes

1. 这里 a checkered reputation 根据上下文可译为"褒贬不一的声誉"。
2. 本句中 be incredibly confusing to remember 译为"记起来让人难以置信地困惑"。
3. 本句有 place ... on what ... 结构。本句可译为：然而，图形用户接口在已经很复杂的系统上增加了另一层不兼容性问题。
   下一句中 closest ... to ... 意为"最接近于……"。
4. 本句主要是翻译问题，可译为：UNIX 销售者以某种羡慕的心情注视着 DOS 世界中可用的大量简装应用程序，并已承诺不同系统的二进制兼容的应用程序即将面世。
5. 本句中 as opposed to assembly language 这短语作状语，可译为"而不用汇编语言"。
6. 本句中 in-house 指"(机构)内部的，自身的"。having the sources in-house 指公司自己有源码……

注：这是一篇短文，简述了 UNIX 的两重性。选取目的是讲解翻译。

4.5　Using Linux in Embedded and Real-time Systems

Why Linux?

Intelligent dedicated systems and appliances used in interface, monitoring, communications, and control applications increasingly demand the services of a *sophisticated*, state-of-the-art operating systems. Many such systems require advanced capabilities like: high resolution and user-friendly graphical user interfaces (GUIs); TCP/IP *connectivity*; substitution of reliable (and low power) flash memory *solid state disk* for conventional disk drives; support for 32-bit *ultra*-high-speed CPUs; the use of large memory arrays; and seemingly infinite capacity storage devices including CD-ROMs and hard disks. [1]

This is not the stuff of *yesteryear's* "*stand-alone*" code, "*roll-your-own*" kernels, or "plain old DOS". No, those days are gone—forever!

Then too, consider the rapidly accelerating pace of hardware and *chipset* innovation—accompanied by extremely rapid *obsolescence* of the older devices. Combine these two, and you can see why it's become an enormous challenge for commercial RTOS vendors to keep up with the constant *churning* of hardware devices[2]. Supporting the newest devices in a timely manner—even just to *stay clear of* the *unrelenting steamroller* of chipset obsolescence—takes a large and constant resource *commitment*. [3] If it's a struggle for the commercial RTOS vendors to keep up, going it alone by writing standalone code or a roll-your-own kernel certainly makes no sense.

With the options narrowing, embedded system developers find themselves faced with a *dilemma*[4]:

- On the one hand, today's highly sophisticated and *empowered* intelligent embedded systems—based on the newest chips and hardware capabilities—demand nothing less than the power, sophistication, and *currency* of support provided by a popular high-end operating system like windows.
- On the other hand, embedded systems demand extremely high reliability (for non-stop, *unattended* operation) plus the ability to customize the OS to match an application's unique requirements.

Here's the *quandary*: general purpose desktop OSes (like Windows) aren't well suited to the unique needs of *appliance-like* embedded systems. However, commercial RTOSes, while designed to satisfy the reliability and configuration flexibility requirements of embedded applications, are increasingly less desirable due to their lack of standardization and their inability to keep pace with the rapid evolution of technology.

What's a Developer to Do?

Fortunately, a new and exciting alternative has emerged: *open-source* Linux. Linux offers powerful and sophisticated system management facilities, *a rich cadre of* device *support*, a *superb* reputation for reliability and *robustness*, and *extensive* documentation. Best of all (say system developers), Linux is available at no charge—and with completely free source code.

Is Linux, like Windows, too large and demanding of system resources to fit the constraints of embedded systems[5]? Unlike Windows, Linux is inherently modular and can be easily scaled into *compact* configurations—*barely* larger than DOS—that can even fit on a single floppy. What's more, since Linux source code is freely available, it's possible to customize the OS according to unique embedded system requirements.

It's not surprising, then, that open-source Linux has created a new OS development and support paradigm *wherein* thousands of developers continually contribute to a constantly evolving Linux code base. In addition, dozens of Linux-oriented software companies have *sprung up*—eager to support the needs of developers building a wide range of applications, ranging from factory automation to intelligent appliances.

Small Foot-print Linux

For many embedded systems, the main challenge in embedding Linux is to minimize system resource requirements in order to fit within constraints such as RAM, solid state disk (SSD), processor speed, and power consumption.[6] Embedded operation may require booting from (and fitting within) a DiskOnChip or CompactFlash SSD; or booting and running without a display and keyboard ("headless" operation); or loading the application from a remote device via an Ethernet LAN connection.

There are many sources of ready-made small *foot-print* Linux[7]. Included among these are a growing number of application-oriented Linux configurations and distributions that are tuned to specific applications. Some examples are routers, firewalls, internet/network appliances, network servers, gateways, etc.

You may also opt to create your own flavor of embedded Linux, starting from a standard distribution and leaving out modules you don't need. Even so, you should consider jump-starting your efforts by beginning with someone else's working configuration[8], since the source code of their version will be available for that purpose. Best of all, this sort of building on the efforts of others in the Linux community is not only completely legal—it's encouraged!

Real-time Linux

Many embedded systems require predictable and *bounded* responses to real-world events. Such "real-time" systems include factory automation, *data acquisition* and control systems, audio/video applications, and many other *computerized* products and devices. What's a "real-time system"? The commonly accepted definition of "real-time" performance is that real-world events must be responded to within a defined, predicable, and relatively short time interval.

Although Linux is not a real-time operating system (the Linux kernel does not provide the required event *prioritization* and preemption functions), there are currently several *add-on* options available that can bring real-time capabilities to Linux-based systems. The most common method is the *dual*-kernel approach. Using this approach, a general purpose (non-real-time) OS runs as a task under a real-time kernel. The general purpose OS provides functions such as disk read/write, LAN/communications, serial/parallel I/O, system initialization, memory management, etc., while the real-time kernel handles real-world event processing. You might think of this as a "have your cake and eat it too"[9] strategy, because it can preserve the benefits of a popular general purpose OS, while adding the capabilities of a real-time OS. In the case of Linux, you can retain full compatibility with standard Linux, while adding real-time functions in a non-interfering manner.

Of course, you could also *dive in* and modify Linux to convert it into a real-time operating system, since its source is openly available. But if you do this, you will be faced with the severe disadvantage of having a real-time Linux that can't keep pace, either features-*wise* or drivers-wise, with mainstream Linux[10]. In short, your customized Linux won't benefit from the continual Linux evolution that results from the *pooled efforts* of thousands of developers world-wide.

Linux is an Operating System, which acts as a communication service between the hardware (or physical equipment of a computer) and the software (or applications which use the hardware) of a computer system. The Linux Kernel (the core, much like a popcorn kernel) contains all of the features that you would expect in any Operating System.

Words and Expressions

sophisticated [sə'fistikeitid] *adj.* 复杂的；高级的；尖端的
connectivity [ˌkɔnek'tiviti] *n.* 连接性，连通性
solid state disk 固体盘
ultra ['ʌltrə] *adj.* 极端的；过分的；超……的
yesteryear ['jestəjiə] *n.* 过去不久，不久以前；去年
stand-alone ['stændəˌləun] *adj.* 独立的 *n.* 单机；独立
roll-your-own 你自己做
chipset ['tʃipset] *n.* 芯片组，芯片集
obsolescence [ˌɔbsə'lesns] *n.* 废弃；逐渐过时
churning ['tʃəːniŋ] *n.* 搅拌；（波浪）翻腾
stay clear of 避开；不接触
unrelenting [ˌʌnri'lentiŋ] *adj.* 不退让的，不屈不挠的；没缓和的
steamroller ['stiːmˌrəulə] *n.* 蒸汽压路机；高压手段
commitment [kə'mitmənt] *n.* 委托；提交；许诺
dilemma [di'lemə] *n.* 窘境，进退两难
empower [im'pauə(r)] *vt.* 授权；准许
currency ['kʌrənsi] *n.* 通用；流行；货币；流通
unattended [ˌʌnə'tendid] *adj.* 没人照顾的，无陪伴的
quandary ['kwɔndəri] *n.* 窘境，为难
appliance-like 类设备；类似于器具的
open-source 开放源码
a rich cadre of 丰富的骨干，丰富的基干
superb [sjuː'pəːb] *adj.* 宏伟的；极好的；超等的

robustness [rəuˈbʌstnis] n. 鲁棒性，坚固性，健壮性
extensive [iksˈtensiv] adj. 广泛的，广博的
compact [ˈkɔmpækt] adj. 紧凑的；紧缩的；紧密的
barely [ˈbɛəli] adv. 几乎没有；仅仅
wherein [wɛərˈin] adv. 在哪里，在什么地方
spring [spriŋ] up 出现，涌现
foot-print 脚印；占地面积
bounded [baundid] adj. 有界限的；受限的 n. 限界；边界
data acquisition 数据采集
computerized [kəmˈpjuːtəraizd] adj. 计算机化的 computerize 的过去式与过去分词
prioritization [praiˈɔritizeiʃən] n. 优先化；列入优先地位
add-on 扩充
dual [ˈdjuːəl] adj. 双的；二重的；对偶的
dive [daiv] vi. 钻研；探究；投入(in)
-wise [后缀][用以构成副词]表示"在……方面"；表示"方向"，"位置"
pooled efforts 共同协力
pool [puːl] vt. 合伙经营；共享；集中(智慧)

Abbreviation

RTOS (Real Time OS) 实时操作系统

Notes

1. advanced capabilities 可译为"高级性能"。本句中的短语 substitution of ... for ... 译为"用可靠的闪存固态盘代替常规的磁盘机"。

   下一句可译为：这不是(不久)以前的"独立"代码、"自己写的"核或"简单的老 DOS"。

2. 本句中 churning 形容硬件设备迅速发展，constant churning of hardware devices 可译为"硬件设备的不断出新"。上一句中 rapidly accelerating pace ... 可译为"（硬件和芯片组）迅速加快的革新步伐"。

3. 本句中，主语是 supporting ...，动名词短语。谓语是 takes，可译为"需要，承担"。stay clear of ... 意为"避开；不接触"。全句可译为：及时地支持最新设备(即使仅是为了避开芯片组过时(造成)的持续高压)需要大量和持续的资源投入。

   下一句中 keep up 指"紧跟硬件发展"。go it alone 即"一个人，单枪匹马"。条件从句中是 it is ... for ... to do ... 结构。全句可译为：如果商用实时操作系统供应商必须奋力紧跟硬件发展的话，那么编写独立的代码或自己的核这种单枪匹马的做法一定是毫无意义的。

4. With the options narrowing 译为"因为选择范围很小"。nothing less than ... 意为"和……一模一样，完全是"。全句可译为：因为选择范围很小，嵌入式系统的开发商面临这样一种困境：一方面，今天高度复杂且授权的智能嵌入式系统(基于最新的芯片和硬件性能)所需要的正是流行的高档操作系统(如 Windows)提供的那种能力、精致性以及通用性。

5. 本句是 too ... to ... 结构，即"太大太多，以致不能满足……"。

下一句中 modular 意思是"模块化的，模块的"。scale into 意思是"缩减成"。全句可译为：不像 Windows，Linux 本来就是模块化的，并且能够很容易地缩减成紧缩配置，这种配置几乎与 DOS 差不多大，甚至能放到一张软盘上。

6. 本句中 fit within constraints ... 意思是"适应……这些约束"。全句可译为：对许多嵌入式系统，为了适应诸如 RAM、固态盘、处理机速度，以及功耗的约束，嵌入的 Linux 的主要任务是，使得系统所需的资源最少。

下一句中 DiskOnChip 可译为"芯片盘，半导体盘"。CompactFlash SSD 可译为"紧凑闪存固态盘"。

7. 本句中 ready-made small foot-print Linux 可译为"现成的小 Linux"。

下一句中 distributions 可译为"发行"，tune to 可译为"适用于"。

8. 本句中 jump-starting 开始的短语可译为"在别人的工作配置基础上开始你的工作"。

9. have your cake and eat it too 意为"两者兼得"。

10. either features-wise or drivers-wise 可译为"不论是特性方面，还是驱动程序方面"。can't keep pace with ... 可译为"不能与……同步前进"。

注：本选文很有说服力地阐述了为何在嵌入式和实时系统中都会选用 Linux。2001 年举行了 Linux 世界大会。

Terms

Real-time system

A computer and/or a software system that reacts to events before the events become obsolete. For example, airline collision avoidance systems must process radar input, detect a possible collision, and warn air traffic controllers or pilots while they still have time to react.

Wizard ['wizəd]（向导）

A wizard—also called a coach, assistant, tutor, expert, and advisor-is a tool to guide you through a step-by-step procedure for completing a standard task, such as prepare a letter, invoice, report, fax, and the like. Wizard do this either by providing directed tutorials or by performing the difficult phases of a task automatically, using the input supplied by the user via some type of dialog box.

Thread

In most traditional operating systems, each process has an address space and a single thread of control. Nevertheless, there are frequently situations in which it is desirable to have multiple thread of control sharing one address space but running in quasi-parallel, as though they were separate processes (except for the shared address space). Thread is sometimes called lightweight process. In many respects, threads are like little mini-processes. Each thread runs strictly sequentially and has its own program counter and stack to keep track of where it is. Threads on a multiprocessor run in parallel.

Hyper-threading(HT，超线程)

Hyper-threading is an Intel-proprietary technology used to improve parallelization of computations (doing multiple tasks at once) performed on PC microprocessors. For each processor core that is physically present, the operating system addresses two virtual processors, and shares the workload between them when possible. Two virtual processors run two threads.

Software

Software is a general term for the various kinds of programs used to operate computers and related devices. Software can be thought of as the variable part of a computer and hardware the invariable part. Software is often divided into application software (programs that do work users are directly interested in) and system software (which includes operating systems and any program that supports application software).

Utility(实用程序)

A difficult-to-classify category of software is the utility, which is a small useful program with limited capability. Some utilities come with operating systems. Like applications, utilities tend to be separately installable and capable of being used independently from the rest of the operating system.

Android (安卓)

Android is an open-source platform founded in October 2003 by Andy Rubin and backed by Google, along with major hardware and software developers (such as Intel, HTC, ARM, Motorola and Samsung) that form the Open Handset Alliance. By 2010, Android became the best-selling smartphone platform.

DOS

DOS /dɔs/, short for Disk Operating System, is an acronym for several closely related operating systems that dominated the IBM PC compatible market between 1981 and 1995, or until about 2000 including the partially DOS-based Microsoft Windows.

Android (operating system)

Android is a mobile operating system developed by Google. It is based on a modified version of the Linux kernel and other open source software, and is designed primarily for touchscreen mobile devices such as smartphones and tablets. In addition, Google has further developed Android TV for televisions, Android Auto for cars and Wear OS for wrist watches, each with a specialized user interface.

Initially developed by Android Inc., which Google bought in 2005, Android was unveiled in 2007, with the first commercial Android device launched in September 2008. The operating system has since gone through multiple major releases, with the current version being 9 "Pie" (the ninth major release and the 16th version of the Android mobile operating system) released in August 2018. The core Android source code is known as Android Open Source Project (AOSP), and is primarily licensed under the Apache License.

Android is also associated with a suite of proprietary software developed by Google, called Google Mobile Services (GMS) that very frequently comes pre-installed in devices, which usually includes the Google Chrome web browser and Google Search and always includes core apps for services such as Gmail, as well as the application store and digital distribution platform Google Play, and associated development platform.

Android has been the best-selling OS worldwide on smartphones since 2011 and on tablets since 2013. As of May 2017, it has over two billion monthly active users, the largest installed base of any operating system, and as of December 2018, the Google Play store features over 2.6 million apps.

iOS (formerly iPhone OS)

iOS is a mobile operating system created and developed by Apple Inc. exclusively for its hardware. It is the operating system that presently powers many of the company's mobile devices, including the iPhone (Originally unveiled in 2007), iPad (January 2010), and iPod Touch (September 2007). It is the second most popular mobile operating system globally after Android.

As of March 2018, Apple's App Store contains more than 2.1 million iOS applications, 1 million of which are native for iPads. These mobile apps have collectively been downloaded more than 130 billion times.

The iOS user interface is based upon direct manipulation, using multi-touch gestures. Interface control elements consist of sliders, switches, and buttons. Interaction with the OS includes gestures such as *swipe*, *tap*, *pinch*, and *reverse pinch*, all of which have specific definitions within the context of the iOS operating system and its multi-touch interface.

Major versions of iOS are released annually. The current version, iOS 12, was released on September 17, 2018. It is available for all iOS devices with 64-bit processors; the iPhone 5S and later iPhone models, the iPad (2017), the iPad Air and later iPad Air models, all iPad Pro models, the iPad Mini 2 and later iPad Mini models, and the sixth-generation iPod Touch. On all recent iOS devices, iOS regularly checks on the availability of an update, and if one is available, will prompt the user to permit its automatic installation.

Network operating system

A network operating system is a specialized operating system for a network device such as a router, switch or firewall.

Historically operating systems with networking capabilities were described as network operating system, because they allowed personal computers (PCs) to participate in computer networks and shared file and printer access within a local area network (LAN). This description of operating systems is now largely historical, as common operating systems include a network stack to support a client-server model.

Exercises

1. Translate the text of Lesson 4.2 into Chinese.
2. Topics for oral workshop.
   - Talk about the basic functions of OS, the main components of OS and their functions.
   - Describe advanced capabilities of OS, such as GUI, window managers, network (TCP/IP connectivity), auto-detection of devices (plug and play).
   - Talk about popular OSs: Windows, UNIX and Linux; their distinguished features; their different application areas (client, server, embedded system, real-time system).
3. Translate the following into English.

软件系统能够分为两大类:应用软件和系统软件。应用软件由执行特定任务的程序组成。与应用软件形成对比,系统软件包含大量的程序。这些程序启动计算机并且作为(function as)所有硬件组成部分和应用软件的主要协调者。没有系统软件装入你的计算机 RAM 中,你的硬件和应用软件就没有用。

系统软件可以分(组)为三个基本部分:操作系统、实用程序软件和语言翻译程序。安装的大多数实用程序软件都是由一些程序组成的,这些程序执行还未包括在 OS 中,但对计算机安装是一些基本的动作(activity)。在某种意义上,实用软件是由扩充 OS 能力的那些软件部件组成的。

计算机的 OS 是管理其活动的程序的主要集合。OS 的主要日常工作(chore)是管理和控制。OS 保证用户请求的所有动作都是正当的,并且以有序的方式被处理。它也管理计算机系统的资源,以便有效、一致地执行这些操作。

应用软件是一种软件,被设计来帮助你解决特定业务的问题或执行特定的业务任务。于是应用软件是最接近于你的软件层。大概有四类应用软件:生产软件、业务(business)/专业软件、娱乐软件和教育/参考软件。

4. Listen to the video "Flash memory" and write down the text.

Unit 5   Computer Networks

5.1   Internet

Many networks exist in the world, often with different hardware and software. People connected to one network often want to communicate with people attached to a different one. This desire requires connecting together different, and frequently *incompatible* networks, sometimes by using machines called *gateways* to make the connection and provide the necessary translation, both in terms of hardware and software. A collection of interconnected networks is called an *internetwork* or just *Internet*.

A common form of Internet is a collection of LANs connected by a WAN. The Internet (note uppercase I) means a specific worldwide internet that is widely used to connect universities, government offices, companies, and of late, private individuals.

By 1995, there were multiple *backbones*, hundreds of mid-level (i.e. regional) networks, tens of thousands of LANs, millions of *hosts*, and tens of millions of users. The size doubles approximately every year. Much of the growth comes from connecting existing networks to the Internet.

The glue that holds the Internet together is the TCP/IP *reference model* and TCP/IP *protocol stack*. TCP/IP makes universal service possible and can be compared to the telephone system or the adoption of standard *gauge* by the railroads in the 19th century.

The Internet is a vast global community of real people who constantly generate more *unadulterated stuff* on more topics than you could ever read in your lifetime. [1] If you haven't tapped into the net yet, here's what you're missing:

E-mail. Let you send messages to Russia, Japan and so on.

File Transfer Protocol (FTP). *Download* files free of charge from thousands of computers around the globe.

Usenet newsgroups. More *banter*, *blather*, and *nuggets* of wisdom than you'll ever be able to read on everything from *archery* to stock. [2]

The World Wide Web (WWW). A fast-growing global network of graphical electronic documents you can browse, interact with, and even create yourself. [3]

New technologies. Global chat, *video conferencing*, fax, free international phone calls and more.

The World Wide Web

The World Wide Web is an *architectural framework* for accessing linked documents spread out over thousands of machines all over the Internet.

Since the Web is basically a client-server system, we discuss both the client (i. e. user) side and the server side.

The Client Side

From the users' point of view, the Web consists of a vast, worldwide collection of documents, usually just called pages for short. Each page may contain links (pointers) to other, related pages[4], anywhere in the world. Users can follow a link (e. g., by clicking on it), which then takes them to the page pointed to. This process can be repeated indefinitely, possibly *traversing* hundreds of linked pages while doing so. Pages that point to other pages are said to use **hypertext**. Web pages can combine color desktop publishing, hypertext linking, *interactive scripting*, sound, video, and even *virtual reality*. When hypertext pages are mixed with other media, the result is called **hypermedia**.

Pages are viewed with a **browser**.

The Server Side

Every *Web site* has a server process listening to TCP port 80 for incoming connections from clients (normally browsers). After a connection has been established, the client sends one request and the server sends one reply. Then the connection is released. The protocol that defines the legal requests and replies is called HTTP. A simple example using it may provide a reasonable idea of how Web servers work. Figure 5-1 shows how the various parts of the Web model fit together.

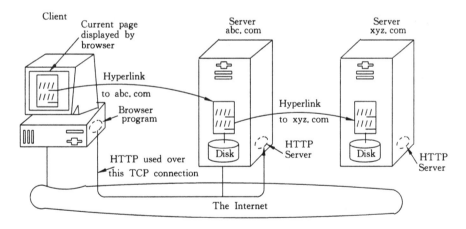

Figure 5-1 The parts of the Web model

***Surfing* the Web**

Once your Web *browser* is open and you are connected to the Internet, the page currently designated as your browser's starting page or home page will be displayed within the browser window. (Usually this page is the home page for your browser's, school's, or ISP's Web site, but it can usually be changed to any page using your browser's Options or Preferences dialog box.)

All browsers have navigational tools to help you move forward or backward through the pages viewed in your current Internet *session*, as well as buttons or menu options to print Web pages when necessary. Figure 5-2 illustrates the most common components of the Microsoft Internet Explorer and Netscape Navigator browsers.

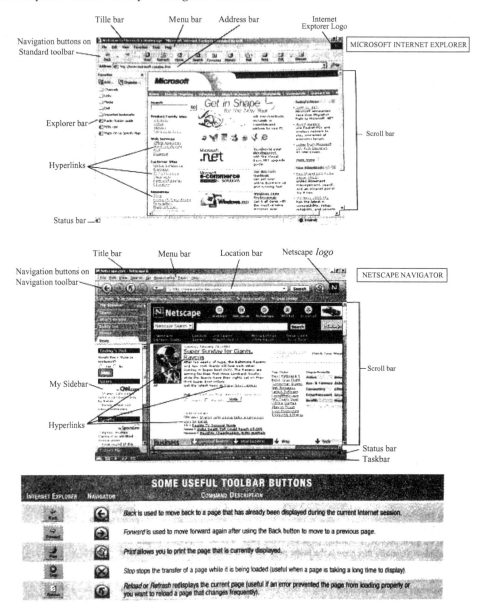

Figure 5-2 Browsers (Microsoft Internet Explorer and Netscape Navigator have emerged as the two most widely used browsers. Both use similar interfaces and commands)

Using URLs and Hyperlinks

To change from the starting Web page to a new Web page, you can type the appropriate URL in the browser's address bar or location bar and press Enter. You can either edit the existing URL or delete it and type a new one. Be sure to match the spelling, *capitalization*, and punctuation exactly. If you don't know the appropriate URL to type, you can search for an appropriate page, as discussed shortly.

If there is a hyperlink displayed for the page you would like to go to, simply click on the link. Remember, hyperlinks can be either text or image-based. If you are not sure if something on a page is a link or not, *rest* the mouse pointer *on* it for a moment. If it is a hyperlink, the pointer should change to indicate that it is a link. The URL for the new page is also displayed on the browser's status bar. Once you click the hyperlink, the appropriate page is displayed. To return to a previous page, click the Back button on your browser's toolbar. To print the current Web page, use the browser's Print button or select Print from the browser's File menu.

Things You May Encounter on a Web Page

You will encounter a variety of different objects on Web pages as you explore the World Wide Web. Though we can't go into an in-depth discussion on the various possible Web-page components here, it is good to be familiar with the most common ones so you'll know how to deal with them as you encounter them. Some common things you may run into are illustrated in Figure 5-3.

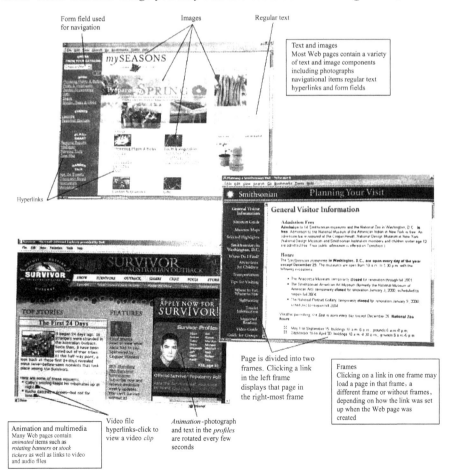

Figure 5-3 Common things that you may encounter on a Web page

Words and Expressions

incompatible ['inkəm'pætəbl] *adj*. 不(可)兼容的，不相容的
gateway ['geitwei] *n*. 网关，入口
internetwork   互联网
Internet ['intənet] *n*. 国际互联网，因特网
backbone ['bækbəun] *n*. 主干(网)，骨干网
host [həust] *n*. 主机，宿主机 *vt*. 托管；做主人，主持
reference model   参考模型
protocol stack   协议栈
gauge [geidʒ] *n*. 轨距；标准尺寸，标准规格；量规，量器，量计，表
unadulterated stuff   不掺杂的资料；真正的资料
download (upload) *vt*. 下载（上传）
Usenet newsgroup   新闻组，用户新闻组
banter ['bæntə] *n*. 逗弄 *vi*. 开玩笑 *vt*. 逗弄
blather ['blæðə] *n*. 胡说；废话 *vi*. 说废话；喋喋不休地说
nugget ['nʌgit] *n*. 块金；小而有价值的东西
archery ['ɑːtʃəri] *n*. 射箭术
video conferencing   开电视会议，视频会议
architectural framework   体系结构框架
traverse ['trævəːs] *vt*. 遍历，通过 *n*. 遍历；穿过；横贯
hypertext ['haipətekst] *n*. 超文本
interactive scripting   交互式文稿
virtual reality (VR)   虚拟现实
hypermedia ['haipəmiːdiə] *n*. 超媒体
Web site   万维网站
surf [səːf] *vt*. 在……冲浪 *vi*. 做冲浪运动 *n*. 海浪，拍岸浪
browser [brauzə] *n*. 浏览器
preference ['prefrəns] *n*. 偏爱；优先选择
session ['seʃən] *n*. 会话期；会话；会议；会晤
capitalization ['kæpitəlai'zeiʃən] *n*. 大写；资本化；股本；资本总额
rest ... on ...   把……放在……上面，搁在；保持(状态)
logo ['ləugəu] *n*. 标识语；Logo教学语言；标识，标志；商标
clip [klip] *n*. 剪辑；剪下来的片段
animation ['æni'meiʃən] *n*. 动画片(制作)；活泼，有生气
animate ['ænimeit] *vt*. 使活泼；用动画片(制作)
rotating banner   旋转的大字标题
stock ticker   股票行情自动收录器
profile ['prəufail] *n*. 配置文件，概要文件；侧面；轮廓，外形；纵剖面(图)，断面(图)

Abbreviations

LAN (Local Area Network)　局域网
WAN (Wide Area Network)　广域网
TCP/IP (Transmission Control Protocol/Internet Protocol)　传输控制协议/互联网协议
FTP (File Transfer Protocol)　文件传输协议
HTTP (Hyper Text Transport Protocol)　超文本传输协议
URL (Uniform Resource Locator)　统一资源定位器
DNS (Domain Name System)　域名系统
ISP (Internet Service Provider)　Internet 服务提供商
IE (Internet Explorer)　微软的网络浏览器名

Notes

1. 本句中 who 引导的从句中用了 more ... than 结构，本意是"比……更多"。全句可译为：因特网是现实世界中人们的巨大的全球性社区。在这个网上，人们在众多主题上不断推出的有价值的资料是你一辈子也读不完的。
  下一句中的 tap into the net 可译为"上网"。tap into 意为"搭上"。
2. 本句也用 more ... than 结构。全句可译为：其内容比你从射箭术到股票行情所能读到的一切更有趣、更热闹和更智慧。
说明：在网上新闻组中，你可随意发表见解，因此可以开玩笑，甚至胡说、凑热闹。
3. 本句中 you can ... 是定语从句，修饰 document。全句可译为：一个快速发展的图形电子文档全球网，你可以浏览其图形电子文档，与之交互，甚至可以建立自己的文档。
4. 本句中 links (pointer) to other, related pages 可译为"指向其他相关页面的连接（指针）"。
  下一句中 which 引导的从句修饰 a link，从句含义为"然后，这个连接把用户带到该连接所指向的页面"。
注：本节内容是对 Internet 的一般介绍，包括 WWW、客户-服务器，以及 TCP/IP、FTP、HTTP、URL、DNS 等术语。

Terms

LAN (lan)

Acronym for **L**ocal **A**rea **N**etwork. A group of computers and other devices dispersed over a relatively limited area and connected by a communications link that enables any device to interact with any other on the network. LANs commonly include microcomputers and shared resources such as laser printers and large hard disks. The devices on a LAN are known as nodes, and the nodes are connected by cables through which messages are transmitted. See also baseband network, broadband network.

MAN

Acronym for **M**etropolitan **A**rea **N**etwork. A high-speed network that can carry voice, data, and images at up to 200 Mbps over distances of up to 75 km. A MAN, which can include one or more LANs as well as telecommunications equipment such as microwave and satellite relay stations, is smaller than a WAN but generally operates at a higher speed.

WAN (Wide Area Network)

A communications network that connects geographically separated areas.

In WAN technologies, a network usually consists of a series of complex computers called packet switches interconnected by communication lines and modems. The size of the network can be extended by adding a new switch and another communication line.

TCP/IP

Acronym for Transmission Control Protocol/Internet Protocol. A protocol developed by the Department of Defense for communications between computers. It is built into the UNIX system and has become the defacto standard for data transmission over networks, including the Internet.

FTP

Acronym for the File Transfer Protocol, the protocol used for copying files to and from remote computer systems on a network using TCP/IP such as the Internet. This protocol also allows users to use FTP commands to work with files, such as listing files and directories on the remote system.

HTTP

Acronym for Hypertext Transfer Protocol. The client/server protocol used to access information on the World Wide Web. By means of URLs, HTTP supports access not only to documents written in HTML, but also to files retrievable through FTP or Gopher, as well as to information posted in newsgroups.

Client/server architecture

By splitting the processing of an application between two distinct components: a "front-end" client and a "back-end" server. The client component is a complete, stand-alone personal computer (not a "dumb" terminal), and it offers the user its full range of power and features for running applications. The server component can be a personal computer, a minicomputer, or a mainframe that provides the traditional strengths offered by minicomputers and mainframes in a time-sharing environment: data management, information sharing between clients, and sophisticated network administration and security features. The client and server machines work together to accomplish the processing of the application being used. Not only does this increase the processing power available over older architectures, but also uses that power more efficiently. The client portion of the application is typically optimized for user interaction, whereas the server portion provides the centralized, multiuser functionality.

URL

Acronym for **U**niform **R**esource **L**ocator. An address for a resource on the Internet. URLs are used by Web browsers to locate these resources. An URL specifies the protocol to be used in accessing the resource (such as http:// for a World Wide Web page or ftp:// for an FTP site), the name of the server on which the resource resides (such as www.whitehouse.gov), and, optionally, the path to a resource (such as an HTML document or a file on that server).

DNS

(1) Acronym for **D**omain **N**ame **S**ystem. The system by which hosts on the Internet have both domain name addresses, such as bluestem.prairienet.org, and IP addresses, such as 192.17.3.4. The domain name address is used by human users and is automatically translated into the numerical IP address, which is used by the packet routing software.

(2) Acronym for **D**omain **N**ame **S**ervice. The Internet utility that implements the Domain Name System (see definition 1). DNS servers, also called name servers, maintain databases containing the addresses and are accessed transparently to the user.

5.2 Extending Your Markup: An XML Tutorial

By now, no doubt, you've heard the acronym XML. You've probably also heard that XML (a) is simple and (b) will solve all your problems. Sounds like magic, doesn't it? But then you look a little deeper and encounter more three-letter acronyms, like DTD, XSL, RDF, and DOM. You begin to doubt XML's simplicity (and you never believed XML could solve all your problems in the first place). In this short tutorial I present what I think are the essential concepts of XML, and hopefully will convince you that despite the *hype*, XML is important for presentation, exchange, and management of information.

More Meaningful Markup

The Standard Generalized Markup Language (SGML) is a rather complicated language that lets you define structure for documents. The Hypertext Markup Language is an application of SGML that has a fixed set of markups. HTML is primarily used for layout on the Web. It tells nothing about the content of the data.

Both SGML and HTML heavily influenced the development of the Extensible Markup Language (XML), a semantic language that lets you meaningfully annotate text. Meaningful annotation is, in essence, what XML is all about. Figure 5-4(a) shows a bibliography entry in HTML; Figure 5-4(b) shows that same entry written in XML. Because the structure of the information in Figure 5-4(b) is more explicit, it is much easier for humans to read and computers to process.

XML Syntax

Syntactically, XML documents look like HTML documents. A *well-formed* XML document—one that *conforms* to the XML syntax—starts with a *prolog* and contains exactly one *element*.[1] Additionally, an arbitrary number of comments and processing instructions (which are not explained in this tutorial) can be included. The prolog looks something like this:

⟨? xml version="1.0" standalone="yes" encoding="UTF-8"?⟩

It tells you that your document follows XML version 1.0, is stand-alone (that is, not accompanied by a document type definition, or DTD), and uses *Unicode* Transformation Format 8-bit (UTF-8) encoding. If the document was accompanied by a DTD, the DTD declaration would also be part of the prolog.

The single element can be viewed as the root of the document. Elements can be *nested*, and attributes can be attached to them. Attribute values must be in quotes, and *tags* must be balanced[2]. Empty element tags must either end with a/> or be explicitly closed.

For a complete description of the XML syntax, see the W3C XML Recommendation.

Defining Structure

DTDs, which are also used in SGML, define the structure of XML documents. It's easiest to think of a DTD as a *context-free grammar*.[3] In particular, DTDs let users specify the set of tags, the order of tags, and the attributes associated with each. A well-formed XML document that conforms to its DTD is called *valid*. Figure 5-5 shows a simple DTD for the bibliography example in Figure 5-4.

⟨UL⟩
⟨LI⟩Aho,A. V.,Sethi,R.,Ullman,J. D. :⟨EM⟩Compilers:Principles,Techniques,and Tools ⟨/EM⟩,
    Addison-Wesley,1985
⟨/UL⟩

(a)

⟨BOOK⟩
  ⟨AUTHOR⟩Aho,A. V.⟨/AUTHOR⟩
  ⟨AUTHOR⟩Sethi,R.⟨/AUTHOR⟩
  ⟨AUTHOR⟩Ullman,J. D.⟨/AUTHOR⟩
  ⟨TITLE⟩Compilers:Principles,Techniques,and Tools⟨/TITLE⟩
  ⟨PUBLISHER⟩Addison-Wesley⟨/PUBLISHER⟩
  ⟨YEAR⟩1985⟨/YEAR⟩
⟨/BOOK⟩

(b)

Figure 5-4　A bibliography entry (a) in HTML and (b) in XML (The HTML description is layout oriented, while the XML description is structure oriented)

A DTD is *declared* in the XML document's prolog using the ! DOCTYPE tag. The DTD can be included within the XML document, or it can be contained in a separate file. If the DTD is in a separate file, say document.dtd, the XML document includes the statement:

⟨! DOCTYPE Document SYSTEM "document. dtd"⟩

You can also *refer to* an external DTD through a URI.

DTD Elements

Elements can be either *nonterminal or terminal*. Nonterminal elements (BIB and BOOK in Figure 5-5) contain subelements, which can be grouped as *sequences or choices*.[4] A sequence defines the order in which subelements must appear. A choice gives a list of alternatives for subelements. Sequences and choices can contain each other.

In Figure 5-5, both BIB and BOOK are sequences. A BOOK element has to have at least one AUTHOR(indicated by +), followed by a TITLE and, *optionally*, by a PUBLISHER and a YEAR(indicated by?). A choice is indicated by the logical operator. The example below shows a choice in which a SECTION can either be a TITLE followed by at least one PARAGRAPH, or a TITLE followed by zero or more PARAGRAPHs (indicated by the *wildcard* *)and at least one SUBSECTION.

⟨! ELEMENT SECTION((TITLE,(PARAGRAPH+))|(TITLE,(PARAGRAPH*), (SUBSECTION+))⟩

⟨! DOCTYPE bib[
　⟨! ELEMENT BIB(BOOK+)⟩
　⟨! ELEMENT BOOK(AUTHOR+,TITLE,PUBLISHER?,YEAR?)⟩
　⟨! ELEMENT AUTHOR(#PCDATA)⟩
　⟨! ELEMENT TITLE(#PCDATA)⟩
　⟨! ELEMENT PUBLISHER(#PCDATA)⟩
　⟨! ELEMENT YEAR (#PCDATA)⟩
]⟩

Figure 5-5　A DTD for the bibliography example (The DTD defines a grammar for documents)

Terminal elements can be declared as parsed character data(#PCDATA) or as EMPTY. Elements can also be declared as ANY. An element declared as ANY is a terminal element in the grammar, but it can contain subelements of any declared type, as well as character data.

Popular applications of DTDs include XHTML and the Chemical Markup Language (CML).

Stylesheets and More

The Extensible Stylesheet Language(XSL)is actually two languages: a *transformation* language (called XSL transformations, or XSLT) and a formatting language (XSL *formatting objects*).

XSLT allows you to transform XML into HTML, thus bypassing the formatting language. It also lets you restructure XML documents so that different kinds of XML representations can be mapped onto one another.[5] This makes XSLT very useful for *electronic commerce* and electronic data interchange.

Figure 5-6 shows XSLT code for turning the bibliography entry in Figure 5-4(b) into an

HTML representation like the one in Figure 5-4(a). [6]

The style sheet element contains a collection of *template* elements and can be included in the document it is to be applied to. [7] The match attribute of a template element *addresses* the document structures, the template can be applied to. Arbitrary XPath expressions can be incorporated into valid values of the match attribute. [8]

The template element indicates the output to be produced. In Figure 5-6, the root template produces an 〈HTML〉…〈/HTML〉 *frame*, the template for BIB elements produces an *unordered list* frame (〈UL〉…〈/UL〉), and the template for BOOK elements produces a list entry frame (〈LI〉…〈/LI〉). These templates all contain an 〈xsl:apply-templates/〉 element that causes the XSLT processor to apply applicable templates on the subelements of the current element *recursively*. [9]

The templates for AUTHOR and TITLE elements contain an 〈xsl:value-of select="."/〉 subelement, which causes the value of the current element (indicated by the path expression ".") to be output. Incidentally, the value of the select attribute can be an arbitrary XPath expression. In the example, the value of the TITLE element is emphasized in the generated HTML output. [10]

For a more detailed discussion of XSL, see http://www.w3.org/Style/XSL/.

```
〈xsl:stylesheet version="1.0"
    xmlns:xsl="http://www.w3.org/1999/XSL/Transform"〉
〈xsl:template match="/"〉
    〈HTML〉〈xsl:apply-templates/〉〈/HTML〉
〈/xsl:template〉

〈xsl:template match="BIB"〉
    〈UL〉〈xsl:apply-templates/〉〈/UL〉
〈/xsl:template〉

〈xsl:template match="BOOK"〉
    〈LI〉〈xsl:apply-templates/〉〈/LI〉
〈/xsl:template〉

〈xsl:template match="AUTHOR"〉
    〈xsl:value-of select="."/〉
〈/xsl:template〉

〈xsl:template match="TITLE"〉
    〈EM〉〈xsl:value-of select="."/〉〈/EM〉
〈/xsl:template〉

〈/xsl:stylesheet〉
```

Figure 5-6   A simple XSL stylesheet for transforming the XML bibliography into an HTML unordered list

Words and Expressions

hype [haip] *n.* 大肆宣传；广告；欺骗

well-formed 格式良好的，形式良好的
conform [kənˈfɔːm] vi. 符合；一致 vt. 使符合；使一致
prolog [ˈprəulɔɡ] n. 序言；声明
Unicode [ˈjuːniˌkəud] n. 统一码，采用双字节对字符进行编码；统一字符编码标准
nest [nest] vi. 嵌套；筑巢 n. (鸟)窝，巢
tag [tæɡ] n. 标记，标签
context-free grammar 上下文无关文法
declare [diˈklɛə] vt. 说明
refer to 参考；涉及；指的是；适用于
nonterminal [nɔnˈtəːminəl] n. 非终结符
terminal [ˈtəːminəl] n. [计] 终结符；终端 adj. 终端的
optionally [ˈɔpʃənəli] adv. 任选，可选择地
wildcard [ˈwaildkɑːd] n. 通配符
transformation [trænsfəˈmeiʃən] n. 转换，变换
formatting object 格式化对象
electronic commerce 电子商务
template [ˈtempleit] n. 模板
address [əˈdres] vt. 处理；对付；向……讲话；写姓名地址，致函；[计]编址；寻址
frame [freim] n. 框架；构架；结构；帧
unordered list 无序表
recursively [riˈkəːsivli] adv. 递归地

Abbreviations

XML (eXtensible Markup Language) 可扩展置标语言
DTD (Document Type Definition) 文档类型定义
XSL (eXtensible Stylesheet Language) 可扩展样式单语言
RDF (Resource Description Format) 资源描述格式
DOM (Document Object Model) 文档对象模型
SGML (Standard Generalized Markup Language) 标准通用置标语言
URI (Uniform Resource Identifier) 统一资源标识符
EDI (Electronic Data Interchange) 电子数据交换

Notes

1. 本句中 one 指 document，prolog 可译为"声明"。全句可译为：格式良好的 XML 文档，即符合 XML 语法的文档，以一个声明开始且只有一个元素。
2. 本句中 balance 译成"平衡"，是指〈标记〉……〈/标记〉这种形式。
3. 句子可译为：最容易的是把 DTD 想象成上下文无关文法。
4. 本句可译为：非终结元素(图 5-5 中的 BIB 和 BOOK)包含多个子元素，多个子元素能够聚集成(顺序)序列或选择。后面两句解释了序列和选择，如选择给出一列子元素可

选择的项。

5. restructure 意为"重新组织,重构"。全句可译为:XSLT 也允许你重构 XML 文档,使得不同种类的 XML 表示能够相互映射(变换)。

6. 本句中 for turning ... into ... 修饰前面的 code,即"把图 5-4(b)中文献目录条目的 XML 表示转换成图 5-4(a)中的 HTML 表示"。

7. 本句中 it is to be applied to 修饰前面的 document。it 指 stylesheet element。这个从句直译是"它将要被应用到"。全句可译为:样式表元素包含一组模板元素,可以把样式表元素包括在将要应用它的文档中。从本句话的意思可看出,样式表通常是一个单独的文档,它可以用于诸 XML 文档。下一句中 the template can be applied to 类似于上面的 it is to be applied to,前面均省略了 that。

8. XPath expression 即"XPath 表达式",属于 XML 链接规范。XML 链接规范有的指明建立资源之间的链接;有的用于指定相对 URL 的绝对路径;有的用于对 XML 文档中的片段或项(fragment,item)定位。XPath 表达式属于最后一种。本句可译为:任意多个 XPath 表达式能够混合到匹配属性的有效值中。

9. 本句中 processor 指"处理器,处理程序"。全句可译为:这些模板都包含〈xsl:apply-template/〉元素,该元素使 XSLT 处理程序把可应用的模板递归地应用于当前元素的诸子元素上。

10. 本句中 emphasize 意为"强调,加强"。〈EM〉...〈/EM〉中间的内容可以是文本、声音等。若是文本,则对文本加粗;若是声音,则对声音放大。全句可译为:在本例子中,TITLE 元素的值在生成的 HTML 输出中被加强。

注:本节通过一个简单的例子介绍了新一代置标语言 XML,包括它的 DTD 和 stylesheet。现在用 schema 来代替 DTD 定义结构。

5.3 Network Protocols

5.3.1 *Protocol Hierarchies*

To reduce their design complexity, most networks are organized as a series of *layers* or levels, each one built upon the one below it. In all networks, the purpose of each layer is to offer certain services to the higher layers, shielding those layers from the details of how the offered services are actually implemented.

Layer n on one machine carries on a conversation with layer n on another machine. The rules and conventions used in this conversation are collectively known as the layer n **protocol**. Basically, a protocol is an agreement between the communicating parties on how communication is to proceed.

The ISO/OSI *model* has seven layers (see Figure 5-7). The entities comprising the corresponding layers on different machines are called **peers**. In other words, it is the peers that communicate using the protocol.

In reality, no data are directly transferred from layer n on one machine to layer n on another machine. Instead, each layer passes data and control information to the layer

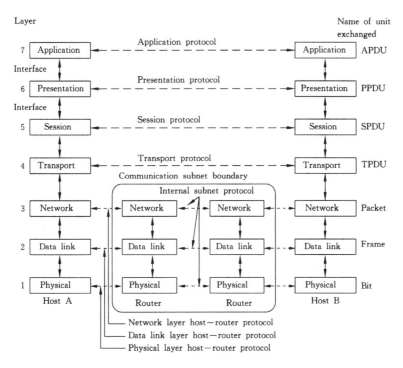

Figure 5-7 The OSI reference model

immediately below it, until the lowest layer is reached.

Between each pair of adjacent layers there is an **interface**. The interface defines which *primitive operations* and services the lower layer offers to the upper one. When network designers decide how many layers to include in a network and what each one should do, one of the most important considerations is defining clean interfaces between the layers. Doing so, in turn, requires that each layer perform a specific collection of well-understood functions.

A set of layers and protocols is called a **network architecture**. The specification of an architecture must contain enough information to allow an implementer to write the program or build the hardware for each layer so that it will correctly obey the appropriate protocol. A list of protocols used by a certain system, one protocol per layer, is called a **protocol stack**.

An analogy may help explain the idea of multilayer communication. Imagine two philosophers (peer processes in layer 3). one of whom speaks *Urdu* and English and one of whom speaks French and Chinese. Since they have no common language, they each *engage* a translator (peer processes at layer 2), each of whom in turn contacts a secretary (peer processes in layer 1). Philosopher 1 wishes to convey his *affection* for *oryctolagus cuniculus* to his peer. To do so, he passes a message (in English) across the 2/3 interface, to his translator, saying "I like rabbits", as illustrated in Figure 5-8. The translators have agreed on a neutral language, Dutch, so the message is converted to "*Ik hou van konijnen*". The choice of language is the layer 2 protocol and is up to the layer 2 peer processes.

The translator then gives the message to a secretary for transmission, by, for example, fax

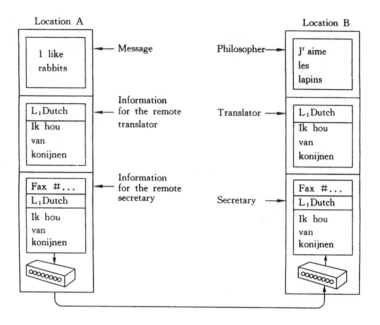

Figure 5-8 The philosopher-translator-secretary architecture

(the layer 1 protocol). When the message arrives, it is translated into French and passed across the 2/3 interface to philosopher 2. Note that each protocol is completely independent of the other ones as long as the interfaces are not changed. The translators can switch from Dutch to say, *Finnish*, at will, provided that they both agree, and neither changes his interface with either layer 1 or layer 3. Similarly the secretaries can switch from fax to e-mail, or telephone without disturbing (or even informing) the other layers. Each process may add some information intended only for its peer. This information is not passed upward to the layer above.

Layering and TCP/IP Protocol

The TCP/IP *layering* model, which is also called the Internet Layering Model or the Internet Reference Model, contains five layers as Figure 5-9 illustrates.

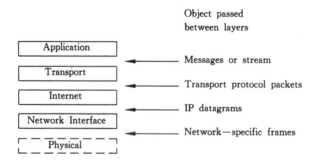

Figure 5-9 The five layers of the TCP/IP reference model

Four of the layers in the TCP/IP reference model correspond to one or more layers in the ISO reference model. However, the ISO model has no Internet Layer. This section summarizes the purpose of each layer.

Layer 1: Physical

Layer 1 corresponds to basic network hardware just as Layer 1 in the ISO 7-layer reference model. For example, the specification of RS-232 belongs in Layer 1, and gives the detailed specification of LAN hardware.

Layer 2: Network Interface

Layer 2 protocols specify how to organize data into *frames* and how a computer transmits frames over a network, similar to Layer 2 protocols in the ISO reference model. A network interface may consist of a device driver (e.g., when the network is a local area network to which the machine attaches directly) or a complex subsystem that uses its own data link protocol (e.g., when the network consists of *packet switches* that communicate with hosts using HDLC).

Layer 3: Internet

Layer 3 protocols specify the format of packets sent across the internet as well as the mechanisms used to forward packets from a computer through one or more *routers* to a final destination. Internet layer encapsulates the packet in an IP *datagram*, fills in the datagram header, uses the routing algorithm (*routing table*) to determine whether to deliver the datagram directly or send it to a router.

Layer 4: Transport

Layer 4 protocols, like layer 4 in the ISO model, specify how to ensure reliable transfer. To do so, transport protocol software arranges to have the receiving side send back *acknowledgments* and the sending side retransmit lost packets. Transport layer provides communication from one application program to another. Such communication is often called end-to-end. A general purpose computer can have multiple application programs accessing the internet at one time. The transport layer must accept data from several user programs. To do so, it adds additional information to each packet, including codes that identify which program sent it and which application program should receive it, as well as a checksum.

Layer 5: Application

Layer 5 corresponds to layers 6 and 7 in the ISO model. Each Layer 5 protocol such as FTP, HTTP, specifies how one application uses the internet. A application program may choose the style of transport needed, which can be either a sequence of individual messages or a continuos stream of bytes.

Words and Expressions

protocol hierarchy 协议分层；协议层次结构

layer (level) ['leiə] *n.* 层；阶层

peer [piə] *n.* 同等层；对等者；同层

primitive operation　基本操作；原始操作
Urdu ['uədu:] *n*. 乌尔都语
engage [in'geidʒ] *vt*. 雇用，聘用　*vi*. 从事，参加（in）
affection [ə'fekʃən] *n*. 感情；慈爱
oryctolagus cuniculus　穴兔
Ik hou van konijnen　我喜欢兔子（荷兰语）
Finnish ['finiʃ] *n*. 芬兰语　*adj*. 芬兰（语）的
layering ['leiəriŋ] *n*. 分层
frame [freim] *n*. 帧；画面；框架；构架；结构
packet switch　包交换（机）；分组交换（机）
packet ['pækit] *n*. 包；[计算机]信息包；数据包；分组（报文）　*vt*. 包装；打包
datagram ['deitəgræm] *n*. 数据报
router ['ru:tə] *n*. 路由器　routing table 路由表
acknowledgment [ək'nɔlidʒmənt] *n*. 确认；认收

Abbreviations

OSI (Open System Interconnect) Model　开放系统互连模型
HDLC (High Level Data Link Communication)　高层数据链路通信

Notes

本节介绍网络协议的分层结构，给出了 ISO/OSI 的七层结构和 TCP/IP 的五层结构。举例说明了为何要分层。

5.3.2　WAP—The Wireless Application Protocol

Preface

The past few years have *witnessed* a radical shift in the way we work, play, and communicate.[1] Today, the Internet and the World Wide Web allow people to exchange messages at the speed of light and access information from any source around the globe. These services are always on, always available, and easy to use.

At the same time, people around the world have jumped to[2] wireless communications at a *frenzied* pace. Today, cellular telephones are critical companions for active consumers and *mobile professionals*.[3] In some countries, as much as 70 percent of the population now uses cellular phones.

These two trends—the expansion of the Internet's reach[4] and the *burgeoning* of mobile communications—are now *converging*. Enter the mobile Internet. The mobile Internet extends the traditional Internet to wireless devices such as cellular phones, personal digital assistants (PDAs), and even automobiles. It brings information and services *to users' fingertips* when and where they need it, anytime, anywhere.[5]

The mobile Internet involves more than simply accessing existing Web pages, however. Mobile users want personalized services that match their individual preferences and needs. They are demanding greater ease of use and immediate results. In this market, services should be tailored to suit the user's current physical location. In addition, services *geared* toward the mobile environment can *push critical* information such as news and stock price alerts *asynchronously to* the user.[6] These new types of Web clients offer the opportunity to integrate Internet access with traditional telephony operations. In summary, the mobile Internet combines data and voice, information and communication, and global reach and personalization.

The realization of the mobile Internet relies on a new set of standards, known as the Wireless Application Protocol (WAP). WAP extends the Internet by addressing the *unique* requirements of the wireless network environment and the unique characteristics of small *handheld* devices. This exciting technology enables efficient access to information, applications, and services from a wide range of mobile devices. It also facilitates the interaction between browsing and telephony services, and it defines how to deliver *pushed* content. The WAP standard provides the necessary network protocols, content types, and run-time application environments to deliver a broad set of new and existing services to consumers and professionals alike.

WAP is an *enabling technology* that is *heralding* a revolution in the way we think about building and deploying Internet services.[7] Some analysts predict that Internet traffic from mobile devices will *outpace traffic* from traditional desktop systems within a few years. Consequently, developers need to keep the mobile environment in mind when they are building and deploying Web site content. A failure to do so could eventually translate into the loss of a significant percentage of available traffic.[8] Conversely, *exploitation* of the mobile Internet as enabled by WAP creates opportunities to reach new customers, provide more personalized service, and sell new applications.

Introduction

WAP is the mobile Internet standard. The wireless environment brings with it several unique challenges that make traditional Internet Web browsing impractical from mobile devices.[9] The WAP standard was defined to address these user interfaces and network challenges. This suite of protocols defines a complete mobile Internet platform, along with the mechanisms needed to bridge the gap between the mobile environment and the wired Internet.

WAP is an attempt to define the standard for how content from the Internet is filtered[10] for mobile communications. Content is now readily available on the Internet and WAP was designed as the (rather than one) way of making it easily available on mobile terminals.

The Wireless Application Protocol takes a client server approach. It incorporates a relatively simple *microbrowser* into the mobile phone, requiring only limited resources on the mobile phone. This makes WAP suitable for *thin clients* and early *smart phones*. WAP puts the intelligence in the WAP Gateways *whilst* adding just a microbrowser to the mobile phones themselves. Microbrowser-based services and applications reside temporarily on servers, not permanently in phones. Someone with a *WAP-compliant* phone uses the *in-built* microbrowser to[11]:

1. Make a request in WML (Wireless Markup Language), a language derived from HTML especially for wireless network characteristics.

2. This request is passed to a WAP Gateway that then retrieves the information from an Internet server either in standard HTML format or preferably directly prepared for wireless terminals using WML.[12] If the content being retrieved is in HTML format, a filter in the WAP Gateway may try to translate it into WML. A WML scripting language is available to format data such as calendar entries and electronic business cards for direct incorporation into the client device.

3. The requested information is then sent from the WAP Gateway to the WAP client, using whatever mobile network *bearer* service is available and most appropriate.[13]

Words and Expressions

witness ['witnis] *n.* 目击者，证人；证明　*vt.* 证明；表明；目击
frenzied ['frenzid] *adj.* 狂乱的，疯狂似的
mobile ['məubail] *adj.* 移动的；流动的；运动的，机动的
professional [prə'feʃənəl] *n.* 专业人员　*adj.* 职业的
burgeon ['bəːdʒən] *vi.* 迅速发展；激增　*vt.* 发芽；生出蓓蕾　*n.* 嫩芽，蓓蕾
converge [kən'vəːdʒ] *vt.* 会合；使会聚；使集中于一点　*vi.* 会合；会聚，集中；收敛
fingertip ['fiŋgətip] *n.* 指尖
to one's fingertips　完全地
gear [giə] *n.* 传动装置；齿轮　*v.* 调整；(使)适合；换挡
push ... to　把……推送给
critical ['kritikəl] *adj.* 紧要的；关键性的；危急的
asynchronously [ei'siŋkrənəsli] *adv.* 不同时地，异步地
realization [riəlai'zeiʃən] *n.* 实现
unique [juː'niːk] *adj.* 独特的
handheld ['hændheld] *adj.* 手持(式)的　*n.* 手持式装置
pushed *adj.* 被推送的
enabling technology　使能技术，支持技术
herald ['herəld] *vt.* 预告；通报；欢呼　*n.* 通报者；预言者；先驱
outpace [aut'peis] *vt.* 从速度上超过；发展快过；胜过
traffic ['træfik] *n.* 流量；通信流量；交通；交通量，运输量
exploitation [ˌeksplɔi'teiʃn] *n.* 开发，宣传；广告
microbrowser ['maikrəub'rauzə] *n.* 微浏览器
thin client　瘦客户机；瘦客户端；精简型计算机
smart phone　智能手机，智能电话
whilst [wailst] *conj.* 和……同时；而；当……时

WAP (Wireless Application Protocol)-compliant　*adj*. 顺应 WAP 的；WAP 功能的
in-built　内置的
bearer ['bɛərə] *n*. 承载(体)，负荷者；运载工具；送信人

Abbreviations

PDA (Personal Digital Assistant)　个人数字助理
WML (Wireless Markup Language)　无线置标语言
SMS (Short Message Service)　短消息服务

Notes

1. 本句中 radical shift 可译为"根本的转变，基本的改变"。本句可译为：过去几年已表明，我们的工作、消遣和通信的方式已有了根本的改变。下一句中 source 意为"信息源"。
2. 本句中 jump to 可译为"转向，转移"。
3. 本句中 active consumer "活跃的消费者"，critical companion 可译为"重要伙伴"。
4. 本句中 reach 指"到达距离，能及的范围"，因此 expansion of the Internet's reach 可译为"因特网覆盖范围的扩大"。
5. 本句中结构为 it brings ... to ... 的短语直译是"移动因特网把信息和服务带到用户的指尖"，因为人们用指尖操作小的移动设备。本句可译为：移动因特网随时随地把信息和服务带给用户。
6. 本句中 geared toward ... 修饰前面的 services；push ... to ... "把……推送给……"。这是 WAP1.2 标准提出的所谓"推送"(push)服务，即相关服务供应商可以根据用户的特定需求，主动把相关信息发送给有关用户。"推"给特定用户的信息可以是语音和数据两种形式。本句可译为：此外，适于移动环境的各种服务能够把紧要的信息，例如新闻和股票价格警报，(异步地)推送给用户。
   下一句中的短语 integrate ... with ... 可译为"集成因特网访问和传统的电话业务"。
7. think about "考虑，回想"。本句可译为：WAP 是一种使能技术，它预示我们在对建立和部署 Internet 服务的思考方式上的一场革命。
8. translate into 这里指"变为，转化"。本句可译为：不这样做，可能最终会失去大量的可用信息。
   下一句中 as enabled by WAP 修饰前面的 mobile Internet，即"因为 WAP 使得移动 Internet 能实现"。本句可译为：相反，通过 WAP 实现的移动因特网的开发创造了各种接触新用户、提供更个性化的服务，以及销售新应用程序的各种机会。
9. make ... browsing impractical 是"使……浏览成为不现实的"。全句可译为：无线环境带来若干独特的挑战，使得从移动设备浏览传统的 Internet 网站成为不切实际的。
10. filter "过滤，用过滤法除去"。因为从 Internet 传来的内容本来是在台式机上显示的，现要在手机一类设备上显示，显然不可能，因此，必须把这些内容进行"过滤"。
    上一句中 wired Internet 指"有线的 Internet"。
11. WAP-compliant phone 可译为"有 WAP 功能的手机"。本句可译为：有 WAP 功能手机的人利用内置的微浏览器(去做……)。

103

12. 本句中 either in standard HTML format or ... 说明前面的 information 用什么格式表示，即"或者以标准的 HTML 格式，或者更可取的是直接为使用 WML 的无线终端准备的"。

13. 本句中短语 using whatever ... 可译为"利用无论哪一种可用且最方便的移动网络承载服务"。

注：本节是 2001 年出版的 *WAP — The Wireless Application Protocol*：*Writing Applications for the Mobile Internet* 的序言，简单介绍了 WAP。2001 年 Comdex 的中心议题是移动 Internet 和 Linux，而移动 Internet 最为热门。WAP 是连接 Internet 和移动通信的桥梁。因此，本节是内容和翻译并重的选文。

5.4 Mobile Internet, Mobile Web

Mobile Internet refers to access to the Internet via a *cellular telephone* service provider. It is wireless access that can *handoff* to another radio tower while it is moving across the service area. It can refer an immobile device that stays connected to one tower,[1] but this is not the meaning of "mobile" here. Wi-Fi and other better methods are commonly available for users not on the move. Cellular base stations are more expensive to provide than a wireless base station that connects directly to an internet service provider, rather than through the telephone system.[2]

A mobile phone, such as a smartphone, that connects to data or voice services without going through the cellular base station is not on mobile Internet. A *laptop* with a broadband modem and a cellular service provider *subscription* that is traveling on a bus through the city is on mobile Internet.[3]

A mobile broadband modem "*tethers*" the smartphone to one or more computers or other end user devices to provide access to the Internet via the protocols that cellular telephone service provider may offer.

According to Buzz-City, mobile internet increased 30% from Q1 to Q2 2011. *As of* July 2012, approximately 10.5% of all Web *traffic* occurs through mobile devices (up from 4% in December 2010).

The mobile Web refers to access to the World Wide Web, i.e. the use of browser-based Internet services, from a handheld mobile device, such as a smartphone or a *feature phone*, connected to a mobile network or other wireless network.

Traditionally, access to the Web has been via fixed-line services on laptops and desktop computers. However, the Web is becoming more accessible by portable and wireless devices. An early 2010 ITU (International Telecommunication Union) report said that with the current growth rates, Web access by people on the go — via laptops and smart mobile devices — is likely to exceed Web access from desktop computers within the next five years. The shift to mobile Web access has been accelerating with the rise since 2007 of larger *multitouch* smartphones, and of multitouch tablet computers since 2010.[4] Both platforms provide better Internet access, screens, and mobile browsers- or application-based user Web experiences than previous generations of mobile devices have done. Web designers may work separately on such pages, or pages may be automatically converted as in Mobile Wikipedia.

The distinction between mobile Web applications and native applications is anticipated to become increasingly *blurred*, as mobile browsers gain direct access to the hardware of mobile devices (including *accelerometers* and GPS chips), and the speed and abilities of browser-based applications improve. *Persistent* storage and access to *sophisticated* user interface graphics functions may further reduce the need for the development of platform-specific native applications.

The mobile Web has also been called Web 3.0, drawing parallels to the changes users were experiencing as Web 2.0 websites *proliferated*.[5]

Mobile Web access today still suffers from *interoperability* and usability problems. Interoperability issues *stem from* the platform *fragmentation* of mobile devices, mobile operating systems, and browsers. Usability problems are centered on the small physical size of the mobile phone *form factors* (limits on display resolution and user input/ operating). Despite these shortcomings, many mobile developers choose to create apps using mobile Web. A June 2011 research on mobile development found mobile Web the third most used platform, trailing Android and iOS.[6]

In an article in *Communications of the ACM* in April 2013, Web technologist Nicholas C. Zakas, noted that mobile phones in use in 2013 were more powerful than Apollo 11's 70 lb (32 kg) Apollo Guidance Computer used in the July 1969 lunar landing. However, in spite of their power, in 2013, mobile devices still suffer from Web performance with slow connections similar to the 1996 stage of Web development. Mobile devices with slower download request/ response times, the latency of over-the-air data transmission, with "high-latency connections, slower CPUs, and less memory" force developers to rethink Web applications created for desktops with "wired connections, fast CPUs, and almost endless memory."[7]

Words and Expressions

cellular telephone　移动电话，蜂窝式电话
handoff ['hændɔf] *n.* 传送；交递；手递手传球
laptop ['læptɔp] *n.* 便携式计算机，笔记本计算机，膝上型计算机
subscription [səb'skripʃən] *n.* 订阅（费），订购（费）
tether ['teðə] *vt.* 系链；（用绳、链）拴；拘束，束缚
as of　到……时为止；在……时；从……时起
traffic ['træfik] *n.* 流量；通信流量；交通；交通量，运输量
feature phone　功能手机（一类低端手机），非智能手机
multitouch ['mʌlti'tʌtʃ] *n.* 多点触摸，多点触控，多重触控；多点感应
blur [blə:] *v.* （使）变模糊；（使）难以区分　*n.* 模糊不清的事物
accelerometer [æk,selə'rɔmitə] *n.* 加速计；加速表
persistent [pə'zistənt] *adj.* 持久的，持续的；持久稳固的
sophisticated [sə'fistikeitid] *adj.* 复杂的；高级的；尖端的
proliferate [prə'lifəreit] *vi.* 激增；扩散；增殖；增生　*vt.* 使激增；使扩散

interoperability ['intərˌɔpərə'biləti] n. 可互操作性；互通性；互用性
stem from 起源于，来自
fragmentation [ˌfrægmen'teiʃən] n. 破碎；错乱；分片
form factor 形状因数；波形因数；形状系数

Abbreviations

Q1 (Quarter 1) 第一季度
ITU (International Telecommunication Union) 国际电信联盟

Notes

1. 本句中 refer 是及物动词，可译为"把……提交，把……转交，把……归属"。前半句可译为：它可能转交给一个一直连接到无线电塔的固定设备。
2. 本句中"more expensive ... than ... "结构的意思是，提供蜂窝基站比提供无线基站更贵。
3. 本句的主要结构是 A laptop is on mobile Internet，其他都是修饰 laptop 的。其中，with a broadband ... subscription 是介词短语，作定语修饰前面的 laptop，可译为"带有宽带调制解调器，并订购了蜂窝服务的 laptop"。后面的 that 从句也修饰前面的 laptop，即"在公共汽车上旅行通过城市的 laptop"。全句可译为：当有人携带一台带有宽带 Modem 并订购了蜂窝网服务的笔记本计算机乘公共汽车经过城市时，这台计算机便连接了移动因特网。
4. 本句中 the rise since 2007 of larger multitouch smartphones = the rise of larger multitouch smartphones since 2007；of multitouch tablet ... = the rise of multitouch tablet ... 。下一句的结构是 ... provide better ... than ... ，即提供了比前几代移动设备更好的 Internet 访问服务。
5. 本句中 parallel 意为"比较；类似的事，相似之处"；draw parallels to ... 可译为"与……做比较"。本句中 user 之前省略了关系代词 that，此 that 引导一个限制性定语从句修饰 the changes。全句可译为：移动 Web 也被称为 Web 3.0，它与用户在 Web 2.0 网站激增时体验到的变化类似。
6. 本句中 trailing Android and iOS 是现在分词短语，用作状语，表示结果。全句可译为：2011 年 6 月的一项关于移动开发的研究发现，移动 Web 是第三大最常用的平台，仅次于 Android 和 iOS。
7. 本句的主要结构是 Mobile devices force developers to rethink ... 。句中 created for ... 修饰前面的 Web applications。
注：本篇解释了什么是移动因特网，即通过蜂窝网无线访问 Internet。仅通过 Wi-Fi 直接访问 Internet 不属于移动因特网，因为如果一个用户在一个 Wi-Fi 区域中迅速移动，一旦离开该 Wi-Fi 区域，他就不可能连续地访问 Internet。此外，虽然在移动中访问 Web 给用户带来极大的方便，但是移动电话的形状因素，如有限的尺寸、输入操作等，仍给用户带来不便。

Terms

PDA
Acronym for Personal Digital Assistant. A lightweight palmtop computer designed to

provide specific functions like personal organization (calendar, note taking, database, calculator, and so on) as well as communications. More advanced models also offer multimedia features. Many PDA devices rely on a pen or other pointing device for input instead of a keyboard or mouse, although some offer a keyboard too small for touch typing to use in conjunction with pen or pointing device. For data storage, a PDA relies on flash memory instead of power-hungry disk drives. *See also* firmware, flash memory, PC Card, pen computer.

Browser plug-in

A browser plug-in is a program that extends the capabilities of your browser. Plug-ins can generally be downloaded for free and installed in your Web browser by simply clicking the appropriate link at the developer's Web site.

Internet content provider

An organization that provides Internet content.

ASP (Application Service Provider)

An organization that manages and distributes software-based services over the Internet.

Portal

Portal is a term, generally synonymous with *gateway*, for a World Wide Web site that is or proposes to be a major starting site for users when they get connected to the Web or that users tend to visit as an anchor site. There are general portals and specialized or niche portals. Some major general portals include Yahoo, Excite, Netscape, Lycos, CNET, Microsoft Network, and America Online's AOL.com. Examples of niche portals include Garden.com (for gardeners).

IM (Instant Messaging)

IM is a type of real time communication service. It is somewhat like e-mail, but much more like a chat room. Both parties are online at the same time, and they "talk" to each other by typing text and sending small pictures in instantaneous time.

IMS (IP Multimedia Subsystem)

IMS represents a 3GPP and 3GPP2 effort to define an all IP based wireless network as compared to the historically disparate voice, data, signaling, and control network elements.

Intranet(内部网)

An Intranet is simply the application of Internet technology within an internal or closed usergroup. Intranets are company specific and do not have to have a physical connection to the Internet.

107

HTML

Acronym for Hypertext Markup Language. HTML allows users to produce Web pages that include text, graphics, and pointers to other Web pages. A markup language describes how documents are to be formatted. It thus contains explicit commands for formatting. For example, in HTML, ⟨B⟩ means start boldface mode, and ⟨/B⟩ means leave boldface mode.

Hub (集线器)

In a network, a device joining communication lines at a central location, providing a common connection to all devices on the network. The term is an analogy to the hub of a wheel.

Switching hub

A central device or switch that connects separate communication lines in a network and routes messages and packets among the computers on the network. The switch functions as a hub, or a PBX, for the network. See also hub.

Bridge (网桥)

1. A device that connects networks using the same communications protocols so that information can be passed from one to the other. Compare gateway.

2. A device that connects two local area networks, whether or not they use the same protocols. A bridge operates at the ISO/OSI data link layer. Compare router.

Switch (交换机)

In communications, a computer or electromechanical device that controls routing and operation of a signal path.

The switching elements in most wide area networks are specialized computers used to connect two or more transmission lines. When data arrive on an incoming line, the switching element must choose an outgoing line to forward them on.

Router (路由器)

An intermediary device on a communications network that expedites message delivery. On a single network linking many computers through a mesh of possible connections, a router receives transmitted messages and forwards them to their correct destinations over the most efficient available route. On an interconnected set of local area networks (LANs) using the same communications protocols, a router serves the somewhat different function of acting as a link between LANs, enabling messages to be sent from one to another. See also gateway.

Gateway (网关)

A device that connects networks using different communications protocols so that information can be passed from one to the other. A gateway both transfers information and converts it to a form compatible with the protocols used by the receiving network. Compare bridge.

Information superhighway

The existing Internet and its general infrastructure, including private networks, online services, etc.

Applet

A small piece of code that can be transported over the Internet and executed on the recipient's machine. The term is especially used to refer to such programs as they are embedded inline as objects in HTML documents on the World Wide Web.

SGML

Acronym for Standard Generalized Markup Language. An information-management standard adopted by the International Organization for Standardization (ISO) in 1986 as a means of providing platform- and application-independent documents that retain formatting, indexing, and linked information. SGML provides a grammar-like mechanism for users to define the structure of their documents, and the tags they will use to denote the structure in individual documents. See also ISO.

Unicode

A 16-bit character encoding standard developed by the Unicode Consortium between 1988 and 1991. By using two bytes to represent each character, Unicode enables almost all of the written languages of the world to be represented in the form of text files. (By contrast, even 8-bit ASCII is not capable of representing even all of the combinations of letters and diacritical marks that are used with the Roman alphabet.) Approximately 28,000 of the 65,536 possible combinations have been assigned to date, 21,000 of them being used for Chinese. The remaining combinations are open for expansion. Compare ASCII.

EDI

Acronym for Electronic Data Interchange. A set of standards for controlling the transfer of business documents, such as purchase orders and invoices, between computers. The goal of EDI is the elimination of paperwork and increased response time. For EDI to be effective, users must agree on certain standards for formatting and exchanging information, such as the X.400 protocol. See also standard, X.400.

URI

Acronym for Uniform Resource Identifier. A character string used to identify a resource on the Internet by type and location. See also relative URL, uniform resource locator.

ISO/OSI model

Acronym for International Organization for Standardization Open Systems Interconnection model. A layered architecture (plan) that standardizes levels of service and types of interaction for computers exchanging information through a communications network. The ISO/OSI model separates computer-to-computer communications into seven layers, or levels, each building upon the standards contained in the levels below it. The lowest of the seven layers deals solely with hardware links; the highest deals with software interactions at the application-program level.

MIME

Acronym for Multipurpose Internet Mail Extensions. A standard that extends the SMTP protocol to permit data, such as video, sound, and binary files, to be transmitted by Internet e-mail without having to be translated into ASCII format first. This is accomplished by the use of MIME types, which describe the contents of a document. A MIME-compliant application sending a file, such as some e-mail programs, assigns a MIME type to the file. The receiving application, which must also be MIME-compliant, refers to a standardized list of documents that are organized into MIME types and subtypes to interpret the content of the file. For instance, one MIME type is text, and it has a number of subtypes, including plain and html. A MIME type of text/html refers to a file that contains text written in HTML. MIME is part of HTTP, and both Web browsers and HTTP servers use MIME to interpret e-mail files they send and receive. See also HTTP, HTTP server, Web browser.

SMTP

Acronym for Simple Mail Transfer Protocol. A TCP/IP protocol for sending messages from one computer to another on a network. This protocol is used on the Internet to route e-mail. See also protocol, TCP/IP.

NNTP

Acronym for Network News Transfer Protocol. The Internet protocol that governs the transmission of newsgroups.

HDLC

Acronym for High-level Data Link Control. A protocol for information transfer developed by the ISO. HDLC is a bit-oriented, synchronous protocol that applies to the data-link (message-packaging) layer of the ISO Open Systems Interconnection (OSI) model for computer-microcomputer communications. Messages are transmitted in units called frames, which can contain differing amounts of data but which must be organized in a particular way. See also frame.

Frame

1. In asynchronous serial communications, a unit of transmission that is sometimes measured in elapsed time and begins with the start bit that precedes a character and ends with

the last stop bit that follows the character.

2. In synchronous communications, a package of information transmitted as a single unit. Every frame follows the same basic organization and contains control information, such as synchronizing characters, station address, and an error-checking value, as well as a variable amount of data. For example, a frame used in the widely accepted HDLC and related SDLC protocols begins and ends with a unique flag (01111110).

RFC

Acronym for Request For Comments. A document in which a standard, a protocol, or other information pertaining to the operation of the Internet is published. The RFC is actually issued, under the control of the IAB, after discussion and serves as the standard. RFCs can be obtained from sources such as InterNIC.

IETF

Acronym for Internet Engineering Task Force. The organization that is charged with studying technical problems facing the Internet and proposing solutions to the IAB. The IETF is managed by the IESG. See also IESG.

Cable modem

A modem that sends and receives data through a coaxial cable television network instead of telephone lines, as with a conventional modem. Cable modems, which have speeds of 500 kilobits per second (Kbps), can generally transmit data faster than current conventional modems. See also coaxial cable, modem.

ADSL

Acronym for Asymmetric Digital Subscriber Line. Technology and equipment allowing high-speed digital communication, including video signals, across an ordinary twisted-pair copper phone line, with speeds up to 9 Mbps downstream (to the customer) and up to 800 kbps upstream. Also called asymmetric digital subscriber loop.

Web browser and IE

A web browser is a software program for retrieving, presenting, and traversing information resources on the World Wide Web. An information resource is identified by a Uniform Resource Identifier (URI) and may be a Web page, image, video, or other piece of content. The most popular web browsers are Microsoft's Internet Explorer (IE), Netscape's Navigator, Opera, Firefox, and Mac Safari.

Wiki (维基)

A wiki is a server program that allows users to collaborate in forming the content of a

Web site. With a wiki, any user can edit the site content, including other users' contributions, using a regular Web browser.

Wikipedia（维基百科全书）

A free encyclopedia built collaboratively using wiki software. Anyone can edit it.

YouTube

YouTube is a video-sharing website on which users can upload, share, and view videos.

SNS and Facebook（社交网服务和Facebook网站）

A social network service focuses on building online communities of people who share interests and/or activities, or who are interested in exploring the interests and activities of others. Most social network services are web based and provide a variety of ways for users to interact, such as e-mail and instant messaging services.

The main types of social networking services are those which contain category divisions (such as former school-year or classmates), means to connect with friends (usually with self-description pages) and a recommendation system linked to trust. Popular methods now combine many of these, with Facebook and Twitter widely used worldwide.

Blog (short for weblog)

A blog is a personal online journal that is frequently updated and intended for general public consumption. Blogs are defined by their format: a series of entries posted to a single page in reverse-chronological order. Blogs generally represent the personality of the author or reflect the purpose of the Web site that hosts the blog. Topics sometimes include brief philosophical musings, commentary on Internet and other social issues, and links to other sites the author favors, especially those that support a point being made on a post.

The author of a blog is often referred to as a blogger. Many blogs syndicate their content to subscribers using RSS, a popular content distribution tool.

Micro-blogging and Twitter（微博和Twitter网站）

Twitter is a free social networking and micro-blogging service that enables its users to send and read messages known as tweets. Tweets are text-based posts of up to 140 characters displayed on the author's profile page and delivered to the author's subscribers who are known as followers. Senders can restrict delivery to those in their circle of friends or, by default, allow open access. Users can send and receive tweets via the Twitter website, Short Message Service (SMS) or external applications. While the service itself costs nothing to use, accessing it through SMS may incur phone service provider fees.

The 140-character limit on message length was initially set for compatibility with SMS messaging, and has brought to the web the kind of shorthand notation and slang commonly used in SMS messages. It is sometimes described as the "SMS of the Internet" since the use of Twitter's application programming interface for sending and receiving short text messages by other applications often eclipses the direct use of Twitter.

Wireless Sensor Network (WSN, 无线传感器网)

A wireless sensor network consists of spatially distributed autonomous sensors to cooperatively monitor physical or environmental conditions, such as temperature, sound, vibration, pressure, motion or pollutants and to cooperatively pass their data through the network to a main location.

Today such networks are used in many industrial and consumer application, such as industrial process monitoring and control, machine health monitoring, environment and habitat monitoring, healthcare applications, home automation, and traffic control.

The WSN is built of "nodes" from a few to several hundreds or even thousands, where each node is connected to one (or sometimes several) sensors. Each such sensor network node has typically several parts: a radio transceiver with an internal antenna or connection to an external antenna, a microcontroller, an electronic circuit for interfacing with the sensors and an energy source, usually a battery.

WeChat (微信)

WeChat (Chinese: 微信; pinyin: Wēixìn; literally: "micro message") is a mobile text and voice messaging communication service developed by Tencent in China, first released in January 2011. It is the largest standalone messaging app by monthly active users.

The app is available on Android, iPhone, BlackBerry, Windows Phone and Symbian phones. As of August 2014, WeChat has 438 million active users; with 70 million outside of China.

WeChat provides text messaging, hold-to-talk voice messaging, broadcast (one-to-many) messaging, sharing of photographs and videos, and location sharing. It can exchange contacts with people nearby via Bluetooth, as well as providing various features for contacting people at random if desired (if these are open to it) and integration with social networking services such as those run by Facebook and Tencent QQ. Photographs may also be embellished with filters and captions, and a machine translation service is available.

Registration is completed through Facebook Connect, email, mobile phone SMS/VM, or Tencent QQ.

WeChat is being promoted in India via gaming site Ibibo, in which Tencent holds a stake.

Web crawler

A Web crawler is an Internet bot that systematically browses the World Wide Web, typically for the purpose of Web indexing. A Web crawler may also be called a Web spider, an ant, an automatic indexer, or a Web scutter.

Web search engines and some other sites use Web crawling or spidering software to update their web content or indexes of others sites' web content. Web crawlers can copy all the pages they visit for later processing by a search engine that indexes the downloaded pages so that users can search them much more quickly.

Crawlers can validate hyperlinks and HTML code. They can also be used for web scraping.

Internet Plus (Chinese: 互联网＋)

Internet Plus, similar to Information Superhighway and Industry 4.0, is proposed by China's Prime Minister Li Keqiang in his Government Work Report on March 5, 2015 so as to keep pace with the Information Trend. "Internet Plus" refers to the application of the internet and other information technology in conventional industries. It is an incomplete equation where various internets (mobile Internet, cloud computing, big data or Internet of Things) can be added to other fields, fostering new industries and business development in China.

5G (from "5th Generation")

5G is the fifth generation cellular network technology (cellular network is also called mobile network). The industry association 3GPP defines any system using "5G NR" (5G New Radio) software as "5G", a definition that came into general use by late 2018. Others may reserve the term for systems that meet the requirements of the ITU IMT-2020. 3GPP will submit their 5G NR to the ITU. It follows 2G, 3G and 4G and their respective associated technologies (such as GSM, UMTS, LTE, LTE Advanced Pro and others). 5G performance targets high data rate, reduced latency, energy saving, cost reduction, higher system capacity, and massive device connectivity.

The first fairly substantial deployments were in April 2019. In South Korea, SK Telecom claimed 38,000 base stations, KT Corporation 30,000 and LG U Plus 18,000. They are using 3.5 GHz (sub-6) spectrum in non-standalone (NSA) mode and tested speeds were from 193 to 430 Mbit/s down.

Seven companies sell 5G radio hardware and 5G systems for carriers: Ericsson, Nokia, Huawei, ZTE, Samsung, Datang Telecom, and Altiostar.

WiMAX

WiMAX is a family of wireless broadband communication standards based on the IEEE 802.16 set of standards, which provide multiple physical layer (PHY) and Media Access Control (MAC) options.

The name "WiMAX" was created by the WiMAX Forum, which was formed in June 2001 to promote conformity and interoperability of the standard. The forum describes WiMAX as "a standards-based technology enabling the delivery of last mile wireless broadband access as an alternative to cable and DSL". IEEE 802.16m or WirelessMAN-Advanced was a candidate for the 4G, in competition with the LTE Advanced standard.

Exercises

1. Translate the text of Lesson 5.3.2 into Chinese.
2. Topics for oral workshop.
   - What is Internet? What are the advantages of using networks?
   - Talk about the services that the Internet provides
   - How do you surf on the Internet?
   - Explain following names: WWW, HTTP, FTP, TCP/IP, DNS, URL, LAN, WAN
3. Translate the following into English.

计算机网络常分类成局域网(LAN)、城域网(MAN)或广域网(WAN)。两个或多个网络的连接称作互联网,世界范围的因特网是互联网的知名例子。

LAN 是一些在单幢建筑或最多(up to)几千米大小的校园内的私有网络,它们被广泛地用来连接公司办公室和工厂里的个人计算机和工作站以共享资源和交换信息。

通常,一种给定的 LAN 将只用一种类型的传输介质。对于 LAN,各种拓扑结构都是可能的。最常用的 LAN 拓扑结构是总线型、环形和星形。

MAN 基本上是 LAN 的更高版本,并且通常使用类似技术。设计 MAN 以延伸至覆盖(over)整个城市。它可以是像有线电视网这样的单个网络,也可以是一种把许多 LAN 连接成一个大网络的手段,以至于资源可以被 LAN 到 LAN 以及设备到设备地共享。例如,一个公司可以使用 MAN 来连接它的遍及(throughout)一个城市的所有办公地点的 LAN。

WAN 跨越(span)很大的地理区域,这个区域可以包括一个国家、一个洲,甚至整个世界。它可在大的地理区域提供数据、声音、图像,以及视频信息的长距离传输。

与 LAN 形成对比,WAN 可以使用公共租用线或私有通信设备,通常两者结合,并且因此能够跨越无限里数。

4. Listen to the video "Web server" and write down the first two paragraphs.

Unit 6 Network Communication

6.1 Two Approaches to Network Communication

Whether they provide connections between one computer and another or between terminals and computers, communication networks can be divided into two basic types: *circuit-switched* (sometimes called connection oriented) and *packet-switched* (sometimes called connectionless). Circuit-switched networks operate by forming a dedicated connection (circuit) between two points.[1] The U. S. telephone system uses circuit switching technology — a telephone call establishes a circuit from the originating phone through the local switching office, across *trunk* lines, to a remote switching office, and finally to the destination telephone. While a circuit is in place, the phone equipment *samples* the microphone repeatedly, encodes the samples digitally, and transmits them across the circuit to the receiver. The sender is guaranteed that the samples can be delivered and *reproduced* because the circuit provides a guaranteed data path of 64 kbps(thousand bits per second), the rate needed to send digitized voice.[2] The advantage of circuit switching lies in its guaranteed capacity: once a circuit is established, no other network activity will decrease the capacity of the circuit. One disadvantage of circuit switching is cost: circuit costs are fixed, independent of traffic[3]. For example, one pays a fixed rate for a phone call, even when the two parties do not talk.

Packet-switched networks, the type usually used to connect computers, take an entirely different approach. In a packet-switched network, data to be transferred across a network is divided into small pieces called packets that are *multiplexed* onto high capacity *intermachine* connections. A packet, which usually contains only a few hundred bytes of data, carries identification that enables the network hardware to know how to send it to the specified destination. For example, a large file to be transmitted between two machines must be broken into many packets that are sent across the network one at a time. The network hardware delivers the packets to the specified destination, where software *reassembles* them into a single file again. The chief advantage of packet-switching is that multiple communications among computers can proceed concurrently, with intermachine connections shared by all pairs of machines that are communicating. The disadvantage, of course, is that as activity increases, a given pair of communicating computers receives less of the network capacity.[4] That is, whenever a packet switched network becomes overloaded, computers using the network must wait before they can send additional packets.

Despite the potential drawback of not being able to guarantee network capacity, packet-switched networks have become extremely popular. The motivations for adopting packet switching are cost and performance. Because multiple machines can share the network hardware, fewer connections are required and cost is kept low. Because engineers have been able to build high speed network hardware, capacity is not usually a problem. So many computer interconnections use packet-switching that, throughout the remainder of this test, the term *network* will refer only to packet-switched networks.

Words and Expressions

circuit-switched　电路交换的，同 circuit-switching
packet-switched　分组交换的，包交换的
trunk [trʌŋk] *n*. 干线，中继线
sample ['sɑ:mpl] *n*. 样本，样值　*vt*. 取样，采样
reproduce [ri:prə'dju:s] *v*. 再生产；复制
multiplex ['mʌltipleks] *adj*. 多路复用
intermachine　机器间
reassemble [ri:ə'sembl] *vt*. 重新组装

Notes

1. 本句后面括号内的 circuit 译为"线路"。全句可译为：电路交换网络运行时在两点之间形成一条专用连接（线路）。
   下一句中 a telephone call 可译为"一次电话呼叫"。全句可译为：美国电话系统采用电路交换技术，即一次电话呼叫建立一条线路，从发起呼叫的电话机通过本地交换局，穿过中继线到一个远程交换局，最后到达目的电话机。
2. 本句中 guaranteed data path 译为"有保障的数据通路（路径）"。the rate needed … 是 64 kbps的同位语。digitized voice"数字化语音"。
3. 本句中 independent of traffic 可译为"与通信（话）量无关"。
4. 本句中 that as activity increases 开始的从句译为"随着网络活动的增加，给定的一对通信的计算机所得到的网络容量会减少"。下一句 before they can send additional packets 译为"在它们能继续发送分组以前"。
   注：IT 技术本身分为计算机和通信，而这两者都有其硬件和软件。本课内容是网络通信硬件方面的基本知识。本节介绍网络通信的两种方法：电路交换和分组交换，这是理解 IP 电话的基础。普通电话用电话交换，一对通话者独占一条电路交换线路。而 IP 电话用分组交换，多对通话者可以共享一条通路（通过下一节讲的多路复用技术），因此 IP 电话比普通电话便宜。

6.2　Carrier Frequencies and Multiplexing

Computer networks that use a *modulated carrier* wave to transmit data are similar to television stations that use a modulated carrier wave to broadcast video. The similarities provide the *intuition* needed to understand a fundamental principle：

Two or more signals that use different carrier frequencies can be transmitted over a single medium simultaneously without interference.

To understand the principle, consider how television transmission works. Each television station is assigned a *channel* number on which it broadcasts a signal. In fact, a channel number is merely shorthand for the frequency at which the station's carrier *oscillates*. [1] To receive a transmission, a television receiver must be tuned to the same frequency as the transmitter. More important, a given city can contain many television stations that all broadcast on separate frequencies simultaneously. A receiver selects one to receive at any time.

Cable television illustrates that the principle applies to many signals traveling across a wire[2]. Although a cable *subscriber* may have only one physical wire that connects to the cable company, the subscriber receives many channels of information simultaneously. The signal for one channel does not interfere with the signal for another.

Computer networks use the principle of separate channels to permit multiple communications to share a single, physical connection. Each sender transmits a signal using a particular carrier frequency. A receiver configured to accept a carrier at a given frequency will not be affected by signals sent at other frequencies. [3] All carriers can pass over the same wire at the same time without interference.

Frequency Division Multiplexing

Frequency division multiplexing(FDM) is the technical term applied to a network system that uses multiple carrier frequencies to allow independent signals to travel through a medium. [4] FDM technology can be used when sending signals over wire, RF, or optical fiber. Figure 6-1 illustrates the concept, and shows the hardware components needed for FDM.

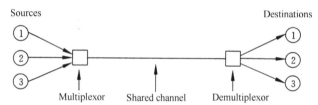

Figure 6-1   The concept of multiplexing(Each pair of source and destination can send data over the shared channel without interference. In practice, each end requires a multiplexor and demultiplexor for 2-way communication, and a multiplexor may need circuitry to generate the carrier waves)

In theory, as long as each carrier operates at a different frequency than the others, it remains independent. In practice, however, two carriers operating at almost the same frequency or at exact multiples of a frequency can interfere with one another. [5] To avoid problems, engineers who design FDM network systems choose a minimum separation between the carriers. The *mandate* for large gaps between the frequencies assigned to carriers means that underlying hardware used with FDM can tolerate a wide range of frequencies. [6] Consequently, FDM is only used on high-bandwidth transmission channels.
In summary,

Frequency division multiplexing (FDM) allows multiple pairs of senders and receivers to communicate over a shared medium simultaneously. The carrier used by each pair operates at a unique frequency that does not interfere with the others.

Time Division Multiplexing

The general alternative to FDM is time division multiplexing (TDM), in which sources sharing a medium take turn. [7] For example, some time-division multiplexing hardware use a *round-robin* scheme in which the *multiplexor* sends a packet from source 1, then sends a packet from source 2, and so on. Figure 6-2 illustrates the idea.

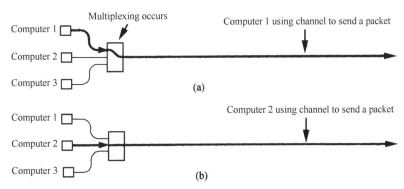

Figure 6-2　Illustration of multiplexing with packets (The sources take turns using the shared communication channel. (a) computer 1 uses the resource to send a packet, and then (b) computer 2 uses the resource to send a packet)

Dividing data into small packets ensures that all sources receive prompt service because it prohibits one source from gaining exclusive access for an arbitrarily long time. [8] In particular, if one source has a few packets to send and another has many, allowing both sources to take turns sending packets guarantees that the source with a small amount of data will finish promptly.

In fact, most computer networks use some form of time division multiplexing.

Words and Expressions

carrier ['kæriə] *n.* 载波；载体；运送者
modulate ['mɔdjuleit] *vt.* 调制
demodulate [diː'mɔdjuleit] *vt.* 解调
intuition [intjuː'iʃən] *n.* 直觉（知识）；直观
channel ['tʃænəl] *n.* 通道；信道；频道
oscillate ['ɔsileit] *vi.* 振荡；振动
cable television　有线电视
subscriber [səb'skraibə] *n.* 用户，订户
mandate ['mændeit] *n.* 授权；命令，训令；委任；要求　*vt.* 托管
round-robin　循环，轮流

multiplexor ['mʌltiplɛksə]　n. 多路复用器
demultiplexor　n. 多路分配器，分路器，信号分离器

Abbreviations

FDM（Frequency Division Multiplexing）　频分多路复用
RF（Radio Frequency）　射频
TDM（Time Division Multiplexing）　时分多路复用

Notes

1. 本句若分两句直译，便译成"其实，频道号只是频率的速记，电视台的载波以这种频率振荡"。全句可译为：其实，频道号只是电视台载波的振荡频率的速记。
   下一句中的 tune to 意为"调谐到"。
2. 本句中 many signals traveling across a wire 译为"许多信号在一根导线上同时传输"。
   下一句中的 cable 指"有线电视"。
3. 本句中 configured to 引导过去分词短语，修饰前面的 receiver。本句译为：一个接收器被设置成接收给定频率的载波，它将不会受其他频率发送的信号的影响（或干扰）。
4. 本句中 applied to … 修饰前面的 term，即"用于计算机网络系统的技术术语"或"计算机网络系统术语"。本句可译为：频分多路复用（FDM）是使用多个载波频率在一个介质中同时传输多个独立信号的计算机网络系统术语。
5. at exact multiples of a frequency "一频率的整数倍"。全句译为：但实际上，两个频率几乎相同或频率成整数倍的载波会彼此干扰。
6. mandate for large gap … 译为"大间隔的要求"。全句译为：在指派给各载波的频率之间有较大间隔的要求意味着与 FDM 一起使用的底层硬件能容纳很宽的频率范围。
7. 本句中 general alternative to FDM 可译为"与 FDM 不同的另一种复用形式"，alternative 是"两者挑一，可供选择的办法（方案）"。句中 sharing a medium 修饰前面的 sources（发送源）。全句译为：与 FDM 不同的另一种复用形式是时分多路复用（TDM），按 TDM，共享通信介质的各发送源依次轮流（发送信息）。
8. 本句中 prompt 意为"敏捷的，及时的"，source 译为"源机"。结构为 prohibit … from … 的短语译为"禁止一台源机任意长时间独占介质"。

注：本节介绍计算机网络传输时所用的两种复用形式：频分多路复用和时分多路复用。采用多路复用可以使多对通信（话）者共享一条通路。

6.3　Internet of Things

The Internet of Things (IoT) is the interconnection of uniquely identifiable *embedded* computing devices within the existing Internet infrastructure. Typically, IoT is expected to offer advanced connectivity of devices, systems, and services that goes beyond machine-to-machine communications (M2M) and covers a variety of protocols, domains, and applications[1]. The interconnection of these embedded devices (including smart objects) is expected to *usher* in automation in nearly all fields, while also enabling advanced applications like a Smart Grid.[2]

Things, in the IoT, can refer to a wide variety of devices such as *heart monitoring implants*, *biochip transponders* on farm animals, automobiles with *built-in* sensors, or *field operation* devices that assist fire-fighters in search and rescue. These devices usually have certain *sensory* capabilities using which they are capable of collecting certain information and passing the same. These devices have identifiable characteristic using which they can be traced (such as RFID). They might also be equipped with *electro-mechanical* capabilities using which they can control the functioning of a system. Current market examples include smart *thermostat* systems and *washers/dryers* that utilize Wi-Fi for remote monitoring.

Internet of Things is actually a *convergence* of multiple technologies. The below individual technologies contributed to the convergence of the capabilities of all of them, which resulted in the Internet of Things[3]:

1. The Internet
2. Wireless communication
3. Embedded systems
4. Electro-Electrical and Electro-Mechanical systems that are *compact*

According to *Gartner*, there will be nearly 26 billion devices on the Internet of Things by 2020. ABI Research estimates that more than 30 billion devices will be wirelessly connected to the Internet of Things (Internet of Everything) by 2020. *As per* a recent *survey* and study done by Pew Research Internet Project, a large majority of the technology experts and *engaged* Internet users who responded — 83 percent — agreed with the notion that the Internet/Cloud of Things, embedded and *wearable* computing (and the corresponding dynamic systems) will have widespread and beneficial effects by 2025.[4] It is, *as such*, clear that the IoT will consist of a very large number of devices being connected to the Internet.

Integration with the Internet implies that devices will utilize an IP address as a unique identifier. However, due to the limited address space of IPv4 (which allows for 4.3 billion unique addresses), objects in the IoT will have to use IPv6 to *accommodate* the extremely large address space required. Objects in the IoT will not only be devices with sensory capabilities, but also provide *actuation* capabilities (e.g., bulbs or locks controlled over the Internet). *To a large extent*, the future of the Internet of Things will not be possible without the support of IPv6; and consequently the global adoption of IPv6 in the coming years will be *critical* for the successful development of the IoT in the future.

The embedded computing nature of many IoT devices means that low-cost computing platforms are likely to be used. In fact, to minimize the impact of such devices on the environment and energy consumption, low-power radios are likely to be used for connection to the Internet. Such low-power radios do not use Wi-Fi, or well established Cellular Network technologies, and remain an actively developing research area. However, the IoT will not be composed only of embedded devices, since higher order computing devices will be needed to perform heavier duty tasks (routing, switching, data processing, etc.).

Besides the *plethora* of new application areas for Internet connected automation to expand into, IoT is also expected to generate large amounts of data from *diverse* locations that is *aggregated* and very high-velocity, thereby increasing the need to better index, store and process such data.[5]

Diverse applications call for different *deployment scenarios* and requirement, which have usually been handled in a *proprietary* implementation. However, since the IoT is connected to the Internet, most of the devices comprising IoT services will need to operate utilizing standardized technologies. *Prominent* standardization bodies, such as the IETF, IPSO Alliance and ETSI, are working on developing protocols, systems, architectures and *frameworks* to enable the IoT.

Words and Expressions

  embedded [im'bedid] *adj.* 嵌入式的
  embed [im'bed] *vt.* 使嵌入，把……嵌入；使插入
  usher ['ʌʃə] *vt.* 迎接，迎来(in)；引，领
  heart monitoring implant 心脏监测植入器
  biochip transponder 生物芯片转发器；生物芯片应答器
  built-in 内置的，嵌入的
  field operation 野外作业
  sensory ['sensəri] *adj.* 感知的，感觉的
  electro-mechanical 电机的，机电的，机电系统的
  thermostat ['θə:məstæt] *n.* 恒温器，温度自动调节器
  washer/dryer *n.* 洗衣机,洗碟机/干燥机
  convergence [kən'və:dʒəns] *n.* 会聚，会合，聚合
  compact ['kɔmpækt] *adj.* 紧凑的；紧缩的；紧密的
  Gartner 美国咨询公司(高德纳)
  as per 根据
  survey [sə:'vei] *vt.* 概述；综观；调查 *n.* 概述；综观；调查
  engaged [in'geidʒd] *adj.* 从事……的，参与的；受雇用的，被占用的
      （engage 的过去式和过去分词）
  wearable ['wɛərəbl] *adj.* 可穿戴式的，可(佩)带的，可穿用的
  as such 照这样；本身
  accommodate [ə'kɔmədeit] *vt.* 提供；供应，供给；使适应
  actuation [ˌæktju'eiʃən] *n.* 激励；启动；开动
  to a large extent 在很大程度上
  critical ['kritik(ə)l] *adj.* 关键性的；紧要的；危急的
  plethora ['pleθərə] *n.* 过多，过剩；多血（症）
  diverse [dai'və:s] *adj.* 多种多样的，各种各样的；形形色色的；不同的

aggregate [ˈæɡrɪɡət] n. 聚集(体)；合计　vt. 使聚集；总计　adj. 聚集的；合计的
deployment scenarios　部署方案
proprietary [prəˈpraɪət(ə)ri] adj. 专用(有)的；所有人的；业主的　n. 所有权；所有人
prominent [ˈprɒmɪnənt] adj. 知名的
framework [ˈfreɪmwɜːk] n. 构架，框架

Abbreviation

RFID (Radio Frequency IDentification)　射频识别

Notes

1. 本句中 that goes beyond … and applications 是定语从句，其中的 that 代表前面的 connectivity。goes beyond "超出"。该从句可译为：(这种高级连接)超越了机器对机器(M2M)的通信，涵盖了各种协议、领域和应用程序。
2. Smart Grid 是指"智能电网"。全句可译为：预料这些嵌入式设备(包括智能对象)的互连将迎来几乎所有领域的自动化，同时也使(诸如)智能电网的先进应用成为可能。
3. 本句中 contribute 意为"(做出)贡献，提供，起一份作用"。which resulted in … 是定语从句，其中的 which 代表前面的 convergence。全句可译为：以下各种技术促成了所有这些技术的能力汇聚，从而产生了物联网。
4. 本句中 As per … Project 是介词短语作状语，表示根据；who responded 的 who 指前面的 experts 和 users；83 percent 是前面 a large majority of 的同位语。全句可译为：根据 Pew Research Internet Project 最近所做的调查和研究，大多数(83%)接受访问的技术专家和受雇用的 Internet 用户都同意这样的看法，在 2025 年前物联网/云、嵌入式和可穿戴式计算(以及对应的动态系统)将会产生广泛和有益的影响。
5. 本句中 to expand into 是不定式，用作前面 automation 的定语；that is aggregated … 是定语从句，其中的 that 代表前面的 large amounts of data；increasing the need … 是现在分词短语，作状语，表示结果。该句中 that 之前的部分可译为：除因特网连接的自动化要扩展过多新应用领域外，预计物联网还将从不同的位置生成大量数据。

注：物联网是多种技术的汇聚，其体系构架可分为感知层、网络层和应用层三个层次。物联网的物一般具有可感知、可标识和可控制的特性。本篇简单介绍什么是物联网，它的物，以及所涉及的一些技术和问题，如低功率无线通信技术、识别技术、IPv6 的支持、标准化等。

6.4　Wireless Network

In the time *span* of just a few years, wireless local area networking went from being a *novelty* to *revolutionizing* the way many organizations connect their computers.[1] Visit any major department store, hospital, or office building, and you will encounter 802.11 cards in all of the PCs and *access points* hanging from ceiling.[2] The speed with which wireless networking has *caught on* is not surprising, as 802.11b offers up to 11 Mbps of bandwidth, and a range of several hundred feet. Newer standards, such as 802.11g, promise five times the speed (54 Mbps). Multiple wireless access points can be easily installed on the same network to increase the coverage area, so that an entire building can be easily connected. Conversely, wiring buildings with Ethernet is expensive and limits the locations from which networked computers can be used. [3]

Most new *laptops* purchased today are *outfitted with* built-in 802.11 networking capabilities, and configuring a home or office wireless network out of the box can take less than 10 minutes. [4] Furthermore, PC card are rapidly coming down in price and increasing in power. The economic forces influencing wireless networking are matched only by the convenience to users. [5] Wide-scale adoption of 802.11 was inevitable, and the general expectation is that it will only increase. Eventually, it is likely that most public areas will offer some sort of wireless connectivity; there are *initiatives* to extend coverage to airplanes and trains, as well as *shopping malls* and airports. [6]

An IEEE 802.11 WLAN is a group of stations (wireless network nodes) located within a limited physical area, where each station is capable of radio communication with a *base station*. There are two WLAN design structures: *ad hoc* and infrastructure networks. The vast majority[7] of installations use infrastructure-based WLANs.

An ad hoc WLAN has no ability to communicate with external networks without using additional *routing protocols*. [8] An ad hoc WLAN is normally created to permit multiple wireless stations to communicate directly with each other, requiring minimal hardware and management.

An infrastructure-based WLAN is composed of one or more Basic Services Set (BSS). Each station has exactly one BSS link connecting it to the infrastructure, the Distribution System (DS), which allows access to external networks. The station's attachment point to the DS, called the Access Point (AP), relays packets from the station within the BSS to the DS as shown in Figure 6-3. [9]

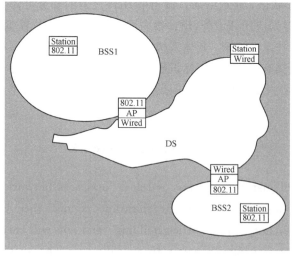

Figure 6-3 DS and BSS

This relaying of traffic means that an *adversary* has additional opportunities to *intercept* traffic. In Figure 6-3, the traffic can be captured by radio receivers in BSS1 or BSS2, as well as by *sniffers* in the wired network. So the *advent* of wireless networking has raised some very unique and compelling issues. The first issue is security. Given the open nature of wireless networks, what threats do they introduce?[10] Other issues are legal and social. Is it right for someone to share with their neighbors the bandwidth for which they are paying an ISP? Do service providers have a right to insist on payment from anyone who obtains connectivity? Is it appropriate for a coffee shop to resell Internet service to wireless users?

Infrared Technology

Infrared (IR) technology has gained popularity in recent years as a way to set up wireless links between office PCs or between an office PC and a handheld device or printer. Infrared technology sends data as infrared light rays. Like your infrared television remote control, infrared technology requires line-of-sight transmission. Because of this limitation, many formerly infrared devices (such as wireless mice and keyboards) now use radio technology instead.

Some applications still using infrared technology include *beaming* data from a handheld PC, notebook, digital camera, or other device to desktop computer, sending documents from a portable PC to a printer, and connecting a portable PC to a company network.

Bluetooth (Wireless PAN, IEEE 802.15)

The *Bluetooth* standard is a low-cost, short-range, wireless radio solution for communications between handheld PCs, mobile phones, and other portable devices, as well as for connecting those devices to home and business equipment, such as PCs, telephones, printers, and more.[11] Bluetooth wireless technology facilitates real-time voice and data transmissions between *Bluetooth-enabled* devices (devices containing a special Bluetooth transceiver chip). For example, a Bluetooth *earpiece* or *headset* can be used *in conjunction with* a cell phone left in a pocket or bag, or a PDA device can be instantly *synchronized* with a desktop PC on entering the office.[12] Since Bluetooth devices automatically recognize each other when they get within transmission range — about 10 meters without an amplifier — handheld PCs, cell phones, and other portable devices can always be networked wirelessly when they are within range. Some industry experts predict that all *household appliances* will be Bluetooth-enabled in the future, resulting in an automatic, always connected *smart home*.

Wireless Ethernet (Wireless LAN, Wi-Fi, IEEE 802.11)

Wireless Ethernet allows the Ethernet standard to be used with wireless network connections. It is also known as *Wi-Fi*, though technically the *Wi-Fi* label can only be used with wireless Ethernet products that are *certified* by the *wireless Ethernet compatibility Alliance*. Users of Wi-Fi *certified* products are assured that their hardware will be compatible with all other Wi-Fi certified hardware.

The IEEE 802.11 standard extends the carrier-sensing multiple access (CSMA) principle employed by Ethernet (IEEE 802.3) technology to suit the characteristics of wireless communication. The 802.11 standard is intended to support communication between computers located within about 150 meters of each other at speed up to 54 Mbps. Wireless Ethernet is a growing choice for organizations wishing to extend their wired Ethernet network.

Words and Expressions

span [spæn] *n.* 跨度；跨径；一段时间 *vt.* 横跨；跨越
novelty ['nɔvəlti] *n.* 新颖，新奇；新奇事物
revolutionize [revə'lu:ʃənaiz] *vt.* 使革命化；彻底改革
catch on   变得流行
laptop ['læptɔp] *n.* 便携式计算机，笔记本计算机，膝上型计算机
outfit with   装备，配备
initiative [i'niʃiətiv] *n.* 主动；积极性
shopping mall   大型购物中心，购物商品区
base station   基站
ad hoc [æd'hɔk] *adv.* 专门；特殊；特定；特别 *adj.* 专门的；特别的
routing protocol   路由协议
adversary ['ædvəsəri] *n.* 对手，敌手
intercept [intə'sept] *vt.* 截取，截获
sniffer ['snifə] *n.* 嗅听者，嗅听器
advent ['ædvənt] *n.* 出现
infrared [,infrə'red] *n.* 红外线 *adj.* 红外线的；使用红外线的
beam [bi:m] *vt.* 播送，定向发出 *n.* 梁；(光线的)束；柱；电波；横梁
bluetooth-enabled   有蓝牙功能的，能够用蓝牙的，实现蓝牙功能的
earpiece ['iəpi:s] *n.* 耳机，耳塞，听筒
headset ['hedset] *n.* 戴在头上的耳机或听筒
in conjunction with   与……共同(或协力)；连同
synchronize ['siŋkrənaiz] *vi.* 同步
household appliances   家用电器
smart home   智能住宅
certify ['sə:tifai] *vt.* 证明；批准 certified *adj.* 被证明的；证明合格的；持有证书的
Wi-Fi ['wai,fai] (Wireless-Fidelity 的缩写)
    n. 一种无线联网技术(WLAN (Wireless LAN) 的同义词)

Abbreviations

AP (Access Point)   接入点；访问点

BSS（Basic Service Set） 基本服务组（集），基本服务单元
DS（Distributed System） 分布式系统
PAN（Personal Area Network） 个人区域网，个人局域网
MAN（Metropolitan Area Network） 城域网
PDA（Personal Digital Assistant） 个人数字助理

Notes

1. 本句中 went from being … to revolutionizing … 译为"从一个新颖事物到彻底改变许多组织连接它们的计算机的方法"。其中 many organization … 是定语从句，修饰前面的 way。全句译为：只有几年的时间，无线局域网已从一个新颖事物发展到彻底改变许多组织连接它们的计算机的方法。
2. 802.11 cards "802.11 网卡"。802 是美国 IEEE（见后面的 Terms）在 20 世纪 70 年代给以太网标准的编号，即 IEEE 802。之后，以太网不断发展，相继出现 IEEE 802.3，IEEE 802.6 等标准。IEEE 802.11 是无线以太网的标准。access points hanging from ceiling 可译为"安装在天花板上的接入点"。

   下一句中，The speed with which … is not surprising 可译为"无线网络变得流行的这种速度并不令人惊奇"。
3. 本句中 wiring building with Ethernet 可译为"用以太网对大楼布线"。limits the locations from … 译为"限制了可以使用联网计算机的位置"。
4. 本句中 out of the box 可译为"从盒子取出"。本句可译为：现在购买的大多数新的膝上型计算机都具有内置的 802.11 联网能力，而配置一个家庭或办公室（用）盒装式无线网络只需要不到 10 分钟。
5. forces 这里指"压力，约束力"。本句可译为：影响无线网络的经济力量只与用户的便利性相匹配。

   下句中 general expectation 意为"普遍预期"。
6. 本句中 to extend coverage … 是不定式短语，修饰前面的 initiative。本句可译为：人们积极地把范围扩大到飞机和火车，以及购物商品区和机场。
7. vast majority "绝大多数"。
8. 本句中 without using … 作条件状语。本句可译为：如果不用附加的路由协议，特定的无线局域网就没有能力与外部网络通信。
9. 本句可译为：工作站到 DS 的连接点（称为接入点，AP）把该 BSS 中的工作站来的包转播到 DS，如图 6-3 所示。
10. 本句中 given the open … 是过去分词短语，作原因状语。全句可译为：由于无线网络的开放特性，它们会带来哪些威胁？
11. 本句中 solution for communications … for connecting … 中两个 for 是并行成分。全句译为：蓝牙标准是手持 PC、移动电话和其他便携式设备之间低成本、短距离的无线通信的解决方案，也是这些设备连接到家里和公司里 PC、电话机、打印机等设备的连接解决方案。
12. 全句译为：例如，蓝牙耳塞或头戴式耳机能够与放在口袋或包里的手机一起使用（无线连接），或者当你走进办公室时，你的 PDA 设备能立刻与台式 PC 同步。

注：笔记本计算机无线连接到 Internet，会像手机无线连接到电话网一样逐步流行。本节简单介绍无线网络的两种设计结构，以及它们所具有的独特问题。

本节介绍的无线网络是 Internet 的一部分。笔记本计算机接入无线网络上 Internet，就像台式计算机接入 LAN 一样直接接入 Internet。

Terms

Cellular network（蜂窝网）

A cellular network or mobile network is a wireless network distributed over land areas called cells, each served by at least one fixed-location transceiver, known as a cell site or base station. In a cellular network, each cell uses a different set of frequencies from neighboring cells, to avoid interference and provide guaranteed bandwidth within each cell.

When joined together these cells provide radio coverage over a wide geographic area. This enables a large number of portable transceivers (e. g., mobile phones, pagers, etc.) to communicate with each other and with fixed transceivers and telephones anywhere in the network, via base stations, even if some of the transceivers are moving through more than one cell during transmission.

Major telecommunications providers have deployed voice and data cellular networks over most of the inhabited land area of the Earth. This allows mobile phones and mobile computing devices to be connected to the public switched telephone network and public Internet.

GSM—2G

GSM (Global System for Mobile Communications) is a standard developed by the European Telecommunications Standards Institute (ETSI) to describe protocols for second-generation (2G) digital cellular networks used by mobile phones. As of 2014 it has become the default global standard for mobile communications-with over 90% market share, operating in over 219 countries and territories.

2G networks developed as a replacement for first generation (1G) analog cellular networks, and the GSM standard originally described a digital, circuit-switched network optimized for full duplex voice telephony. This expanded over time to include data communications.

3G

3G is the third generation of mobile telecommunications technology. This is based on a set of standards used for mobile devices and mobile telecommunications use services and networks that comply with the International Mobile Telecommunications-2000 (IMT-2000) specifications by the International Telecommunication Union. 3G finds application in wireless voice telephony, mobile Internet access, fixed wireless Internet access, video calls and mobile TV.

3G telecommunication networks support services that provide an information transfer rate of at least 200 kbit/s. Later 3G releases, often denoted 3.5G and 3.75G, also provide mobile broadband access of several Mbit/s to smartphones and mobile modems in laptop computers.

The following standards are typically branded 3G:
- the UMTS system, first offered in 2001, standardized by 3GPP, used primarily in Europe, Japan, China (however with a different radio interface) and other regions predominated by GSM 2G system infrastructure. The cell phones are typically UMTS and GSM hybrids. Several radio interfaces are offered, sharing the same infrastructure:
  - The original and most widespread radio interface is called W-CDMA.
  - The TD-SCDMA radio interface was commercialized in 2009 and is only offered in China.
  - The latest UMTS release, HSPA+, can provide peak data rates up to 56 Mbit/s in the downlink in theory and 22 Mbit/s in the uplink.
- the CDMA2000 system, first offered in 2002, standardized by 3GPP2, used especially in North America and South Korea, sharing infrastructure with the IS-95 2G standard. The cell phones are typically CDMA2000 and IS-95 hybrids. The latest release EVDO Rev B offers peak rates of 14.7 Mbit/s downstream.

4G

4G is the fourth generation of mobile telecommunications technology, succeeding 3G and preceding 5G. A 4G system, in addition to the usual voice and other services of 3G, provides mobile broadband Internet access, for example to laptops with wireless modems, to smartphones, and to other mobile devices. Potential and current applications include amended mobile web access, IP telephony, gaming services, high-definition mobile TV, video conferencing, 3D television, and cloud computing.

Two 4G candidate systems are commercially deployed: the Mobile WiMAX standard (first used in South Korea in 2007), and the first-release Long Term Evolution (LTE) standard (in Oslo, Norway and Stockholm, Sweden since 2009).

The consumer should note that 3G and 4G equipment made for other continents are not always compatible, because of different frequency bands.

UMTS

The Universal Mobile Telecommunications System (UMTS) is a third generation mobile cellular system for networks based on the GSM standard. UMTS originally uses wideband code division multiple access (W-CDMA) radio access technology to offer greater spectral efficiency and bandwidth to mobile network operators.

NIC(Network Interface Card, 网络接口卡)

NIC is a computer hardware component designed to allow computers to communicate over a computer network.

RFID(Radio Frequency Identification,射频识别)

A technology similar in theory to bar code identification. With RFID, the electromagnetic or electrostatic coupling in the RF portion of the electromagnetic spectrum is used to transmit signals. An RFID system consists of an antenna and a transceiver, which read the radio frequency and transfer the information to a processing device, and a transponder, or tag, which is an integrated circuit containing the RF circuitry and information to be transmitted.

RFID systems can be used just about anywhere, from clothing tags to missiles to pet tags to food — anywhere that a unique identification system is needed. The tag can carry information as simple as a pet owners name and address or the cleaning instruction on a sweater to as complex as instructions on how to assemble a car. Some auto manufacturers use RFID systems to move cars through an assembly line. At each successive stage of production, the RFID tag tells the computers what the next step of automated assembly is.

One of the key differences between RFID and bar code technology is RFID eliminates the need for line-of-sight reading that bar coding depends on. Also, RFID scanning can be done at greater distances than bar code scanning. High frequency RFID systems (850 MHz to 950 MHz and 2.4 GHz to 2.5 GHz) offer transmission ranges of more than 90 feet, although wavelengths in the 2.4 GHz range are absorbed by water (the human body) and therefore has limitations.

FDDI

Acronym for **F**iber **D**istributed **D**ata **I**nterface. A standard developed by the American National Standards Institute (ANSI) for high-speed fiber-optic local area networks. FDDI provides specifications for transmission rates of 100 megabits (100 million bits) per second on networks based on the token ring standard. FDDI II, an extension of the FDDI standard, contains additional specifications for the real-time transmission of analog data in digitized form.

IEEE or I.E.E.E.

Institute of **E**lectrical and **E**lectronics **E**ngineers.

IEEE 802 standards

A set of standards developed by the IEEE to define methods of access and control on local area networks. The IEEE 802 standards correspond to the physical and data-link layers of the ISO Open Systems Interconnection model, but they divide the data-link layer into two sublayers. The logical link control (LLC) sublayer applies to all IEEE 802 standards and covers station-to-station connections, generation of message frames, and error control. The media access control (MAC) sublayer, dealing with network access and collision detection, differs from one IEEE 802 standard to another: IEEE 802.3 is used for bus networks that use CSMA/CD, both broadband and baseband, and the baseband version is based on the Ethernet standard. IEEE 802.4 is used for bus networks that use token passing; and IEEE 802.5 is used for ring networks that use token passing (token ring networks). In addition, IEEE 802.6 is an emerging standard for Metropolitan Area Networks, which transmit data, voice, and video over distances of more than five kilometers. See also bus network, ISO/OSI model, ring network, token passing, token ring network.

CSMA/CD

Acronym for **C**arrier **S**ense **M**ultiple **A**ccess with **C**ollision **D**etection. A network protocol for handling situations in which two or more nodes (stations) transmit at the same time, thus causing a collision. With CSMA/CD, each node on the network monitors the line and transmits when it senses that the line is not busy. If a collision occurs because another node is using the same opportunity to transmit, both nodes stop transmitting. To avoid another collision, both then wait for differing random amounts of time before attempting to transmit again. Compare token passing.

ATM (Asynchronous Transfer Mode)

ATM is a switching technique for telecommunication networks. It uses asynchronous time-division multiplexing, and it encodes data into small, fixed-sized cells. Each cell has 53 bytes. This differs from networks such as Ethernet LANs that use variable sized packets or frames. ATM provides data link layer services that run over OSI Layer 1 physical links. ATM has functional similarity with both circuit switched networking and small packet switched networking. This makes it a good choice for a network that must handle both traditional high-throughput data traffic (e.g., file transfers), and real-time, low-latency content such as voice and video. ATM uses a connection-oriented model in which a virtual circuit must be established between two endpoints before the actual data exchange begins.

Exercises

1. Translate the text of Lesson 6.2 into Chinese.
2. Topics for oral workshop.
- Talk about the differences between traditional networks and wireless networks.
- Explain the differences between circuit-switched networks and packet-switched networks.
- Discuss the differences between wireless networks and wireless communication.
3. Translate the following into English.

传输介质用来在网络上传输消息。例如，网络中使用的传输介质可以是一组私有的电缆、公共电话线或卫星系统。传输介质可以是有线的或者无线的。

三类最常用来承载消息的有线介质是双绞线(twisted-pair wire)、同轴电缆和光纤电缆(光缆)。最近几年在传输介质中最成功的研发之一是光纤。光缆通常用于网络的高速主干线或Internet基础设施。

近几年无线传输介质已经变得特别流行。它们支持在物理布线不现实或者不方便的环境中的通信，具备便于(facilitate)移动性。无线介质通常用来把设备连接到网络，共享计算机之间的信息，把无线鼠标连接到计算机，并且用于手持式 PC、无线电话和其他移动设备。通过空中传播的无线电信号传输数据是大多数无线介质的核心(heart)。除常规的广播电台应用之外，微波、蜂窝和卫星传输介质也利用无线电信号传输数据。

无线电传输需要使用发射器通过空气发送无线电信号。接收器(通常包含某种类型的天线)在另一端接收数据。当一个设备既作为接收器又作为发射器时，它常称为收发器(transceiver)或发送接收器。

4. Listen to the video "Web server" and write down the last paragraph.

Unit 7 Database

7.1 An Overview of a Database System

Let us consider an enterprise, such as an airline, that has a large amount of data kept for long periods of time in a computer. This data might include information about passengers, flights, aircraft, and personnel, for example. Typical relationships that might be represented include bookings (which passengers have seats on which flights?) flight *crews* (who is to be the pilot, copilot, etc., on which flights?), and service records (when and by whom was each aircraft last serviced?).

Data, such as the above, that is stored more-or-less permanently in a computer we term a *database*. [1]

The software that allows one or many persons to use and/or modify this data is a database management system (DBMS). The primary goal of a DBMS is to provide an environment that is both convenient and efficient to use in *retrieving* information from and storing information into the database. [2]

Data Abstraction

It should be obvious that between the computer, dealing with bits, and the ultimate user dealing with abstractions such as flights or assignment of personnel to aircraft, there will be many levels of abstraction. A fairly standard viewpoint regarding levels of abstraction is shown in Figure 7-1. There we see a single database, which may be one of many databases using the same DBMS software, at three different levels of abstraction.

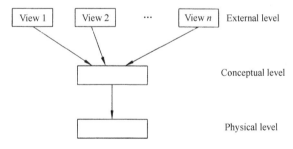

Figure 7-1 Data abstraction levels in a DBMS

The lowest level, i.e., the *physical level* has the data stored on hardware devices. [3] User programs cannot access them directly. They have to go through the logical level to access the data. The *external level* defines the different views of the database as required by the external or user programs. One user program may not require all the data in the database. Hence the user/application programs view only the required information from the database.

That means different programs will have different views of the database depending on their requirement of data. Such views are external to the database and are specified at the external level. Also it is not necessary that different views should contain altogether different data.[4] There can be common information in different views.

The *conceptual level* describes the entire database. It is used by database administrators, who must decide what information is to be kept in the database.

Data Models

A data model is a collection of conceptual tools for describing data, data relationships, data *semantics* and data constraints. The data models are divided into three classes, viz., object-based logical models, record-based logical models and physical data models.

Object-based logical models are used for describing data at the conceptual and view levels. They are very close to human logic. Many different models are available to describe object-based logical models. The most important among them are semantic data model and *entity-relationship model*. Semantic data model provides a *facility* for expressing meaning about the data in the database. The Entity-Relationship model (E-R model) is based on a *perception* of a real world which consists of a collection of objects called entities and relationships among these objects.[5] An entity is an object, which can be uniquely distinguished from other objects. For instance, the *designation*, physical *dimensions* and weight per unit length uniquely describe a particular rolled steel section. The set of all entities of the same type and relationships of the same type are termed as entity set and relationship set respectively. Examples of entity set are:

      all rooms in a building
      all elements in a finite element *mesh*
      all bearings in a machine

Entities and relationships are to be distinguished and a database model should specify how this can be carried out. This is achieved using the concept of *primary key*. An entity-relationship model may define certain constraints to which the contents of a database must confirm. One important constraint is the number of entities to which another entity can be associated via a relationship. For relationships involving two entity sets, there can be relationships like one-to-one, one-to-many, many-to-one and many-to-many. Schematic representation of these relationships are shown in Figure 7-2.

Record-based logical models define the overall logical structure of the database as well as higher level description of its implementation. Three different record-based logical models are widely used. They are:

1. Hierarchical Model
2. Network Model
3. Relational Model

Physical data models are used to describe data at the lowest level. There are very few[6] physical data models in use. Some of the widely known ones are:

1. *Unifying model*
2. Frame memory

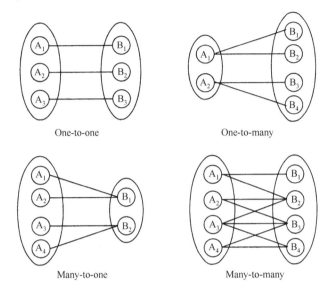

Figure 7-2 Different types of entity relationships

Words and Expressions

database ['deitəbeis] *n.* 数据库
crew [kru:] *n.* 全体乘务员
retrieve [ri'tri:v] *vt.* 检索;取回;重新得到;恢复 *n.* 检索;恢复;找回
physical level 物理层
external level 外部层
conceptual level 概念层
semantics [si'mæntiks] *n.* 语义学
entity-relationship model 实体关系模型
facility [fə'siliti] *n.* 便利,方便;容易;灵巧;熟练,敏捷;[常用复数] 设备,工具
perception [pə'sepʃən] *n.* 感知,感觉
designation [ˌdezig'neiʃən] *n.* 名称;指定,选派
dimension [di'menʃən] *n.* 尺寸;尺度;维(数);度(数);元
mesh [meʃ] *n.* 网格
primary key 主关键字;主键
unifying model 统一模型

Abbreviation

DBMS (DataBase Management System) 数据库管理系统

Notes

1. 这是倒装句,we 是主语,term 是谓语动词,data 是宾语,a database 是宾语补足语。全句可译为:我们把上述那种不同程度地长久存储在计算机中的数据称为数据库。
2. 全句可译为:DBMS 的主要目标是提供这样的环境,在从数据库中检索信息和把信息存储在数据库中时使用起来既方便,又高效。
3. has 这里含意为"让"。the data stored 构成"名词+分词"形式的复合宾语。全句可译为:最低层,也即物理层,把数据存储在硬件设备上。
4. altogether 这里含意为"完全"。全句可译为:不同的视图也无必要包含完全不同的数据。
5. 关系代词 which 引导的限制性定语从句修饰 a real world。全句可译为:实体关系模型(E-R 模型)基于这样的感知:现实世界是由一组称为实体的对象和这些对象之间的关系组成的。
6. very few 意为"很少很少"或"几乎没有"。请注意它与 a very few(极少数)、a good few 及 quite a few(相当多,不少)的区别。

注:一个数据库系统一般由下列四部分组成:数据库、数据库管理系统、数据库管理员与应用程序。

数据库是数据的汇集,这些数据以一定的组织形式存储在存储介质上;数据库管理系统对数据库进行管理,实现数据库系统的各项功能;数据库管理人员进行数据库的规划、设计、协调、维护和管理;应用程序则实现背景问题所预期的功能,它通过数据库管理系统访问数据库。

7.2 Introduction to SQL

Ideally, a database language must enable us to create the database and table structures; it must enable us to perform basic data-management *chores* (add, delete, and modify data); and it must enable us to perform complex *queries* designed to transform the raw data into useful information. Moreover, it must perform such basic functions with minimal user effort, and its *command* structure and *syntax* must be easy to learn. Finally, it must be portable; that is, it must conform to some basic standard so that we do not have to learn from scratch when we move from one RDBMS to another.

SQL meets these ideal database language requirements well. First, SQL coverage fits into three categories:

1. Data definition creates the database and its table structures.
2. Data management uses a set of commands to enter, correct, delete, and update data within the database tables.
3. Data query uses a set of commands to explore the database contents and allows the user to convert the raw data into useful information.

Second, SQL is relatively easy to learn: It performs the required database functions by using a basic *vocabulary* of about thirty commands. Better yet, SQL is a *nonprocedural language*: You merely have to command *what* is to be done; *you don't have to worry about how it is to be done*.[1]

Finally, the American National Standards Institute (ANSI) does prescribe a standard SQL. Never mind that the ANSI standard is so limited that all commercial SQL products exceed it.[2] In fact, some vendors already meet the proposed ANSI SQL 2 standard, to be implemented in 1993.[3] Consequently, it is seldom possible to take a SQL-based application from one RDBMS to another without making some changes. Nevertheless, the different SQL dialects share the same basic command set and structure, thus allowing us to conclude that there is a useable standard. We will use this standard as the basis for our presentation. But we will also present a few SQL enhancements, especially when those enhancements are widely shared among the many RDBMS vendors.[4]

Don't become discouraged over the fact that several SQL dialects exist. Since the differences between the various SQL dialects are minor, you should have little trouble adjusting to your software requirements.[5] Whether you use XDB, ORACLE, dBASE IV, DB2, R:BASE for DOS, IBM's OS/2 Database Manager, or any other well-established RDBMS software, our experience is that a few hours spent with a software manual will be sufficient to get you up to SQL speed if you know the material presented in this chapter.[6] In short, the knowledge you gain in this lesson is portable.

There are some very good reasons for studying SQL basics:

1. The ANSI *standardization* effort has led to a de facto query standard for *relational databases*. In fact, many relational database experts are inclined to argue that, "If it's not SQL, it's not relational."
2. SQL has become the basis for present and expected future DBMS integration efforts, allowing us to link hierarchical, network, and relational databases.
3. SQL has become the *catalyst* in the development of *distributed databases* and database client/server architectures.

Words and Expressions

chore [tʃɔː] *n.* 日常零星工作
query ['kwiəri] *n.* 查询；询问　*v.* 询问
command [kə'mɑːnd] *n.* 命令
syntax ['sintæks] *n.* 语法
vocabulary [və'kæbjuləri] *n.* 词汇表，词典；词汇；词汇量
nonprocedural language　非过程型语言
standardization [ˌstændədai'zeiʃən] *n.* 标准化
relational database　关系型数据库
catalyst ['kætəlist] *n.* 催化剂
distributed database　分布式数据库

Abbreviations

RDBMS (Relational Database Management System)　关系数据库管理系统
SQL (Structured Query Language)　结构式查询语言
ANSI (American National Standards Institute)　美国国家标准协会
DOS (Disk Operating System)　磁盘操作系统

Notes

1. what 是关系代词，引导一个从句作 command 的宾语。to be done 是不定式短语，作从句中的表语。全句可译为：你必须做的仅仅是发出要做什么的命令，不必操心它是如何做的。
2. 本句是祈使句，表示一种建议。全句可译为：不必介意 ANSI 标准如此受限制，以致所有的商用 SQL 产品都胜过它。
3. to be implemented … 是不定式短语，作非限制性同位语，修饰前面的 standard。全句可译为：事实上，某些厂家已经符合要在 1993 年实施的所提议 ANSI SQL 2 标准。
4. present 在这里读音为[pri'zent]，而非['preznt]，含意为"介绍"。when 为连词，这里含意为"考虑到"。全句可译为：然而我们也将介绍几个 SQL 增强特性，主要是考虑到这些增强特性广泛地为很多 RDBMS 供应商所共享。
5. adjusting 引导的分词短语作状语。全句可译为：由于各种 SQL 方言版本之间的差异是微小的，适应你的软件需求几乎没有什么麻烦。
6. well-established 这里含意为"一致公认的"。whether … or … 含意为"不管……还是……"。全句可译为：不管你使用的是 XDB、ORACLE、dBASE Ⅳ、DB2、DOS 下的 R:BASE、IBM 的 OS/2 数据库管理程序，还是任何其他一致公认的 RDBMS 软件，我们的经验是：如果你了解本章中介绍的材料，那么花几个小时了解软件手册将足以让你快速掌握 SQL。

Term

SQL

Acronym for structured query language.

A database sublanguage used in querying, updating, and managing relational databases; the de facto standard for database products.

7.3　Object-relational Database

Object-relational database management systems span object and relational technology.

Object Database

In an object oriented database information is represented in the form of objects as used in

Object-Oriented Programming. When database capabilities are combined with object programming language capabilities, the result is an object database management system (ODBMS). An ODBMS makes database objects appear as programming language objects in one or more object programming languages. An ODBMS extends the programming language with transparently *persistent* data, concurrency control, data recovery, associative queries, and other capabilities.

Some object-oriented databases are designed to work well with object-oriented programming languages such as Python, Java, C#, Visual Basic. NET, C++ and Smalltalk. Others have their own programming languages. ODBMSs use exactly the same model as object-oriented programming languages. Object databases are generally recommended when there is a business need for high performance processing on complex data.

Why "Object-relational?"

Essentially, the SQL:1999 object model has all the same features as the object model used by object database management systems (ODBMSs), but in a way different from most ODBMSs. This was caused by the *mandate* that SQL:1999 be *backward compatible* to SQL-92. That mandate required adapting the SQL:1999 object model to the SQL-92 relational model. As a result, the SQL:1999 object model does not match the object model used by object programming languages. The term object-oriented cannot be used to describe this model because that would imply that the database models match object programming models.[1] Instead, the term *object-relational* is used.

SQL:1999 expands the data complexity and performance capability of relational databases with the object structures that were added. Nevertheless, more complex structures that[2] can be used by an object programming language will result in *impedance mismatch* and require mapping. The mapping, however, is not as complicated as it is for SQL-92 RDBMSs. The data mapping is needed to move data from a database to an object programming language (or an application server using an object programming language) and additional mapping to move the data back to the database.[3] (ODBMSs that use the same object model as object programming languages do not need this type of mapping.)

A common *initialism* used for an object-relational database management systems is ORDBMS.

Object-Relational Database (ORD)

An object-relational database (ORD) or object-relational database management system (ORDBMS) is a relational database management system that allows developers to integrate the database *with* their own custom data types and methods. The term object-relational database is sometimes used to describe external software products running[4] over traditional DBMSs to provide similar features; these systems are more correctly referred to as object-relational mapping systems.

Whereas RDBMS or SQL-DBMS products focused on the efficient management of data drawn from a limited set of data types (defined by the relevant language standards), an object-relational DBMS allows software developers to integrate their own types and the methods that apply to them into the DBMS.[5] The goal of ORDBMS technology is to allow developers to raise the level of abstraction at which they view the problem domain.

Object-relational database management systems *grew out of* research that occurred in the early 1990s. That research extended existing relational database concepts by adding object concepts. The idea was to retain a declarative query language based on *predicate calculus* as a central component of the architecture. Probably the most notable research project was Postgres (UC Berkeley). Two products trace their *lineage* to that research: Illustra and PostgreSQL.

Many of the ideas of early object-relational database efforts have largely been added to SQL:1999. In fact, any product that *adheres to* the object-oriented aspects of SQL:1999 could be described as an object-relational database management product. For example, IBM's DB2, Oracle database, and Microsoft SQL Server, make claims to support this technology and do so with varying degrees of success.[6]

Words and Expressions

object-relational database 对象关系数据库
object database 对象数据库
persistent [pə'sistənt] *adj.* 持久的,持续的;持久稳固的
mandate ['mændeit] *n.* 命令,训令;委任;授权 *vt.* 托管
backward compatible 向后兼容的
impedance [im'pi:dəns] *n.* [电]阻抗,全电阻;[物]阻抗
impedance mismatch 阻抗失配
initialism [i'niʃəliz(ə)m] *n.* [语]词首字母缩略词
integrate with 使结合
integrate into 使并入,使成一体
grow out of 产生自……;变得不适合于;长大得与……不再相称
predicate calculus 谓词演算
lineage ['liniidʒ] *n.* 血统,世系
adhere to 坚持;追随;依护;胶着

Abbreviations

ODBMS (Object DataBase Management System) 对象数据库管理系统
ORDBMS (Object-Relational DataBase Management System) 对象关系数据库管理系统

Notes

1. 本句可译为:面向对象这一术语不能用来描述该模型,因为那将意味着数据库模型匹配对象程序设计模型。

2. 该 that 是关系代词，引导一个限制性定语从句修饰其前的 structures，意为"能为对象程序设计语言所使用的"。
3. 本句中 using 是现在分词，引导的短语修饰其前的 server。全句可译为：需要有数据映像把数据从数据库移到对象程序设计语言（或者说使用对象程序设计语言的应用程序服务器），以及另外的映像把数据移回到数据库。
4. running 是现在分词，引导的短语修饰其前的 products，意为"在提供一些类似特性的各传统 DBMS 上运行的"。
5. 这是由连接词 Whereas 连接的并列句，其中的 drawn 是过去分词，引导的短语修饰其前的 data；后面的 that 引导的定语从句修饰其前的 methods。全句可译为：尽管各种 RDBMS 或 SQL-DBMS 产品集中于对从一组有限的（由相关的语言标准定义的）数据类型提取的数据进行有效管理，对象关系 DBMS 使得软件开发人员能把他们自己的类型和应用于这些数据类型的方法并入 DBMS 中。
6. 本句可译为：例如，IBM 的 DB2、Oracle 数据库和微软 SQL Server 声称支持这一技术，并且已取得不同程度的成功。

Terms

Relational database

Relational database is a database that conforms to the relational model, and refers to a database's data and schema (the database's structure of how that data is arranged). Common usage of the term "Relational database management system" technically refers to the software used to create a relational database, but sometimes mistakenly refers to a relational database.

A relational database management system (RDBMS) is a database management system (DBMS) that is based on the relational model. Relational databases are the most common kind of database in use today (assuming one does not count a file system as a database).

A short definition of a RDBMS may be a DBMS in which data is stored in the form of tables and the relationship among the data is also stored in the form of tables.

Distributed database

A distributed database is a database in which storage devices are not all attached to a common processing unit such as the CPU, controlled by a distributed database management system (together sometimes called a distributed database system). It may be stored in multiple computers, located in the same physical location; or may be dispersed over a network of interconnected computers. Unlike parallel systems, in which the processors are tightly coupled and constitute a single database system, a distributed database system consists of loosely-coupled sites that share no physical components.

Intelligent database

In the late 1980s the concept of an intelligent database was put forward as a system that manages information (rather than data) in a way that appears natural to users and which goes beyond simple record keeping.

This concept postulated three levels of intelligence for such systems: 1. high level tools, 2. the user interface and, 3. the database engine. The high level tools manage data quality and automatically discover relevant patterns in the data with a process called data mining. This layer often relies on the use of artificial intelligence techniques. The user interface uses hypermedia in a form that uniformly manages text, images and numeric data. The intelligent database engine supports the other two layers, often merging relational database techniques with object orientation.

In the twenty-first century, intelligent databases have now become widespread.

NoSQL database

A database that features fewer consistency restrictions than conventional relational databases. While most often called NoSQL databases, these databases are sometimes also referred to as "Not only SQL" databases to reflect that SQL queries can be used in some cases.

NoSQL databases feature highly optimized key-value data designed for quick retrieval and appending operations, and they can offer substantial performance benefits over relational databases in terms of latency and throughput. As a result they've become quite popular recently, particularly in use with Big Data, cloud computing and real-time web applications.

Sensor database system

When multiple sensors work in a certain range, they form a sensor network. Sensor networks consist of sensors bound by carriers and servers that receive and process data sent back by sensors. In sensor networks, sensor data is generated by signal processing functions in sensors. The signal processing function measures and classifies the data detected by the sensor, and marks the classified data with a timestamp, which is then sent to the server and processed by the server. In sensor networks, a large number of data queries have to deal with the data flow between sensors or between sensors and front-end servers. Data flow engine and data flow operators are the main methods to control this large flow of data.

One of the functions of the sensor database system is to access the sensor network and feedback the queried results to the application users. Sensor database systems need to consider the existence of a large number of sensor devices, as well as their mobility and decentralization. Sensor databases must make use of all sensors in the system, and can manage data in sensor databases as conveniently and concisely as traditional databases; establish a mechanism to obtain and distribute source data; establish a mechanism to adjust data flow according to sensor networks; configure, install and restart components in sensor database conveniently, etc.

Real-time database (RTDB)

Real-time database is a branch of the development of database system, which is produced by the combination of database technology and real-time processing technology. Real-time database system is the supporting software for developing real-time control system, data acquisition system and CIMS system. Real-time database has become the basic data platform of enterprise informatization, which can directly collect and acquire all kinds of data in the process of enterprise operation in real time, and transform it into effective public information for all kinds of business. An important feature of real-time database is real-time, including data real-time and transaction real-time. Transaction real-time refers to the speed at which the database processes its transactions. It can be event triggering or timing triggering. In general the data of real-time database mainly come from DCS control system, control system established by configuration software + PLC, data acquisition system (SCADA), relational database system, direct connecting hardware equipment and manual input data through man-machine interface.

By real-time database, applications such as product planning, maintenance management, expert system, laboratory information system, simulation and optimization can be integrated, and it plays a bridge role between enterprise management and real-time production.

7.4 Data Warehouse

7.4.1 Data Warehouse

Getting a *handle* on corporate data isn't a new problem, but there's a new solution that many corporations are turning to: the *data warehouse*.[1]

As its name implies, a data warehouse is a storage place—although it consists of silicon rather than mortar and bricks. Data warehouses can contain *archives* of information going back 10 years, but they can also consists of recently gathered data.

The concept of the data warehouse embodies a new architectural approach to structuring diverse sources of corporate data with the goal of giving workers widespread and easy access.

Although they prove popular with retail, banking, utility, and telecommunications industries, data warehouses can be useful to most any enterprise.

That's because the data warehouse addresses the fundamental problem every business faces.

The data warehouse allows you to better manage and utilize corporate knowledge, which *in turn* helps a business become more competitive, better understand customers, and more rapidly meet market demands.[2]

The term "data warehouse" was coined a few years ago by William Inmon, vice president of technology for Prism Solutions Inc., a Sunnyvale, Calif., maker of warehouse management systems. Having authored several books on the subject, Inmon is generally referred to as the father of the concept, although he points out that aspects of data warehousing have existed for years.

According to Inmon, the data warehouse separates an organization's decision support, or informational data, from its *day-to-day* operational data.

Four characteristics distinguish the material found in the data warehouse. Warehouse data is:

- subject oriented;
- integrated;
- time-variant;
- *nonvolatile*.

For example, the operational data of a financial institution would focus on transactional functions such as loans, savings, and bank cards, while information on customers would be maintained separately — and inconsistently — in numerous other databases.

But data within the warehouse is organized around subjects such as customer, vendor, and product, which end-users can readily understand. Warehouse data is often gathered from sources throughout an enterprise, including different applications, databases, and computer systems, and is likely to be fragmented and inconsistent. [3]

Warehouse data must be cleaned up to conform to consistent formats, *naming conventions*, and access methods.

As for time variance, data is accurate *as of* a specific moment in time, excluding the present. [4]

In the day-to-day transactional world, a single data record, such as a customer's account *balance*, can change constantly. But in the data warehouse, that customer's balance will consist of a long string of single *snapshots* over time. And because the snapshots don't change, they are nonvolatile.

Most organizations are implementing their data warehouse in stages.

First, an organization must create a data model, as well as a physical design for the data warehouse. It must also establish a system of record, which is Inmon's term for the source of the data. [5]

The data then has to be extracted from the operational databases, transformed or cleaned up, and loaded into the data warehouse. This task is typically handled by data warehouse management software. Companies providing such software include Prism Solutions, Evolutionary Technologies, and Carleton Corp. of Birlington, Mass.

Data warehouse management software runs on a workstation and allows an IS *staff* member to specify particular transformations, such as data mapping, *conversion*, and summarization.

Data warehouse management software also maintains the *meta data*, which in essence is data about the data. Meta data defines the relationships between operational databases and the data warehouse and is stored within the warehouse. When end-users access the data warehouse, they can use the meta data to find definitions of the data or to look up subject areas.

Of course, along with getting the data into the warehouse, organizations also have to choose a DBMS for the warehouse.

All of the popular relational database management systems (RDBMS) — DB2/Oracle, Ingres, Informix, Sybase SQL Server, and Hewlett-Packard Co.'s Allbase/SQL — are suitable for data warehousing. Additionally, there are databases made expressly for data warehousing, such as Red Brick Warehouse.

Proponents of this latter category say a DBMS such as Red Brick offers the advantage of faster query processing, because it doesn't have a built-in *overhead* for handling transactional processes.

Opponents argue that these products offer performance advantages only to medium-size warehouse containing as much as 20 gigabytes of data.

Organizations must also choose the appropriate data access and *reporting tools* and database connectivity software that let end-users to store information in the data warehouse.

Trinzic Corp.'s Forest & Trees and Metaphor Inc.'s Data Interpretation System (DIS) are examples of data access tools, while Microsoft Corp.'s Open Database Connectivity (ODBC) and the SQL Access Group API are examples of *middleware* that link different databases and front-end applications.

Warehouse builders may also want to evaluate data analysis and presentation products that allow data gathered from the warehouse to be analyzed and graphically represented. [6] This last category of products can include anything from spreadsheets to executive information systems.

Corporations that are already *immersed* in the process say implementing a data warehouse becomes easier and less expensive as their efforts continue.

And, of course, the users are happy with the result of the data warehouse. Sometimes it's like a kid in a candy store, they are so *thrilled* to be getting what they want. [7]

Words and Expressions

data warehouse　数据仓库
handle ['hændl] *n.* 处理；柄，把手；句柄　　*vt.* 处理，操纵
archive ['ɑːkaiv] *n.* 档案文件　*vt.* 存档
in turn　依次，轮流
day-to-day　日常的，逐日的
nonvolatile ['nɔn'vɔlətail] *adj.* 非易失性的
naming convention　命名法约定
as of　从……时起；到……时为止；在……时
balance ['bæləns] *n.* 余额；收付平衡；平衡；天平　　*vt.* 使平衡；结算；用（天平等）称
snapshot ['snæpʃɔt] *n.* 快照；瞬像；抽点打印
staff [stɑːf] *n.* 全体工作人员
conversion [kən'vəːʃən] *n.* 变换，转换
meta data　元数据
overhead ['əuvəhed] *n.* 开销，企业一般管理费用
reporting tool　报表工具

middleware ['midlwɛə] *n.* 中间件，媒件；中间设备
immerse [i'mə:s] *vt.* 沉浸；投入；专心于
thrill [θril] *vt.* 使激动　*vi.* 激动　*n.* 一阵激动

Abbreviations

EIS (Executive Information System)　执行信息系统，主管信息系统
DIS (Data Interpretation System)　数据解释系统
ODBC (Open DataBase Connectivity)　开放数据库互连
API (Application Programming Interface)　应用程序设计接口
IS (Information System)　信息系统

Notes

1. that 引导 a new solution 的同位语。全句可译为：对公司数据的处理不是一个新问题，然而现在很多公司正致力于数据仓库这一新的解决办法。
2. in turn 这里含意为"依次，轮流"。which 引导的非限制性定语从句修饰前面的整个一句话。全句可译为：数据仓库使得人们能更好地管理和利用企业知识，这反过来又有助于企业变得更有竞争性，更好地理解顾客，更快速地满足市场需求。
3. including 引导 sources 的同位语。全句可译为：数据仓库数据经常是从整个企业的各种来源收集的，包括不同的应用、数据库与计算机系统等，因而可能是支离破碎和不一致的。
4. as of 含意为"从……时起；到……时为止；在……时"。全句可译为：至于时间上的变化，从任一特定时刻起的数据，除当前的以外，都是准确的。
5. which 引导的非限制性定语从句修饰 a system of record。全句可译为：它也必须建立一个记录系统，此记录系统是 Inmon 为数据源定的术语。
6. allow 后跟"名词＋不定式（被动）"形式的复合宾语。全句可译为：数据仓库建立者们可能也想对数据分析和演示产品进行评价，这些产品能对数据仓库收集来的数据进行分析并用图形表示出来。
7. to be getting 引导一个状语，表示原因。全句可译为：有时这像是小孩在糖果店里，因为得到想要的东西而如此激动。

7.4.2　What is Data Mining?

Overview

Generally, *data mining* (sometimes called data or knowledge discovery) is the process of analyzing data from different perspectives and summarizing it into useful information — information that can be used to increase *revenue*, cuts costs, or both.[1] Data mining software is one of a number of analytical tools for analyzing data. It allows users to analyze data from many different dimensions or angles, categorize it, and summarize the relationships identified[2]. Technically, data mining is the process of finding *correlations* or patterns among dozens of fields in large relational databases.

Continuous Innovation

Although data mining is a relatively new term, the technology is not. Companies have used powerful computers to *sift through* volumes of supermarket scanner data and analyze market research reports for years. However, continuous innovations in computer processing power, disk storage, and statistical software are dramatically increasing the accuracy of analysis while *driving down* the cost.

Data, Information, and Knowledge

- **Data**

Data are any facts, numbers, or text that can be processed by a computer. Today, organizations are accumulating vast and growing amounts of data in different formats and different databases. This includes:

  * Operational or transactional data such as, sales, cost, inventory, payroll, and accounting.

  * Nonoperational data, such as industry sales, forecast data, and macro economic data.

  * *Meta data*—data about the data itself, such as logical database design or data dictionary definitions.

- **Information**

The patterns, associations, or relationships among all this data can provide information. For example, analysis of retail point of sale transaction data can yield information on which products are selling and when. [3]

- **Knowledge**

Information can be converted into knowledge about historical patterns and future trends. For example, summary information on retail supermarket sales can be analyzed *in light of* promotional efforts to provide knowledge of consumer buying behavior. Thus, a manufacturer or retailer could determine which items are most *susceptible* to promotional efforts.

Data Warehouses

Dramatic advances in data capture, processing power, data transmission, and storage capabilities are enabling organizations to integrate their various databases into data warehouses. Data warehousing is defined as a process of centralized data management and retrieval. Data warehousing, like data mining, is a relatively new term although the concept itself has been around for years. Data warehousing represents an ideal vision of maintaining a central repository of all organizational data. Centralization of data is needed to maximize user access and analysis. Dramatic technological advances are making this vision a reality for many companies. And, equally dramatic advances in data analysis software are allowing users to access this data freely. The data analysis software is what supports data mining. [4]

What Can Data Mining Do?

Data mining is primarily used today by companies with a strong consumer focus—retail, financial, communication, and marketing organizations. It enables these companies to determine relationships among "internal" factors such as price, product positioning, or staff skills, and "external" factors such as economic indicators, competition, and customer *demographics*. And, it enables them to determine the impact on sales, customer satisfaction, and corporate profits. Finally, it enables them to "drill down" into summary information to view detail transactional data.

With data mining, a retailer could use point-of-sale records of customer purchases to send targeted *promotions* based on an individual's purchase history. [5] By mining demographic data from comment or warranty cards, the retailer could develop products and promotions to appeal to specific customer segments.

How Does Data Mining Work?

While large-scale information technology has been evolving separate transaction and analytical systems, data mining provides the link between the two. Data mining software analyzes relationships and patterns in stored transaction data based on *open-ended* user queries. Several types of analytical software are available: statistical, machine learning, and neural networks. Generally, any of four types of relationships are sought:

- Classes: Stored data is used to locate data in predetermined groups. For example, a restaurant chain could mine customer purchase data to determine when customers visit and what they typically order. This information could be used to increase traffic by having daily specials.
- *Clusters*: Data items are grouped according to logical relationships or consumer preferences. For example, data can be mined to identify market segments or consumer affinities.
- Associations: Data can be mined to identify associations. The beer-diaper example is an example of associative mining.
- Sequential patterns: Data is mined to anticipate behavior patterns and trends. For example, an outdoor equipment retailer could predict the likelihood of a backpack being purchased based on a consumer's purchase of sleeping bags and hiking shoes. [6]

Data mining consists of five major elements:

- Extract, transform, and load transaction data onto the data warehouse system.
- Store and manage the data in a multidimensional database system.
- Provide data access to business analysts and information technology professionals.
- Analyze the data by application software.
- Present the data in a useful format, such as a graph or table.

Words and Expressions

data mining　数据挖掘
revenue ['revənjuː] *n.* 收入，收益；[复]总收入；税收
correlation [ˌkɔri'leiʃən] *n.* 相互关系；相关(性)

sift [sift] *vi.* 被筛下；通过(through, into)；细查　　*vt.* 详审　　*vt. & vi.* 筛；精选
drive down　压低
meta data　元数据
in light of　按照，根据
susceptible [sə'septəbl] *adj.* 易受影响的；易感动的；容许……的
demographic [ˌdemə'ɡræfik] *n.* 人口统计数据；人口统计学　　*adj.* 人口统计(学)的；人口的
promotion [prə'məuʃn] *n.* 推销，(商品等的)宣传；促进，增进；发扬；提升
open-ended *adj.* 不固定的，随便的；无限制的，无尽头的
cluster ['klʌstə] *n.* 一簇；一组；一串；一束　　*vt.* 使成群　　*vi.* 丛生；群集

Notes

1. 本句中的 analyzing 和 summarizing 引导两个并列的动名词短语作 of 的宾语。此后的 that 是关系代词，引导一个限制性定语从句，修饰其前的 information。全句可译为：一般，数据挖掘(有时称为数据或知识发现)是从各种不同的视角分析数据，并把它概括成有用信息的过程——这些信息能用于增加收入、削减成本，或者既增加收入又削减成本。
2. identified 是过去分词，修饰其前的 relationships，可译为"所标识的"。
3. 本句中的 which 是形容词，含意为"哪一个(些)"。全句可译为：例如，对销售交易数据之零售要点的分析，能产生关于哪些产品正在销售和哪些产品何时销售的信息。
4. what 是关系代词，引导的从句作主句的表语。全句可译为：数据分析软件是支持数据挖掘技术的软件。
5. 本句中的 based 引导的过去分词短语作状语。全句可译为：使用数据挖掘，零售商能基于个人的购买历史，使用销售点的顾客购买记录来安排一些有针对性的促销活动。下一句中的 to appeal 是不定式，引导的短语作定语，修饰其前的 promotions。全句可译为：通过挖掘评论意见或保证卡中的人口统计数据，零售商能开发一些产品并向专门的顾客群进行推销。
6. 本句中 being purchased 与其前的 a backpack 一起构成复合宾语；based 引导一个分词短语作状语。全句可译为：例如，户外设备零售商可依据消费者购买睡袋和徒步旅行鞋的情况预测购买背包的可能性。

7.5　Big Data

According to a recent study by IBM, 90 percent of all the information ever created[1] has been produced in the last two years. That's Big Data, much of it so large that it's beyond the capacity of conventional database management systems to process efficiently.[2] It includes data produced by social media, email, video, audio, text, and web pages. The data, in turn, comes from smart phones, digital cameras, Global Positioning Systems (GPS), industrial sensors, social networks, and even public surveillance and traffic monitoring systems, among other sources. Every time you use a cell phone, do a Google or Baidu search, use your *credit rating*, or buy something from *Amazon*, you are expanding your digital *footprint* and creating more data.

Such is the scale of this wave of information that we need new ways to comb through it in order find the value hidden inside. [3]

As Internet companies such as Google, Baidu, and Amazon have sought ways to improve their network planning, advertising placement, and customer care, they have created innovative new applications of Big Data analysis. Today, the biggest generators of data are Twitter, *Facebook*, LinkedIn, and other *emerging* social networking services.

The enormous growth of data has coincided with the development of the "cloud", which is really just a new alias for the global Internet. As data *proliferates*, especially with the growth of mobile data applications using[4] Apple and Android smartphones and operating systems, the requirements for data transmission and storage also grow. No single site server configuration offers sufficient capacity for the enormous amounts of data derived from sites such as Microsoft.com or FedEx.com, which serve global communities of customers, with transactions numbering in the millions[5]. Users want fast response times and immediate results, especially in the largest and most complex web-based business environments, operating[6] at Internet speed.

Big Data analysis requires a cloud-based solution — a networked approach capable of[7] keeping up with the accelerating volume and speed of data transmission. This means that clusters of servers and software such as Hadoop and Enterprise Control Language (ECL), which[8] help find hidden meaning in large amounts of data, are distributed throughout the cloud on high-bandwidth cables. Using the cloud-based approach, a Google marketing analyst may be sitting in Silicon Valley inputting[9] information requests, but the data is being processed simultaneously in Tokyo, Amsterdam and Austin, Texas.

The largest companies making the most sophisticated industrial equipment are now able to go from merely detecting equipment failures to predicting them. [10] This allows the equipment to be replaced before a serious problem develops.

Large Internet commerce companies such as Amazon and Taobao use Big Data analytics to predict buyer activity as well as to understand warehousing requirements and geographic positioning.

Government medical agencies and medical scientists use Big Data for early discovery and tracking of potential *epidemics*. A sudden increase in *emergency room* visits, or even increased sales of certain *over-the-counter* drugs, can be early warnings of a *communicative disease*, allowing doctors and emergency response officials to activate control and containment procedures.

Many Big Data applications were created to help web companies struggling with unexpected volumes of data. Only the biggest companies had the development capacity and budgets for Big Data. But the continuing *decline* in the costs of data storage, bandwidth, and computing means that Big Data is quickly becoming a useful and affordable tool for medium-sized and even some small companies. Big Data analytics are already becoming the basis for many new, highly specialized business models. [11]

E-commerce is expected to *account for* eight percent of retail sales in China by 2015. Many younger Chinese in the 25-35 age group have already started to buy much of their clothing and consumer goods online. This means that Chinese companies like Taobao and *360buy.com* will need ever more sophisticated Big Data analytics to run their businesses effectively. Currently, most Chinese companies use open-source software, particularly Hadoop, to manage their Big Data applications. But the scale of potential Big Data suggests that China's own IT experts are likely to develop *indigenous* analytical products. It is possible that China could develop a globally *branded* Big Data product in the next five years.

Whether or not Chinese companies develop their own products, the scale of China's Big Data market will have a major impact on the data analytics sector and international product and service development.

In the meantime, emerging global Chinese corporations like Huawei and Haier are using variants of Big Data to define hidden or previously unrecognized buyer behavior and market trends. And just as major corporations such as General Electric, *Walmart* and Google use data analytics today, small companies will use the same processes on a smaller scale to improve their competitive positions and use scarce resources more effectively.

Words and Expressions

credit rating 信用评级，信用等级；信誉评价，信誉等级；客户信贷分类
Amazon ['æməzən] *n.* 亚马逊(公司)；亚马孙河(南美洲大河)
footprint ['futprint] *n.* (计算机)占用的空间；脚印，足迹
Facebook ['feisbuk] *n.* 脸谱(网)(一个社交网络服务网站)
emerging [i'mə:dʒiŋ] *adj.* 新兴的 *vi.* 出现；浮现(emerge 的现在分词)
proliferate [prə'lifəreit] *vi.* 激增；增生，增殖；扩散 *vt.* 使激增；使扩散
epidemic [ˌepi'demik] *n.* 流行病，时疫；(流行病的)传播
emergency room 急诊室
over-the-counter ['əuvəðə'kauntə] *adj.* 非处方的，不需处方可以出售的
communicative disease 传染病
decline [di'klain] *n.* 下降，下倾；衰退 *v.* (使)下降，(使)下倾
account for (在数量、比例上)占；对……负责；说明(原因、理由等)
360buy.com 京东商城
indigenous [in'didʒinəs] *adj.* 本土的；土生土长的；生来的，固有的
branded ['brændid] *adj.* 名牌的，有品牌的；打有烙印的
Walmart ['wɔlmɑ:t] 沃尔玛(世界连锁零售企业)

Abbreviations

GE (General Electric) 通用电气公司
GPS (Global Positioning System) 全球定位系统

Notes

1. 过去分词 created 作为定语修饰其前的名词(information)，这种语法现象将在后面多处出现，请注意。information ever created 可译为"不断创建的信息"。
2. 此处的 it 是先行代词，代替后面的不定式短语 to process efficiently。全句可译为：那就是大数据，其中许多(数据量)如此之大，以致要高效地处理已超出常规数据库管理系统的能力。
3. that 引导状语从句，表示目的或结果。全句可译为：这波信息的规模如此之大，我们需要一些新的方式去梳理，以发现隐藏于其中的价值。
4. using 引导的现在分词短语作定语，修饰其前的 applications，意为"使用……的应用程序"。
5. numbering 引导的现在分词短语作定语，修饰其前的 transactions。transactions numbering in the millions 可译为"数以百万计的事务"。前面的 which 引导一个非限制性定语从句，修饰 sites。
6. operating 引导一个现在分词短语，作谓语动词 want 的宾语。
7. capable of 引导的短语作定语，修饰其前的 approach。capable of keeping up with … 意为"能跟上……的"。
8. which 引导一个非限制性定语从句，修饰其前的 software。help 后面的不定式 to find 中省略了 to。Hadoop 是一个能够对大量数据进行分布式处理的软件框架。Hadoop 是可靠的，因为它假设计算元素和存储会失败，因此它维护多个工作数据副本，确保能够针对失败的节点重新分布处理。Hadoop 是高效的，因为它以并行的方式工作，通过并行处理加快处理速度。Hadoop 还是可伸缩的，能够处理 PB 级数据。此外，Hadoop 依赖于社区服务器，因此它的成本比较低，任何人都可以使用。
9. inputting 是现在分词，引导分词短语作状语，表示主语正在进行的另一动作。
10. 注意本句中现在分词的各种不同用法。making 引导一个现在分词短语作定语，修饰其前的 companies，而 detecting 与 predicting 分别作为 from 与 to 的宾语。全句可译为：一些制造最先进工业设备的最大的公司，现在已从仅能检测设备故障发展为预测故障。
11. allowing 引导一个动名词短语作状语，表示结果。全句可译为：急诊室就诊人数突然增加，甚至某些非处方药的销售量增加，可能是某种传染病的早期预警信号，医生和应急部门可据此启动一些控制和遏制程序。

Exercises

1. Translate the text of Lesson 7.3 into Chinese.
2. Topics for oral workshop.
- Explain the DBS. What does a DBS consist of?
- How many types of DBMS do you know? What features does each type of DBMS have respectively?
- Why are so many corporations turning to the data warehouse to handle their corporate data?

3. Translate the following into English.

(a) 因为很多数据库系统用户没受过计算机训练(be not trained)，开发者通过几级抽象——物理层、逻辑层与视图层——对用户隐藏复杂性，简化用户与系统的交互。

(b) 数据库模式(schema)用称为数据定义语言(DDL)的专门语言表达的一组定义来详细说明。诸 DDL 语句的编译结果是存储在一个称为数据字典或数据目录的专门文件中的一组表。

(c) 结构化查询语言(SQL)是关系数据库管理系统使用最广泛且标准的查询语言，它是一种非过程型(non-procedural)语言。

(d) 实体是现实世界中可与其他对象区分开的"东西"或"对象"。例如，每个人是实体，银行账户也能被认为是一些实体。实体在数据库中用一组属性描述。

(e) 数据仓库是 IT 领域和商务环境中最新且最热门的时髦技术词语(buzzword)和概念之一。数据仓库是信息的逻辑汇集(collection)——从很多不同操作中的数据库收集的——它支持各种商务分析活动和决策(decision-making)任务。

4. Listen to the video "Wi-Fi" and write down the text.

Unit 8 Multimedia

8.1 Introduction

Multimedia means, from the user's *perspective*, that computer information can be represented through audio and/or video, in addition to text, image, graphics and animation. For example, using audio and video, a variety of dynamic situations in different areas, such as sport or *ornithology lexicon*, can often be presented better than just using text and image alone.

The integration of these media into the computer provides additional possibilities for the use of computational power currently available (e. g., for interactive *presentation* of huge amounts of information). Furthermore, these data can be transmitted through computer and telecommunication networks, which implies applications in the areas of information distribution and cooperative work. Multimedia provides the possibility for *a spectrum of* new applications, many of which are in place[1] today.

8.1.1 Main Properties of a Multimedia System

If we *derive* a multimedia system *from* the meaning of the words in the American *Heritage* Dictionary, then a multimedia system is any system which supports more than a single kind of media. This characterization is insufficient because it only deals with a *quantitative* evaluation of the system. For example, each system processing text and graphics would be classified as a multimedia system according to this narrow definition. Such systems already existed before the multimedia notion was used in a computer environment. Hence, the notion multimedia implies a new quality in a computer environment.

We understand multimedia more in a *qualitative* rather than a quantitative way. Therefore, the kind rather than the number of supported media should determine if a system is a multimedia system. It should be pointed out that this definition is controversial. Even in the standardization *bodies*, e. g., ISO, a weaker interpretation is often used.

A multimedia system distinguishes itself from other systems through several properties. We elaborate on the most important properties such as combination of the media, *media-independence*, computer control and integration.

Combination of Media

Not every arbitrary combination of media *justifies* the usage of the term multimedia. A simple text processing program with incorporated images is often called a multimedia application because two media are processed through one program. But one should talk about multimedia only when both continuous and *discrete media* are utilized. A text processing program with incorporated images is therefore not a multimedia application.

Independence

An important aspect of different media is their level of independence from each other. In general, there is a request for independence of different media, but multimedia may require several levels of independence. On the one hand, a computer-controlled *video recorder* stores audio and video information, but there is an inherently tight connection between the two types of media. Both media are coupled together through the common storage medium of the tape. On the other hand, for the purpose of presentations, the combination of DAT recorder (Digital Audio Tape) signals and computer-available text satisfies the request for media-independence.[2]

Computer-supported Integration

The media-independence *prerequisite* provides the possibility of combining media in arbitrary forms. Computers are the ideal tool for this purpose. The system should be capable of computer-controlled media processing. Moreover, the system should be programmable by a system programmer or even a user. Simple input or output of different media through one system (e. g., a video recorder) does not satisfy the requirement for a computer-controlled solution. Computer-controlled data of independent media[3] can be integrated to accomplish certain functions. For such a purpose, timing, spatial and semantic synchronization relations will be included. A text processing program that supports text, table calculations and video clips does not satisfy the demand for integration if program supporting the connection between the data cannot be established.[4] A high integration level is accomplished if changing the content of a table row causes corresponding video *scene*[5] and text changes.

8.1.2 Multimedia

A multimedia system is characterized by computer-controlled, integrated production, manipulation, presentation, storage and communication of independent information,[6] which is encoded at least through a continuous (time-dependent) and a discrete (time independent) medium.

Multimedia is very often used as an attribute of many systems, components, products, ideas, etc., without satisfying the above presented characteristics. From this viewpoint our definition is *deliberately* restrictive.

Thus, two notions of multimedia can be distinguished:
- "Multimedia", strictly speaking:
  In this context, continuous media will always be included in a multimedia system. At the same time important timely marginal conditions (through the continuous media) for the processing of discrete media will be introduced. They have barely been considered in computer use until now.
- "Multimedia", in the broader sense:
  Often the notion multimedia is used to describe the processing of individual images and text, although no continuous medium is present. Many of the processing tasks in this environment will also be necessary in the multimedia system according to the

restrictive definition. In any case, if more media are processed together, one can talk about multimedia according to this second notion.

Words and Expressions

perspective [pə'spektiv] n. 观点，看法；视角；远景；透视；透视图；观察
ornithology lexicon 鸟学词典（字典）
presentation [ˌprezən'teiʃən] n. 演示，表演；介绍；播放；赠送
a spectrum of 一系列
derive [di'raiv] ... from 从……取得；导出；推知
heritage ['heritidʒ] n. 传统；遗产，继承物
quantitative ['kwɔntitətiv] adj. 定量的
qualitative ['kwɔlitətiv] adj. 定性的；性质上的
body ['bɔdi] n. 本体；主要部分；机构，团体，组织
media-independence 媒体独立性，媒体无关性
justify ['dʒʌstifai] vt. 证明……是正当的（合法的）
discrete media 离散媒体
video recorder 录像机
prerequisite [priːˈrekwizit] adj. 必要的 n. 先决条件
scene [siːn] n. 景色；情景；场面；片段
deliberately [di'libəritli] adv. 审慎地

Notes

1. in place 这里指"已存在，正在开发，到位"。
2. 全句译为：另一方面，为了演示，把数字音频磁带录音机的信号与计算机可用的文本结合起来满足媒体独立性的要求。
3. Computer-controlled data of independent media 译为"计算机控制的独立媒体数据"。
4. 本句主句的主要结构为 program ... satisfy ... demand；从句的主要结构为 program ... be established。全句译为：如果不能建立支持这些数据之间连接的程序，则支持文本、表格计算和视频剪辑的文本处理程序不满足集成要求。
5. 本句中的 scene 和上一句中的 clips 都是指（一段）视频片段，但前者是播放出来的一段情景，后者是存放在介质上的一段视频数据（就像电影胶卷的一段剪辑）。
6. is characterized by 意为"以……为特点，以……为特性，以……标志"。主句译为：多媒体系统以独立信息的计算机控制、集成制作、处理、演示、存储和通信为特征。

注：本节描述什么是多媒体，以及它应具有的特性。

Terms

ISO
The International Organization for Standardization (often incorrectly identified as an

acronym for International Standards Organization), an international association of countries, each of which is represented by its leading standard-setting organization—for example, ANSI (American National Standards Institute) for the United States. The ISO works to establish global standards for communications and information exchange. Primary among its accomplishments is the widely accepted ISO/OSI model, which defines standards for the interaction of computers connected by communications networks. ISO is not an acronym; rather, it is derived from the Greek word isos, which means "equal" and is the root of the prefix "iso-".

8.2 Audio

Sound is produced by the vibration of matter. During the *vibration, pressure variations* are created in the air surrounding it. The pattern of the *oscillation* is called a waveform (see Figure 8-1).

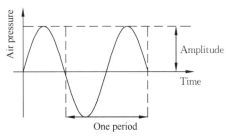

Figure 8-1 Oscillation of an air pressure wave

The waveform repeats the same shape at *regular intervals* and this portion is called a period. Since sound waves occur naturally, they are never perfectly smooth or uniformly periodic.[1] However, sounds that display a recognizable *periodicity* tend to be more musical than those that are nonperiodic.[2] Examples of periodic sound sources are musical instruments, *vowel sounds*, the whistling wind and bird songs. Nonperiodic sound sources include *unpitched percussion* instruments, coughs and sneezes and rushing water.

Frequency

The frequency of a sound is the *reciprocal* value of the period; it represents the number of periods in a second and is measured in hertz (Hz) or cycles per second (cps). A convenient abbreviation, kHz (kilohertz), is used to indicate thousands of oscillations per second: 1 kHz equals 1 000 Hz. The frequency range is divided into:

Infra-sound	from 0 to 20 Hz
Human hearing frequency range	from 20 Hz to 20 kHz
Ultrasound	from 20 kHz to 1 GHz
Hypersound	from 1 GHz to 10 THz

Multimedia systems typically make use of sound only within the frequency range of human hearing. We will call sound within the human hearing range audio and the waves in this frequency range acoustic signals. [3] For example, speech is an *acoustic signal* produced by humans; music signals have a frequency range between 20 Hz and 20 kHz. Besides speech and music, we denote any other audio signal as noise.

Amplitude

A sound also has an *amplitude*, a property *subjectively* heard as *loudness*. [4] The amplitude of a sound is the measure of the *displacement* of the air pressure wave from its *mean*, or *quiescent state*.

8.2.1 Computer Representation of Sound

The smooth, continuous curve of a sound waveform is not directly represented in a computer. A computer measures the amplitude of the waveform at regular time intervals to produce a series of numbers. Each of these measurements is a sample. Figure 8-2 illustrates one period of a digitally sampled waveform.

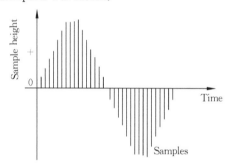

Figure 8-2 Sampled waveform

The mechanism that converts an audio signal into digital samples is the Analog-to-Digital Converter (ADC). The reverse conversion is performed by a Digital-to-Analog Converter (DAC). The *AM79C30A* Digital Subscriber Controller chip is an example of an ADC and is available on *SPARC stations*. DAC is also available as a standard UNIX device.

Sampling Rate

The rate at which a continuous waveform (Figure 8-1) is sampled is called the *sampling rate*. Like frequencies, sampling rates are measured in Hz. The CD standard sampling rate of 44 100 Hz means that the waveform is sampled 44 100 times per second. This seems to be above the frequency range the human ear can hear. However, the bandwidth (which in this case is 20 000 Hz-20 Hz=19 980 Hz) that digitally sampled audio signal can represent, is at most equal to half of the CD standard sampling rate (44 100 Hz). This is an application of the *Nyquist sampling theorem*. ("For *lossless* digitization, the sampling rate should be at least twice the maximum frequency responses.") Hence, a sampling rate of 44 100 Hz can only represent frequencies up to 22 050 Hz, a boundary much closer to that of human hearing.

Quantization

Just as a waveform is sampled at discrete times, the value of the sample is also discrete. The resolution or *quantization* of a sample value depends on the number of bits used in measuring the height of the waveform. An 8-bit quantization yields 256 possible values; 17-bit CD-quality quantization results in over 65 536 values.

Figure 8-3 presents a 3-bit quantization. The sampled waveform with a 3-bit quantization results in only eight possible values: 0.75, 0.5, 0.25, 0, -0.25, -0.5, -0.75 and -1. The shape of the waveform becomes less *discernible* with a lowered quantization, i. e., the lower the quantization, the lower the quality of the sound[5] (the result might be a *buzzing* sound).

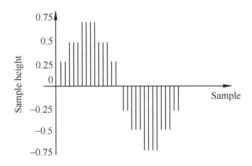

Figure 8-3　Three-bit quantization

Sound Hardware

Before sound can be processed, a computer needs input/output devices. Microphone *jacks* and built-in speakers are devices connected to an ADC and DAC, respectively for the input and output of audio.

8.2.2　Audio Formats

The AM79C30A Digital Subscriber Controller provides voice-quality audio. This converter uses an 8-bit μ-*law* encoded quantization and a sampling rate of 8 000 Hz. This representation is considered fast and accurate enough for telephone-quality speech input.

CD-quality audio is generated if the stereo DAC operates at 44 100 samples per second with a 17-bit linear PCM (Pulse Code Modulation) encoded quantization.

The above examples of telephone-quality and CD-quality audio indicate that important format parameters for specification of audio are: sampling rate (e. g., 8 012.8 samples/second) and sample quantization (e. g., 8-bit quantization).

8.2.3　MP3 Compression

The MP3 format is a *compression* system for music. The MP3 format helps reduce the number of bytes in a song without hurting the quality of the song. MP3 is officially MPEG Audio Layer-3, a MPEG (Moving Picture Experts Group) compression standard. Each layer uses a different sampling rate to obtain different compression results. Layer 3—the *norm* for music today—typically compress a CD-quality song by a factor of 10 to 12 while still retaining CD-quality sound.[6] For example, the 51 MB Queen song shown in the accompanying illustration compresses to less than 5 MB after conversion to the MP3 format. Higher compression levels can be used, but the quality of the song is affected. For example, near-CD quality (with a 15∶1 compression ratio) stores about 1 minute of music in 700 KB of space, instead of about 1 MB for 11∶1 compression. Compressing it to the much lower FM radio quality (22∶1) would store the same one minute of music in about 400 KB of space. Compared to the 10 MB required to store the one minute in the CD's native WAV format, MP3 compression is pretty impressive.[7]

The MP3 format became widely used when downloading songs from the Internet began to become a popular *pastime*. MP3 files can be downloaded in minutes rather than hours, and the MP3 format lets you store hundreds of songs on a single storage medium.

To reduce the file size, MP3 compression Layer 3 uses *perceptual coding techniques* to remove all *superfluous* information (the parts of the song that the human ear wouldn't hear anyway). A coding scheme called *Huffman coding* is then used to substitute shorter strings of bits for frequently used larger strings.[8] This process is typically repeated until an appropriate level of compression is reached. The result-saved with the extension .MP3[9]—can then be played on a PC with MP3 software or using any MP3-compatible device (such as an MP3, CD, or DVD player).

Words and Expressions

vibration [vaiˈbreiʃən] *n.* 振动
pressure variation 压力变化
oscillation [ˌɔsiˈleiʃən] *n.* 振动,振荡
regular interval 固定间隔
periodicity [piəriəˈdisiti] *n.* 周期性
vowel sound 元音
unpitched percussion 未定调的打击(敲打)
reciprocal [riˈsiprəkəl] *adj.* 倒数
infra-sound 次声
ultrasound [ˈʌltrəsaund] *n.* 超声(波)
hypersound [ˌhaipəsaund] *n.* 特超声
GHz (GigaHerts) 千兆赫兹; THz (TeraHerts) 兆兆赫兹
acoustic signal 声信号
amplitude [ˈæmplitjuːd] *n.* 振幅
subjectively [səbˈdʒʌktivli] *adv.* 主观上
loudness [ˈlaudnis] *n.* 响度;音量
displacement [disˈpleismənt] *n.* 位移(量)

mean [mi:n] *n.* 中间；平均数
quiescent state 静止状态，不动状态
AM79C30A Digital Subscriber Controller 一种数字用户控制器
SPARC station SPARC 工作站
sampling rate 取样率，采样率
Nyquist sampling theorem 尼奎斯特抽样定理
lossless ['lɔsləs] *adj.* 无损（耗）的
quantization [kwɔnti'zeiʃən] *n.* 量化
discernible [di'sə:nəbl] *adj.* 看得清的，可分辨的
buzz [bʌz] *n.* 嗡嗡叫
jack [dʒæk] *n.* 插孔，插头，插口
μ-law μ-定律
compression [kəm'preʃən] *n.* 压缩
norm [nɔ:m] *n.* 标准；规范
pastime ['pɑ:staim] *n.* 消遣，娱乐
perceptual [pə'septjuəl] *adj.* 感知的；（五官所）知觉的，感性的
perceptual coding technique 感知编码技术
superfluous [ˌsju:'pə:fluəs] *adj.* 多余的，过剩的，过量的
Huffman coding 哈夫曼编码

Abbreviations

ADC (Analog-to-Digital Converter) 模/数转换器
DAC (Digital-to-Analog Converter) 数/模转换器
PCM (Pulse Code Modulation) 脉码编码调制

Notes

1. 全句译为：因为声波自然地发声，所以它们不可能完全平滑或有均匀的周期。
2. 本句的主语和谓语动词分别是 sounds … 和 tend。全句译为：然而，在波形上可看出周期性的声音比非周期性的声音更好听。
   下一句译为：乐器、元音、呼啸的风和鸟唱的歌都是周期性声源的例子。
3. 本句中 call 意为"把……叫作"，有两个宾语：一是 (call) sound … audio；二是 (call) the waves … acoustic signals。全句译为：我们把人类听觉范围内的声音称为音频，把这个频率范围内的波称为声信号。
4. 本句中 a property 引导的短语是 an amplitude 的同位语，补充说明振幅。heard 修饰 property。全句译为：声音也有振幅，反映人们听到的音量（或响度）。
   下一句译为：声波的振幅是空气压力波距离中间位置或（即）静止状态的位移量的量度。
5. 本句中 lowered quantization 意为"量化精度低"。全句译为：在量化精度低的情况下，波形的形状可分辨性差，即量化精度越低，声音质量越差。例如，值为 0.729 和 0.721

的两个振幅,若取小数后两位的量化,则都变为 0.72,因此没有区别。但若取小数点后三位,则它们就有区别。

6. 本句中 by a factor of 10 to 12 意为"以 10 到 12 的(比例)因子"。全句可译为:层 3,现今音乐的标准,典型地以 10:1～12:1 压缩率压缩 CD 质量的歌曲,而仍然保持 CD 质量的声音。

7. 本句中 one minute 即 one minute of music。native "本地的,出生的",这里指"本来的,原来的"。全句译为:与用 CD 原来的 WAV 格式存储一分钟的音乐所需要的 10 MB 空间相比,MP3 压缩是相当令人赞叹的。

8. substitute A for B "用 A 代 B"。全句可译为:然后,一种称为哈夫曼编码的编码方案被用来把频繁用的较长的(字符)串替换成较短的位串。

9. result-saved with the extension .MP3 译为"用扩展名.MP3 保存的结果"。

注:本节介绍声音数字化的原理:采样和样值的量化。文中有不少术语。

8.3 Video

8.3.1 Video Compression

To understand video on the PC today, it is important to know the history, and a little bit about the state-of-the-art video technology.

Performance Anxiety

Like many new technologies, digital video has a somewhat checkered past: products released before their time, *CPU-hungry* technologies running on under-powered machines, and *jerky*, *fuzzy*, postage-stamp video touted as the second coming of the GUI.[1] But video on the PC has a unique *Achilles' heel* best *typified* by the question "Why does video look better on my $ 200 TV?". Well, it is simple: Video is *analog*; computers are digital.

Computers can't handle analog information directly—they have to *simulate* it. Computers start by *digitizing* the video, or by dividing the analog frame into individual picture elements (pixels). A 640-by-480 frame requires 307,200 pixels, each of which takes one byte to store. Add the 24-bit *color depth* necessary for high-quality video and you've *tripled* the storage requirement to 921,600 bytes per frame. At 30 frames per second, you're up to 9.2 million pixels and over 27 MB of data—each second! This not only presents tremendous storage and bandwidth problems, it's more data than even the most powerful desktop computer can handle in real time.[2]

A number of simple "scaling" techniques are used to reduce the load. The first is to *shrink* the image size. *Quarter-screen* video (320-by-240) requires one quarter the bandwidth of full-screen video (640-by-480), which brings you down to 6.912 megabytes per second. The second is to cut the frame rate, which is the number of frames displayed per second. Most video is acquired at 30 frames per second (fps), the same rate as television. Dropping from 30 to 15 fps cuts your data rate in half, to about 3.5 MB per second. The third technique is to reduce the color depth. Dropping from 24-bit to 8-bit reduces the data rate by another two thirds, taking you down to just over 1.1 MB per second. Depending on the type of video, you'll know which approach to take.

Unfortunately, even a data rate of 1.1 MB per second is still too high for even *quadruple-speed* CD-ROM drives and most non-SCSI hard disks. So scaling the image size, frame rate, and color depth still doesn't get the job done. That's where video compression comes in.

Video *Squeeze*

Video compression is a collection of techniques used to shrink video files. *Embodied* by products called *codecs* (compression/*decompression*), these methods fall into two general categories: *interframe* and *intraframe* compression.

Interframe compression uses *a system of key* and *delta frames* to eliminate redundant information between frames. Key frames store an entire frame, while delta, or "difference", frames record only the interframe changes. During decompression, the CPU builds frames from the key frames and *accumulated* deltas.

Intraframe compression is performed entirely within individual frames. During intraframe compression, codecs use a variety of techniques to convert pixels to more compact mathematical formulas. The simplest technique is called run length encoding (RLE), in which rows of adjacent identical pixels are grouped together[3].

Intraframe technologies range from simple RLE to documented standards such as JPEG to *exotic* mathematical disciplines such as *wavelets* and *fractal transform*.[4] Not all codecs use both interframe and intraframe techniques — some use only intraframe. Those that use both apply intraframe compression on key frames and the information remaining in delta frames after removing interframe redundancies.[5]

All video sequences have differing amounts of interframe and intraframe redundancy, which means that different sequences will compress with various degrees of success. Today's codecs can generally guarantee that you'll reduce your data rate to a somewhat more reasonable level. The question is, at what price?

Lunch Money Required

The saying "there is no such thing as a free lunch" holds true for codecs. The price of a codec's reduction in data rate is *diminished* video quality.

All relevant codecs are "lossy" in nature. During compression, pixels are replaced with more compact formulas designed to approximate the original detail rather than recreate it exactly.[6] As you would expect, the greater the compression, the greater the loss of video quality.

Compression aside, there are the formidable operational aspects of video to consider. Video I/O and decompression take lots of CPU cycles. Local-bus architectures ease the load by speeding data transfer. And with Pentium processors, we finally have systems up to the task.

Postage-stamp video is dead. Long live quarter-screen video.

Figure 8-4 shows the procedure of video capture and editing.

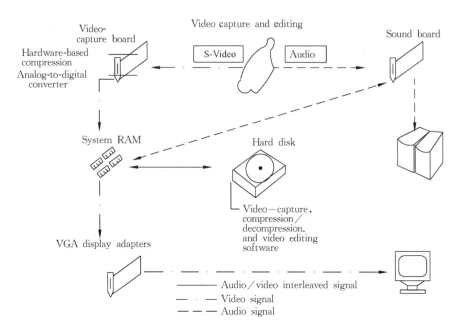

Figure 8-4 Procedure of video capture and editing

Video's Route Through the PC

A **standard video source**, such as a *camcorder*, VCR, or *laserdisk player*, transmits the analog video signal to the *video-capture board*, while analog audio is sent to a sound board inside the PC.

The capture board utilizes analog-to-digital converters (ADCs) to transform the analog video signal into binary code. The video *footage* can be captured as a raw sequence of video frames[7], which is sent to and held in system RAM where software compression is performed.

Meanwhile, **the audio signal** undergoes analog-to-digital conversion by the sound board's converters. This information is also sent to the PC's main system RAM.

After the **video and sound** tracks have been captured, the captured signal can be stored directly to the hard disk or software-based compression can be applied. Generally, the digital video and audio signals are stored as a synchronized, or interleaved, AVI file on the hard disk.

Because the video-capture boards in this *roundup*[8] do not provide **hardware-base decompression support**, the CPU was responsible for decompressing the files in system memory. The decompressed digital video portion of the AVI file is sent from memory to a conventional VGA or SVGA graphics adapter, where it is converted into a noninterlaced analog signal.

8.3.2 MP4

MP4 is a *container* format. A container format allows you to combine different multimedia streams (mostly audio and video) into one single file. Multimedia containers are for example the well known AVI (.avi), MPEG (.mpg,.mpeg), Matroska (.mkv,.mka), OGM (.ogm), Quicktime (.mov) or Realmedia (.rm,.rmvb). MP4 supports all kinds of multimedia content (multiple audio streams, video streams, *subtitle* streams, pictures) and advanced content (officially called "Rich Media" or "BIFS") like 2D and 3D animated graphics, user *interactivity*, DVD-like menus.

MP4 is specified as a part of the ISO/IEC MPEG-4 international standard. MPEG-4 (ISO 14496) is a broad Open Standard developed by the Moving Picture Experts Group (MPEG), a working group of the International Organization for Standardization (ISO) which also did the well known MPEG-1 (MP3, VCD) and MPEG-2 (DVD, SVCD) Standards, standardizing all sorts of audio/video compression formats and much more. By its nature the MPEG-4 Standard doesn't aim at standardizing one potential product (e.g., something comparable to DVD) but covers a broad range of Sub-Standards, which Product Providers can choose from to follow, according to what they need for their product. The MPEG-4 Standard, as mentioned, is divided into many different sub-standards, where for our users on Sharewareguide.net the following parts might be of major interest:

- ISO 14496-2 (Video), e.g., Advanced Simple Profile (ASP), as followed by XviD, DivX5, 3ivx…
- ISO 14496-10 (Video), Advanced Video Coding (AVC), also known as H.264
- ISO 14496-14 (Container), MP4 container format (uses the .MP4 extension)

Words and Expressions

CPU-hungry 拼命用 CPU 的,需大量 CPU 时间的
jerky [dʒəːki] *adj.* 急动的;不平稳的
fuzzy [ˈfʌzi] *adj.* 模糊的;(录音等)失真的
Achilles' heel 唯一致命的弱点
typify [ˈtipifai] *vt.* 作为……典型,代表
analog [ˈænəlɔg] *n.* 模拟;类似物
simulate [ˈsimjuleit] *vt.* 模拟
digitize [ˈdidʒitaiz] *vt.* 数字化
color depth 颜色浓度
triple [ˈtripl] *vt.* 使增至三倍,三倍于 *vi.* 增至三倍 *n.* 三元组;三倍数;三个一组
shrink [ʃriŋk] *v.* (使)缩小
quarter-screen 1/4 屏幕
quadruple-speed 四倍速
decompression [diːkəmˈpreʃn] *n.* 解压
squeeze [skwiːz] *vt.* 使缩减 *v.* 压榨;挤 *n.* 压榨;挤
embody [imˈbɔdi] *vt.* 体现;使具体化;包含
codec [ˈkəudek] *n.* 压缩解压;编码解码器
interframe [ˈintəːfreim] *n.* 帧间
intraframe [ˈintrəfreim] *n.* 帧内
a system of ……体系,方法

key frame 主要帧，关键帧
delta frame δ 帧
accumulate [əˈkjuːmjuleit] v. 累加，累积
exotic [igˈzɔtik] adj. 特殊的；奇异的
wavelet [ˈweivlit] n. 小波
fractal transform 分形变换
diminish [diˈminiʃ] vt. 降低；减小，减少；缩减 vi. 变少，缩小
camcorder [ˈkæmkɔːdə] n. 可携式摄像机
laserdisk player 激光盘唱机，光碟播放机
video-capture board 视频捕获卡
footage [ˈfutidʒ] n. 尺长（电影胶卷的总长度）；以尺计算长度
roundup [ˈraundʌp] n. 综述；摘要；围捕
container [kənˈteinə] n. 容器
subtitle [ˈsʌbtaitl] n. 副标题（书本中的）；说明对白的字幕
interactivity [ˌintərˈæktiviti] n. 交互性，互动性

Abbreviations

RLE (Run-Length Encoding)　行程长度编码
JPEG (Joint Photographic Experts Group)　联合摄影专家组，联合图像专家组
VCR (Video Cassette Recorder) 盒式磁带录像机
SVGA (Super Video Graphics Array)　超视频图形阵列

Notes

1. 本句后半句中有三个并列短语。全句译为：像许多新技术一样，数字视频有一段波折的经历：产品提前发布，需大量 CPU 时间的技术运行在低功能的机器上，以及不平稳的、模糊的、邮票大小的视频图像被吹捧为 GUI 的第二次到来。

2. 本句中 it's more data than ... 译为"其数据量超过了功能最强大的台式计算机所能实时处理的量"。

3. 本句中 in which 引导的从句译为"按这种编码法，各行中相邻的相同像素组合（或分组）在一起"。例如，未压缩的数据若为 ABCCCCCCCCCDEFGGG，则其行程长度编码为 ABC！8DEFG！3。

4. 本句中有结构：range from ... to ... to ...。documented 意为"已成文档的"。全句译为：帧内压缩技术从简单的 RLE 到已成文档的诸标准（如 JPEG），到一些特殊数学方法（如小波和分形变换）都有。

5. 本句中 that use both ... 为定语从句，修饰 those。本句译为：那些使用两种压缩技术的 codecs 在消除了帧间冗余后，对关键帧和 δ 帧内的信息进行帧内压缩。

6. 本句中 designed to 引导的短语修饰前面的 formulas。全句译为：在压缩期间，使用更紧凑的公式来替换诸像素，这些公式旨在近似原来的细节，而不是精确地重建。

7. video footage 这里指捕获卡得到的所有视频信号。主句可译为：这些视频信号能被捕

获作为连续的原始视频帧。
8. in this roundup 这里指"在本综述中"。

注：本节描述视频压缩的基本原理、为何要压缩以及压缩的代价，文中有不少术语。视频压缩技术在不断发展中。

Terms

JPEG

1. Acronym for Joint Photographic Experts Group. An ISO/ITU standard for storing images in compressed form using a discrete cosine transform. JPEG trades compression off against loss; it can achieve a compression ratio of 100∶1 with significant loss and possibly 20∶1 with little noticeable loss.

2. A graphic stored as a file in the JPEG format.

MPEG

1. Acronym for Motion Pictures Experts Group. An ISO/ITU standard for storing motion pictures in compressed digital form. Individual periodic key frames (called I-frames) are encoded using the JPEG method and then further compressed by recording only the differences between them. As a result, the method is lossy, and the linear interpolation required to decode the images calls for a very fast CPU or a dedicated MPEG board. The MPEG-1 standard is used in CD-ROMs and video CDs; it provides a resolution of 352 by 240 pixels (somewhat lower than VCR quality) at 30 frames per second with 24-bit color and CD-quality sound. MPEG-2, used in higher-capacity DVD media, provides full-screen images.

2. A video/audio file in the MPEG format. Such files generally have the extension .mpg. See also JPEG.

DVI

Acronym for Digital Video Interface. A hardware-based compression/decompression technique for storing full-motion video, audio, graphics, and other data on a computer or on a CD-ROM. DVI technology was developed by RCA in 1987 and acquired by Intel in 1988. Intel has since developed a software version of DVI, called Indeo. Also called digital video-interactive.

VGA

Acronym for Video Graphics Array. A video adapter that duplicates all the video modes of the EGA (Enhanced Graphics Adapter) and adds several more.

SVGA

Acronym for Super Video Graphics Array. A video standard established by the Video Electronics Standards Association (VESA) in 1989 to provide high-resolution color display on IBM-compatible computers. Although SVGA is a standard, compatibility problems can occur with the video BIOS. SVGA offers resolutions up to 1024×768 pixels and color formats up to 24 bits per pixel.

AVI

Acronym for Audio Video Interleaved. A Windows multimedia file format for sound and moving pictures that uses the Microsoft RIFF (Resource Interchange File Format) specification.

8.4 Synchronization

Advanced multimedia systems are characterized by the integrated computer-controlled generation, storage, communication, manipulation and presentation of independent time-dependent and time-independent media.[1] The key issue which provides integration is the digital representation of any data and the synchronization of and between various kinds of media and data.

The word synchronization refers to time. Synchronization in multimedia systems refers to the *temporal* relations between media objects in the multimedia system. In a more general and widely used sense some authors use synchronization in multimedia systems as comprising content, spatial and temporal relations between media objects.[2] We differentiate between time-dependent and time-independent media objects. A time-dependent media object is presented as a media stream. Temporal relations between *consecutive units* of the media stream exist. If the *presentation duration* of all units of a time-dependent media object are equal, it is called continuous media object. A video consists of a number of ordered frames; each of these frames has a fixed presentation duration. A time-independent media object is any kind of traditional media like text and images. The semantic of the respective content does not depend upon a presentation according to the *time domain*.[3]

Synchronization between media objects comprises relations between time-dependent media objects and time-independent media objects. A daily example of synchronization between continuous media is the synchronization between the visual and acoustical information in television. In a multimedia system, the similar synchronization must be provided for audio and moving pictures. An example of temporal relations between time-dependent media and time-independent media is a *slide show*. The presentation of slides is synchronized with the *commenting audio stream*. To realize a slide show in a multimedia system, the presentation of graphics has to be synchronized with the appropriate units of an audio stream.

Synchronization is *addressed* and supported by many system components including the operating system, communication system, databases, documents and even often by applications. Hence, synchronization must be considered at several levels in a multimedia system.

The operating system and lower communication layers handle single media streams with the objective[4] to avoid *jitter* at the presentation of the units of one media stream. For example, users will be annoyed if an audio presentation is interrupted by pauses or if clicks result in short gaps in the presentation of the audio clip.

On top of this level is the run-time support for the synchronization of multiple media streams is located. The objective at this level is to maintain the temporal relations between various streams. In particular the *skew* between the streams must be restricted. For example, users will be annoyed if they notice that the movement of the lips of a speaker does not correspond to the presented audio.

The next level holds the run-time support for the synchronization between time dependent and time-independent media together with the handling of user interactions.[5] The objective is to start and stop the presentation of the time-independent media within a *tolerable* time interval, if some previously defined points of the presentation of a time-dependent media object are reached. The audience of a slide show is annoyed if a slide is presented before the audio comment introduces a new picture.[6] A short delay after the start of the introducing comment is tolerable or even useful.

The temporal relations between the media objects must be specified. The relations may be specified *implicitly* during capturing of the media objects, if the goal of a presentation is to present the media in the same way as they were originally captured. This is the case of audio/video recording and *playback*.

The temporal relations may also be specified *explicitly* in the case of presentations that are composed of independently captured or otherwise created media objects.[7] In the slide show example, a presentation designer selects the appropriate slides, creates an audio object and defines the units of the audio presentation stream where the slides have to be presented. Also, the user interactivity may be part of a presentation and the temporal relations between media objects and user interactions must be specified. The tools that are used to specify the temporal relations are located on top of the previous levels.

Words and Expressions

temporal ['tempərəl] *adj.* 时间的，时序的
consecutive unit   连续单元(元素)
presentation duration   演示(的持续)时间
time domain   时域，时间范畴
slide show   放映幻灯片
commenting audio stream   解说的音频流
address   *vt.* 处理；对付；向……讲话；写姓名地址，致函；[计]编址；寻址
jitter ['dʒitə] *vi.* 抖动，跳动   *n.* 振动，剧跳；紧张不安
skew [skjuː] *adj.* 歪斜的；偏的；斜交的   *n.* 偏离；歪斜；斜交
tolerable ['tɔlərəbl] *adj.* 可容许的，可容忍的
implicitly [im'plisitli] *adv.* 隐式地，隐含地
explicitly [ik'splisitli] *adv.* 显式地，明显地
playback ['pleibæk] *n.* 重放；读出

Notes

1. 本句中前一个 independent "独立的"，后一个 independent 与 time 构成复合词，译为 "时间无关的"。time-dependent 译为 "时间相关的，时间有关的"。
   本句下一句中 the synchronization of and between various kinds of media and data 译为 "各类媒体和数据之间的同步"，其中 of, between 是两个并列介词。
2. 本句译为：在更普遍、更广泛使用的意义上，一些作者认为多媒体系统中的同步包括内容、空间和媒体对象之间的时序关系。
3. 本句译为：各个内容的语义与演示的时间没有关系。
4. 本句中 with the objective ... 译为 "其目标是……"
5. 全句可译为：下一层包含对时间有关和时间无关媒体之间同步，以及对用户交互的处理的运行时刻支持。
6. 本句直译是：如果在讲解、介绍一张新图片之前就放映幻灯片，则看幻灯节目的观众是不满意的。本句也可译为：如果幻灯片跟不上讲解，则看幻灯节目的观众是不满意的。而下一句可译为：解说开始后的短暂延迟是可容忍的，甚至是有用的。
7. 本句中 that are composed of ... 是定语从句修饰 presentations。全句译为：在演示是由独立捕获或用其他方法（独立）制作的媒体对象所组成的情况下，这些时序关系也可以显式指定。

注：本节介绍多媒体演示时各种媒体之间的同步概念。

Terms

T. 120 standard

A family of International Telecommunications Union (ITU) specifications for multipoint data communications services within computer applications, such as conferencing and multipoint file transfer.

HDTV (High-Definition TeleVision)

HDTV is a type of television that supports high-resolution, digital broadcast signals. HDTVs can display digital broadcast images many times clearer and sharper than analog TVs. Many can also support the wider format commonly used with motion pictures.

GIF (Graphics Interchange Format), PNG (Portable Network Graphics)

Common graphic formats for line art images.

MIDI

Acronym for **M**usical **I**nstrument **D**igital **I**nterface. A serial interface standard that allows for the connection of music synthesizers, musical instruments, and computers. The MIDI standard is based partly on hardware and partly on a description of the way in which music and sound are encoded and communicated between MIDI devices. The information transmitted between MIDI devices is in a form called a MIDI message, which encodes aspects of sound such as pitch and volume as 8-bit bytes of digital information. MIDI devices can be used for creating, recording, and playing back music. Using MIDI, computers, synthesizers, and sequencers can communicate with each other, either keeping time or actually controlling the music created by other connected equipment. See also synthesizer.

Exercises

1. Translate the text of Lesson 8.4 into Chinese.
2. Topics for oral workshop.
   - What is multimedia?
   - Talk about the devices that a computer needs to support multimedia.
   - Describe three areas in which multimedia is being used. (training and education, entertainment ...).
   - Discuss the digitalization of sound.
3. Translate the following into English.

动画片是用来描述一系列图像的术语,这些图像被一张接一张地显示以模拟运动。电视上的卡通片是动画片的一个例子。

视频不同于动画片,当视频被记录时,它通常开始作为连续的视觉信息流,该信息流被分成(be broken into)个别的图像或帧。当放映(project)这些帧时——典型地以每秒30帧的速率——其效果(effect)是平稳地重新构造原始的连续信息流。像你可以想象的那样,以每秒30帧,在多媒体演示期间,显示一个视频所涉及的数据量可能需要大量的存储空间。因而,视频数据——如音频数据——经常被压缩。存在各种压缩标准。一些最常见的视频文件格式是.avi、.mpeg、.mov、.rm。

为了多媒体演示,可以用标准的(模拟)摄像机记录视频,然后当它(视频)输入计算机时转换成数字形式。另一种选择是,可以用数码摄像机数字地记录影片。Web页面上经常使用流式视频以缩减文件的大小。类似于流式音频,一旦视频的一部分已经下载,流式视频文件就可以开始播放。

4. Listen to the video "Camcorder" and write down the first three paragraphs (before step).

Unit 9　Artificial Intelligence

9.1　Overview of Artificial Intelligence

What is Artificial Intelligence (AI) exactly? As a beginning we offer the following definition:

AI is a branch of computer science concerned with the study and creation of computer systems that exhibit some form of intelligence: systems that learn new concepts and tasks, systems that can *reason* and draw useful conclusions about the world around us, systems that can understand a natural language or perceive and comprehend a *visual* scene, and systems that perform other types of *feats* that require human types of intelligence.

Like other definitions of complex topics, an understanding of AI requires an understanding of related terms such as intelligence, knowledge, reasoning, thought, *cognition*, learning, and a number of computer-related terms. While we lack precise scientific definitions for many of these terms, we can give general definitions of them. And, of course, one of the objectives of this text is to *impart* special meaning to all of the terms related to AI, including their operational meanings.

Dictionaries define intelligence as the ability to acquire, understand and apply knowledge, or the ability to exercise thought and reason. Of course, intelligence is more than this. It embodies all of the knowledge and feats, both conscious and unconscious, which we have acquired through study and experience[1]: highly refined sight and sound perception; thought; imagination; the ability to converse, read, write, drive a car, memorize and recall facts, express and feel emotions; and much more.

Intelligence is the integrated sum of those feats which gives us the ability to remember a face not seen for thirty or more years, or to build and send rockets to the moon. It is those capabilities which set *Homo sapiens* apart from other forms of living things[2]. And, as we shall see, the food for this intelligence is knowledge.

Can we ever expect to build systems which exhibit these characteristics? The answer to this question is yes! Systems have already been developed to perform many types of intelligent tasks, and expectations are high for near term development of even more impressive systems.[3] We now have systems which can learn from examples, from being told, from past related experiences, and through reasoning. We have systems which can solve complex problems in mathematics, in scheduling many diverse tasks, in finding optimal system configurations, in planning complex strategies for the military and for business, in diagnosing medical diseases and other complex systems, to name a few.[4] We have systems which can "understand" large parts of natural languages. We have systems which can see well enough to "recognize" objects from photographs, video cameras and other *sensors*. [5] We have systems which can reason with incomplete and uncertain facts. Clearly, with these developments, much has been accomplished since the advent of the digital computer.

In spite of these impressive achievements, we still have not been able to produce *coordinated*, *autonomous* systems which possess some of the basic abilities of a three-year-old child. These include the ability to recognize and remember numerous *diverse* objects in a scene, to learn new sounds and associate them with objects and concepts, and to adapt readily to many diverse new situations. These are the challenges now facing researchers in AI. And they are not easy ones. They will require important breakthroughs before we can expect to equal the performance of our three-year old.

To gain a better understanding of AI, it is also useful to know what AI is not. AI is not the study and creation of conventional computer systems. Even though one can argue that all programs exhibit some degree of intelligence, an AI program will go beyond this in *demonstrating* a high level of intelligence to a degree that equals or exceeds the intelligence required of a human in performing some task. [6] AI is not the study of the mind, nor of the body, nor of languages, as customarily found in the fields of *psychology*, *physiology*, cognitive science, or *linguistics*. To be sure, there is some overlap between these fields and AI. All seek a better understanding of the human's intelligence and sensing processes. But in AI the goal is to develop working computer systems that are truly capable of performing tasks that require high levels of intelligence. The programs are not necessarily meant to imitate human senses and thought processes. Indeed, in performing some tasks differently, they may actually exceed human abilities. The important point is that the systems all are capable of performing intelligent tasks effectively and efficiently.

Finally, a better understanding of AI is gained by looking at the component areas of study that make up the whole. These include such topics as *robotics*, memory organization, knowledge *representation*, storage and recall, learning models, *inference* techniques, commonsense reasoning, dealing with uncertainty in reasoning and decision making, understanding natural language, *pattern recognition* and machine *vision* methods, search and matching, *speech recognition* and *synthesis*, and a variety of AI tools.

How much success have we realized in AI to date? What are the next big challenges? The answers to these questions form a large part of the material covered in this text. We shall be studying many topics which *bear* directly or indirectly *on* these questions in the following Lessons. We only mention here that AI is coming of an age where practical commercial products are now available including a variety of robotic devices, vision systems that recognize shapes and objects, expert systems that perform many difficult tasks as well as or better than their human expert counterparts, *intelligent instruction systems* that help pace a student's learning and monitor the student's progress[7], "intelligent" editors that assist users in building special knowledge bases, and systems which can learn to improve their performance.

Words and Expressions

reason ['ri:zn] vt. 推理,推论;辩论　vi. 推理,推论;思考　n. 理由
visual ['viʒuəl] adj. 视觉的;看得见的;凭视力的
feat [fi:t] n. 技艺
cognition [kɔg'niʃən] n. 认识,认知
impart [im'pɑ:t] vt. 给予(尤指抽象事物),把……分给;告知,透露
Homo sapiens ['həuməu'sæpiənz] 人类,智人(现代人的学名)
sensor ['sensə] n. 传感器,灵敏元件
coordinate [kəu'ɔ:dinit] vt. 协调;同等　n. 坐标(用复数)　adj. 同等的
autonomous [ɔ:'tɔnəməs] adj. 自治的,自立的
diverse [dai'və:s] adj. 多种多样的;形形色色的;多变化的;(和……)不一样的(from)
demonstrate ['demənstreit] vt. 表明,表示;(用实例、实验等)说明,表演;论证,证实
psychology [sai'kɔlədʒi] n. 心理学;心理状态
physiology [ˌfizi'ɔlədʒi] n. 生理学
linguistics [liŋ'gwistiks] n. 语言学
robotics [rəu'bɔtiks] n. 机器人学,机器人技术
representation [ˌreprizen'teiʃən] n. 表示法;表现
inference ['infərəns] n. 推理,推论,推断;推断的结果;(逻辑上的)结论
pattern recognition　模式识别
vision ['viʒən] n. 视觉,视力;眼光;想象力　vt. 想象;幻见
speech recognition　语音识别
synthesis ['sinθisis] n. 综合,合成
bear on　有关;对准,瞄准
intelligent instruction system　智能教育系统

Notes

1. both conscious and unconscious 是插入语。全句可译为：它具体体现了有意识地和无意识地通过学习和经验获得的所有知识和技艺。
   冒号后面的 highly refined sight and sound perception 可译为"高度精练的视觉和听觉感知"。
2. set Homo sapiens ... living things 可译为"使人类区别于其他形式的生命体"。
3. 本并列句第二句中的 near term development 含意为"近期开发,短期开发"。该简单句可译为：对近期开发给人印象更为深刻的系统寄予了很高的期望。
4. 本句谓语动词 have 有两个宾语：systems 和 other complex systems。前者有一个 which 引导的很长的定语从句修饰。最后的 to name a few 是插入语,意为"仅举几个例子"。
5. 句中 can see well enough to "recognize" objects 可译为"能看得很好,足以识别物体"。全句可译为：我们有些系统能看得很清楚,足以"识别"照片上、摄像机和其他传感器拍摄的图像上的物体。

6. 全句可译为：即使有人可能争辩说，所有程序都显示出某种程度的智能，但 AI 程序将超过它，AI 程序表现出的高水平智能达到或超过了人在完成某个任务中所需智能的程度。

7. pace 含意为"为……定步速（或速度），调整，控制"。intelligent instruction systems that help pace ... 可译为"帮助调整学生的学习，并监控学生学习进度的智能教育系统"。

9.2 About Expert System

An expert system is a set of programs that manipulate encoded knowledge[1] to solve problems in a specialized domain that normally requires human *expertise*. An expert system's knowledge is obtained from expert sources and coded in a form suitable for the system to use in its inference or reasoning processes. The expert knowledge must be obtained from specialists or other sources of expertise, such as texts, journal articles, and data bases. This type of knowledge usually requires much training and experience in some specialized field such as medicine, geology, system configuration, or engineering design. Once a sufficient body of expert knowledge[2] has been acquired, it must be encoded in some form, loaded into a *knowledge base*, then *tested*, and *refined* continually throughout the life of the system.

Characteristic Features of Expert Systems

Expert systems differ from conventional computer systems in several important ways.

1. Expert systems use knowledge rather than data to control the solution process. "In the knowledge lies the power"[3] is a theme repeatedly followed and supported throughout this book. Much of the knowledge used is *heuristic* in nature rather than *algorithmic*.

2. The knowledge is encoded and maintained as an entity separate from the control program. As such, it is not compiled together with the control program itself. This permits the *incremental* addition and modification (*refinement*) of the knowledge base without recompilation of the control programs. Furthermore, it is possible in some cases to use different knowledge bases with the same control programs to produce different types of expert systems. Such systems are known as expert system *shells* since they may be loaded with different knowledge bases.

3. Expert systems are capable of explaining how a particular conclusion was reached, and why requested information is needed during a *consultation*. This is important as it gives the user a chance to assess and understand the system's reasoning ability, thereby improving the user's confidence in the system.

4. Expert systems use *symbolic* representations for knowledge (rules, networks, or frames) and perform their inference through symbolic computations that closely resemble manipulations of natural language. (An exception to this is the expert system based on *neural network* architectures.)

5. Expert systems often reason with *metaknowledge*; that is, they reason with knowledge about themselves, and their own knowledge limits and capabilities.

The Reality and Future of Expert Systems

Companies will not truly use expert systems (ES) for business operations until ES applications can be fully integrated with other conventional software and data bases, can be developed and operated cost-effectively, and can be *validated* to protect against errors that can have a *deleterious* impact on business operations.[4] The AI industry is making significant efforts toward defining software standards and development methodology, including *verification* and validation techniques—the ES technology issues identified by CIOs as major roadblocks hindering ES progress.

The two problems with the current knowledge encoding process are (1) current ES techniques do not facilitate modeling of commonsense knowledge[5], and (2) the knowledge base does not learn from its experience. *Case-based reasoning* (CBR) and the technologies that are emerging from the CYC project at MCC Corp., address both of these problems.

CBR is a theory that deals with memory, learning, planning, and problem solving. It assumes that the human expert derives knowledge from experience (called cases) and can better *articulate* knowledge as experience than as rules.[6] CBR has been modeled in new AI tools to improve their problem-solving techniques and to enable them to adapt to new situations. A case is described to the computer in terms of its features. Determining the appropriate case features is the main task of knowledge engineering. CBR does not capture commonsense knowledge that a human expert may use in solving a problem. NCC Corp. seeks to resolve this issue by developing an extremely large knowledge base of commonsense. Its CYC project has developed the following special tools for efficiently capturing and processing commonsense knowledge:

- Representation language, CYCL, which handles *prepositional attitudes*, such as beliefs, goals, and purpose;
- knowledge base, which exists at two levels: (1) easy user and external system *interacting*, (2) efficient processing, a facility for automatically translating sentences from the user level into the most appropriate representation for processing.[7]
- Dealing with *imprecision* has been a problem for AI. And inference mechanism that uses *fuzzy* logic provides a way to process imprecise rules and data easily and efficiently. Fuzzy logic performs an approximation of possible outcomes, clusters these outcomes as overlapping *gradations*, and processes these as series of overlapping sets.[8]

The architecture of the next generation of expert systems provides for full integration of expert systems with external systems and data bases. Although AI researchers are addressing most of the technological limitations of early expert systems, companies are still waiting for software standards and development methodologies. The success of neural networks, CASE technology, and client/server computing in the last two years indicates that US companies realize they must modernize their information systems. The new generation of information systems will most likely use AI and other state-of-the-art technologies.

Words and Expressions

expertise [ˌekspəˈtiːz] n. 专门知识（或技能、意见等），专长；鉴定
knowledge base 知识库
test [test] vt. 测试；试验；检验 n. 测试；试验；检验
refine [riˈfain] vt. 精化，精炼，提纯；精制
heuristic [hjuəˈristik] adj. 启发式的
algorithmic [ˈælgəriðəmik] adj. 算法的
incremental [ˌinkriˈmentəl] adj. 增量（式）的；增加的
refinement [riˈfainmənt] n. 精化，细化；改进
shell [ʃel] n. 外壳；（房屋的）框架、骨架 vt. 设定命令行解释器的位置
consultation [ˌkɔnsəlˈteiʃən] n. 咨询；磋商
symbolic [simˈbɔlik] adj. 符号的；象征的
neural network 神经网络
metaknowledge [ˌmetəˈnɔlidʒ] n. 元知识
validate [ˈvælideit] vt. 确认，证实；验证
deleterious [ˌdeliˈtiəriəs] adj. 有害的，有毒的
verification [ˌverifiˈkeiʃən] n. 验证；证实；核对
case-based reasoning 基于实例的推理，基于案例的推理
articulate [ɑːˈtikjulit] vt. 明确表达
prepositional attitude 前置看法；前置态度
interact [ˌintərˈækt] vi. 交互；互相作用；互相影响
imprecision [ˌimpriˈsiʒən] n. 不精确；不明确；含糊不清
fuzzy [ˈfʌzi] adj. 模糊的；（录音等）失真的 fuzzy logic 模糊逻辑
gradation [grəˈdeiʃən] n. 层次，等级，分次，分等

Abbreviations

ES (Expert System) 专家系统
CIO (Chief Information Officer) 信息部门主管
CBR (Case-Based Reasoning) 基于实例的推理，基于案例的推理

Notes

1. encoded knowledge 意为"经过编码的知识"。
2. a sufficient body of expert knowledge 可译为"足够多的专家知识"。
3. in the knowledge lies the power 意为"知识就是力量"。
4. protect against 含意为"防止，不受……的侵害"。后半句可译为"并确认能防止对业务运作产生有害影响的错误（时）"。
5. 第(1)点可译为"当前的专家系统技术不便于常识性知识的建模"。

6. 后半句中用了 better ... than 结构,可译为"经验比规则能更好地明确表达知识"。
7. 本句可译为:知识库,它分两个层次:(1)简便的用户和外部系统交互层;(2)高效处理层,即一个把用户层的句子自动翻译成最适于处理的表示的设施。
8. cluster ... as ... 含意为"群集成,把……集成一组"。全句可译为:模糊逻辑估算可能结果,把这些结果群集成一些重叠的层次,再把它们作为重叠集合系列进行处理。

9.3 Deep Learning

Deep learning (also known as deep structured learning or hierarchical learning) is part of a broader family of machine learning methods based on learning data representations, as opposed to task-specific algorithms[1]. Learning can be supervised, semi-supervised or unsupervised.

Deep learning architectures such as deep neural networks, *deep belief networks* and recurrent neural networks (RNN) have been applied to fields including[2] computer vision, speech recognition, natural language processing, audio recognition, social network filtering, machine translation, *bioinformatics*, drug design, medical image analysis, material inspection and board game programs, where they have produced results comparable to and in some cases superior to human experts.

Deep learning models are *vaguely* inspired by information processing and communication patterns in biological nervous systems yet have various differences from the structural and functional properties of biological brains (especially human brains), which make them incompatible with neuroscience evidences.[3]

Definition

Deep learning is a class of machine learning algorithms that:

- use a *cascade* of multiple layers of nonlinear processing units for feature extraction and transformation. Each successive layer uses the output from the previous layer as input.
- learn in supervised (e.g., classification) and/or unsupervised (e.g., pattern analysis) manners.
- learn multiple levels of representations that correspond to different levels of abstraction; the levels form a hierarchy of concepts.

Overview

Most modern deep learning models are based on an artificial neural network, although they can also include propositional formulas or *latent* variables organized layer-wise in *deep generative models* such as the nodes in deep belief networks and deep Boltzmann machines.[4]

Why is it called deep learning and not just learning? The answer is because it works at deeper and deeper levels or layers. In deep learning, each level learns to transform its input data into a slightly more abstract and composite representation.[5] In an image recognition application, the raw input may be a matrix of pixels; the first representational layer may abstract the pixels and encode edges; the second layer may compose and encode arrangements of edges; the third layer may encode a nose and eyes; and the fourth layer may recognize that the image contains a face. Importantly, a deep learning process can learn which features to optimally place in which level on its own. (Of course, this does not completely *obviate* the need for hand-tuning; for example, varying[6] numbers of layers and layer sizes can provide different degrees of abstraction.)

The "deep" in "deep learning" refers to the number of layers through which the data is transformed. More precisely, deep learning systems have a substantial credit assignment path (CAP) depth. The CAP is the chain of transformations from input to output. CAPs describe potentially causal connections between input and output. For a *feedforward* neural network, the depth of the CAPs is that of the network and is the number of hidden layers plus one (as the output layer is also parameterized). For recurrent neural networks, in which a signal may propagate through a layer more than once, the CAP depth is potentially unlimited. No universally agreed upon *threshold* of depth divides *shallow learning* from deep learning, but most researchers agree that deep learning involves CAP depth > 2. CAP of depth 2 has been shown to be a universal approximator in the sense that it can emulate any function.[7] Beyond that more layers do not add to the function approximator ability of the network. Deep models (CAP > 2) are able to extract better features than shallow models and hence, extra layers help in learning features.

Deep learning architectures are often constructed with a greedy layer-by-layer method. Deep learning helps to *disentangle* these abstractions and pick out which features improve performance.

For supervised learning tasks, deep learning methods obviate feature engineering, by translating the data into compact intermediate representations *akin* to principal components, and derive layered structures that remove redundancy in representation.[8]

Deep learning algorithms can be applied to unsupervised learning tasks. This is an important benefit because unlabeled data are more *abundant* than labeled data. Examples of deep structures that can be trained in an unsupervised manner are *neural history compressors* and deep belief networks.

Interpretations

Deep neural networks are generally interpreted in terms of the *universal approximation theorem* or probabilistic inference.

The universal approximation theorem concerns the capacity of feedforward neural networks with a single hidden layer of finite size to approximate continuous functions.[9] In 1989, the first proof was published by George Cybenko for Sigmoid activation functions[10] and was generalised to feed-forward multi-layer architectures in 1991 by Kurt Hornik.

The probabilistic interpretation derives from the field of machine learning. It features inference, as well as the optimization concepts of training and testing, related to fitting and generalization, respectively. [11] More specifically, the probabilistic interpretation considers the activation nonlinearity as a cumulative distribution function. [12] The probabilistic interpretation led to the introduction of *dropout* as *regularizer* in neural networks. The probabilistic interpretation was introduced by researchers including Hopfield, Widrow and Narendra and popularized in surveys such as the one by Bishop. [13]

Applications

- **Automatic speech recognition**

Large-scale automatic speech recognition is the first and most convincing successful case of deep learning. LSTM[14] RNNs can learn "Very Deep Learning" tasks that involve multi-second intervals containing speech events separated by thousands of discrete time steps, where one time step corresponds to about 10 ms. LSTM with forget gates is competitive with traditional speech recognizers on certain tasks.

All major commercial speech recognition systems (e. g. , Microsoft Cortana, Xbox, Skype Translator, Amazon Alexa, Google Now, Apple Siri, Baidu and iFlyTek voice search, and *a range of* Nuance speech products, etc.) are based on deep learning.

- **Image recognition**

A common evaluation set for image classification is the MNIST database data set. MNIST is composed of handwritten digits and includes 60,000 training examples and 10,000 test examples.

Deep learning-based image recognition has become "superhuman", producing more accurate results than human *contestants*. This first occurred in 2011.

Deep learning-trained vehicles now interpret 360° camera views. Another example is Facial Dysmorphology Novel Analysis (FDNA) used to analyze cases of human *malformation* connected to a large database of *genetic syndromes*.

- **Natural language processing**

Neural networks have been used for implementing language models since the early 2000s. LSTM helped to improve machine translation and language modeling.

Other key techniques in this field are negative sampling[15] and word embedding. Word embedding can be thought of as a representational layer in a deep learning architecture that transforms an atomic word into a positional representation of the word relative to other words in the dataset; the position is represented as a point in a vector space. Using word embedding as an RNN input layer allows the network to parse sentences and phrases using an effective compositional vector grammar[16]. Recursive auto-encoders built *atop* word embeddings can assess sentence similarity and detect paraphrasing. Deep neural architectures provide the best results for *constituency* parsing[17], *sentiment* analysis, information retrieval, spoken language understanding, machine translation, contextual entity linking, writing style recognition, text classification and others.

Recent developments generalize word embedding to sentence embedding. Google Translate (GT) uses a large end-to-end long short-term memory network. Google Neural Machine Translation (GNMT) translates "whole sentences at a time, rather than pieces". The network encodes the "semantics of the sentence rather than simply memorizing phrase-to-phrase translations".

- **Drug discovery and *toxicology***

A large percentage of candidate drugs fail to win *regulatory approval*. These failures are caused by insufficient efficacy (on-target effect), undesired interactions (*off-target* effects), or unanticipated *toxic* effects. Research has explored use of deep learning to predict the *biomolecular* targets, off-targets, and toxic effects of environmental chemicals in *nutrients*, household products and drugs.

- **Financial *fraud* detection**

Deep learning is being successfully applied to financial fraud detection and *anti-money laundering*. Deep anti-money laundering detection system can *spot* and recognize relationships and similarities between data and, further *down the road*, learn to detect *anomalies* or classify and predict specific events. The solution *leverages* both supervised learning techniques, such as the classification of suspicious transactions, and unsupervised learning, e.g. anomaly detection.

- **Military**

The Department of Defense applied deep learning to train robots in new tasks through observation.

Words and Expressions

deep learning   深度学习
deep belief network   深度信念网络
bioinformatics [ˌbaiəuˌinfəˈmætiks] *n.* 生物信息学；生物资讯
vaguely [ˈveigli] *adv.* 含糊地；模糊地；不明确地
cascade [kæsˈkeid] *n.* 级联；串联；小瀑布；瀑布状物；
　　　　　　　　　　vi. & vt. (使)阶式地连接；(使)瀑布似地落下
latent [ˈleitənt] *adj.* 隐而不见的；潜在的，潜伏的　*n.* 隐约的指印
deep generative model   深度生成模型
obviate [ˈɔbvieit] *vt.* 避免；排除，消除；预防
feedforward [ˈfiːdfɔːwəd] *n.* (控制系统中的)前馈
threshold [ˈθreʃhould] *n.* 阈值；临界值；界限；门槛；入门；开端
shallow learning   浅层学习
disentangle [ˈdisinˈtæŋgl] *vt.* 解开，清理；解决；使解脱　*vi.* 解开；解决
akin [əˈkin] *adj.* 相近似的,同类的(to)；同族的，有血缘关系的
abundant [əˈbʌndənt] *adj.* 丰富的，充裕的；充分的

neural history compressor 神经历史压缩器
universal approximation theorem 通用逼近定理,通用近似定理,万能近似定理
dropout ['drɔpaut] n. 随机失活;磁盘信息失落;漏失信息,丢失信息
regularizer ['regjuləraizə] n. 正则项,规则化项;正则化因子;规管人员
a range of 一套;一系列
contestant [kən'testənt] n. 竞争者;参加比赛者
dysmorphology [diz'mɔːfələdʒi] n. 畸形学
malformation ['mælfɔː'meiʃən] n. 畸形(性);畸形物;畸形体
genetic syndrome [dʒi'netik 'sindrəum] 遗传综合征
atop [ə'tɔp] adv. 在顶上 prep. 在……的顶上
constituency [kən'stitjuənsi] n. 选区;全体选民;(一批)赞助者
sentiment ['sentimənt] n. 情绪;感情;情操;意见;观点;感伤
toxicology [ˌtɔksi'kɔlədʒi] n. 毒理学,毒物学
regulatory approval ['regjulətəri ə'pruːvəl] 监管部门的批准
off-target 脱靶,偏离目标
toxic ['tɔksik] adj. 有毒的,有毒性的;中毒的
biomolecular [ˌbaiəuməu'lekjulə] adj. 生物分子的
nutrient ['njuːtriənt] n. 营养素,营养物,营养品 adj. 营养的,滋养的
fraud [frɔːd] n. 欺诈,欺骗;欺诈行为;骗子;假货
anti-money laundering 反洗钱
spot [spɔt] vt. 认出,发现;把……弄脏;玷污
          n. 地点;场所;污点;斑点;少量;插播节目
          adj. 现场的;现货的;插播的
down the road 在将来,今后;沿着这条路走
anomaly [ə'nɔməli] n. 异常;不规则;反常事物
leverage ['liːvəridʒ; 'levəridʒ] vt. 利用;举债经营 n. 力量;影响;杠杆作用;杠杆率

Abbreviations

RNN (Recurrent Neural Network) 循环神经网络,递归神经网络
CAP (Credit Assignment Path) 信度分配路径
LSTM (Long Short-Term Memory) 长的短期记忆
MNIST (Mixed National Institute of Standards and Technology) 国家标准与技术混合研究所
FDNA (Facial Dysmorphology Novel Analysis) 面部畸形学异常分析

Notes

1. 本句中,based 引导分词短语作定语,修饰其前的 methods;as opposed to … 含意为"与……截然相反,对照,与……相对,不是……",此短语可译为"而不是任务特定的算法"。
2. 本句中,including 引导的是 fields 的同位语,即解释 fields 包括哪些领域。

3. 全句可译为：深度学习模型的灵感隐约地来自生物神经系统中的信息处理和交流模式，但与生物大脑(尤其是人脑)的结构和功能特性有各种各样的差异，这使它们与神经科学证据不相容。

4. 本句中，organized 引导的分词短语作定语，修饰其前的 variables。全句可译为：大多数现代的深度学习模型都是基于人工神经网络的，然而它们也可以包括命题公式或如深度信念网络和深度玻尔兹曼机中的节点这样的、在深度生成模型中按层组织的隐变量。

5. 本句中，不定式 to transform 引导的短语作宾语。

6. 本句中，varying 引导动名词短语作主语。

7. 本句中，that 是连接词，引导名词性从句作其前之 sense 的同位语。全句可译为：深度为 2 的 CAP 可以模拟任何函数，从这个意义上说，表明它是一个通用的逼近器。

8. 本句可译为：对于有监督学习任务，深度学习方法通过把数据转换成与主要组成部分近似的紧凑中间表示来避免特征工程，并导出消除表示中的冗余的分层结构。

9. 本句中，the capacity of 后跟的是名词＋不定式类型的复合宾语。全句可译为：通用逼近定理涉及具有有限大小单隐层的前馈神经网络逼近连续函数的能力。

10. Sigmoid activation function 译为"Sigmoid 激活函数"。所谓激活函数，就是在人工神经网络的神经元上运行的函数，负责将神经元的输入映射到输出端。引入激活函数是为了增加神经网络模型的非线性，使得神经网络可以任意逼近任何非线性函数，这样神经网络就可以应用到众多的非线性模型中。由于其单调增加以及反函数单调增加等性质，Sigmoid 激活函数常被用作神经网络的阈值函数，将变量值映射到(0,1)之间。

11. 本句中，fitting 与 generalization 的含意分别为"拟合"与"泛化"。拟合指的是：一组观测结果的数字统计与相应数值组的吻合。泛化指的是：由具体的、个别的扩大为一般的。全句可译为：它的特征是推理，以及分别与拟合和泛化相关的训练和测试的优化概念。

12. cumulative distribution function(缩写为 CDF)译为"累积分布函数"。它能完整描述一个实数随机变量 x 的概率分布，是概率密度函数的积分。对于所有实数 x，累积分布函数与概率密度函数(probability density function，缩写为 PDF)相对。全句可译为：更明确地说，概率解释把激活非线性看作累积分布函数。下一句可译为：概率解释导致了在神经网络中引进随机失活作为正则项。

13. 本句中，including 后跟的是 researchers 的同位语。全句可译为：概率解释是由包括霍普菲尔德、威德罗和纳伦德拉在内的一些研究人员引入的，并在一些调查(如由毕肖普主持的调查)中得到普及。

14. LSTM 是 Long Short-Term Memory 的缩写，含意为"长的短期记忆"。Long Short Term 网络(一般称为 LSTM)是 RNN 的一种特殊类型。LSTM 通过刻意的设计来避免神经网络中的长期依赖问题。记住长期的信息在实践中是 LSTM 的默认行为，而非需要付出很大代价才能获得的能力。LSTM 的关键是记忆单元状态。信息从上一个单元传递到下一个单元，LSTM 通过"门"(gate)来控制丢弃或者增加信息，从而实现遗忘或记忆的功能。"门"是一种使信息选择性通过的结构，由一个 Sigmoid 函数和一个向量内积操作组成。一个 LSTM 单元有三个这样的门，分别是遗忘门（forget gate）、输入门（input gate）和输出门（output gate）。遗忘门用来控制上一单元状态被遗忘的程度：Sigmoid 函数的输出值在[0,1]区间，0 代表完全丢弃，1 代表完全通过。

15. 本句中，negative sampling 含意为"负采样技术"。在自然语言处理领域，判断两个单词是不是一对上下文词（context）与目标词（target），如果是一对，则是正样本；如果不是一对，则是负样本。

16. 本句中，第一个 using 引导动名词短语作主语，第二个 using 引导分词短语作状语。全句可译为：使用单词嵌入作为 RNN 输入层使得网络能使用有效的组合向量文法去分析句子和短语。下一句中的 built 是过去分词，引导分词短语作定语，修饰其前的 Recursive auto-encoders。

17. constituency parsing 含意为"选区分析"。一个选区分析树将句子分解成子短语。树中的非终结符号是短语的类型，如名词短语，终结符号是句子中的单词，边是无标记的。与 constituency parsing 相对照的是依赖分析（dependency parsing）。

Term

Neural network

Neural networks are a type of artificial intelligence that attempts to imitate the way a human brain works. Rather than using a digital model, in which all computations manipulate zeros and ones, a neural network works by creating connections between processing elements, the computer equivalent of neurons. The organization and weights of the connections determine the output. Neural networks are particularly effective for predicting events when the networks have a large database of prior examples to draw on. Strictly speaking, a neural network implies a non-digital computer, but neural networks can be simulated on digital computers. Neural networks are currently used prominently in voice recognition systems, image recognition systems, industrial robotics, medical imaging, data mining and aerospace applications.

Neural network computers can learn from experience. Their inherent ability to learn 'on the fly' is one of the primary reasons researchers are excited and optimistic about their future. Deep learning architectures can be a variety of different neural networks such as deep neural networks, deep belief networks and recurrent neural networks.

9.4 Robot Sophia

Sophia is a social *humanoid* robot developed by Hong Kong based company Hanson Robotics, see Figure 9-1. Sophia was activated on February 14, 2016 and made her first public appearance at South by Southwest Festival (SXSW) in mid-March 2016 in Austin, Texas, United States. It is able to display more than 50 facial expressions.

Sophia has been *covered* by media around the globe and has participated in many *high-profile* interviews. In October 2017, Sophia became the first robot to receive citizenship of any country. In November 2017, Sophia was named the United Nations Development Programme's first ever Innovation Champion, and is the first non-human to be given any United Nations *title*.

Figure 9-1 Sophia at ITU's AI for Good Global Summit in Geneva in May 2018

History

Sophia was activated on February 14, 2016. The robot, modeled after actress Audrey Hepburn, is known for her human-like appearance and behavior compared to previous robotic variants. According to the manufacturer, David Hanson, Sophia uses artificial intelligence, visual data processing and facial recognition. Sophia also imitates human *gestures* and facial expressions and is able to answer certain questions and to make simple conversations on predefined topics (e. g. on the weather). Sophia uses voice recognition (speech-to-text) technology from Alphabet Inc. (parent company of Google) and is designed to get smarter over time. Sophia's intelligence software is designed by Hanson Robotics. The AI program analyses conversations and extracts data that[1] allows her to improve responses in the future.

Hanson designed Sophia to be a suitable companion for the *elderly* at *nursing homes*, or to help crowds at large events or parks. He has said that he hopes that the robot can ultimately interact with other humans sufficiently to gain social skills.

Sophia has seven robot humanoid "siblings" who were also created by Hanson Robotics. Fellow Hanson robots are Alice, Albert Einstein Hubo, BINA48, Han, Jules, Professor Einstein, Philip K. Dick Android, Zeno, and Joey Chaos.

Features

Sophia smiles *mischievously*, *bats* her eyelids and tells a joke. Without *the mess of* cables that make up the back of her head, see Figure 9-2, you could almost mistake her for a human. Cameras within Sophia's eyes combined with computer algorithms allow her to see. It can follow faces, *sustain* eye contact, and recognize individuals. It is able to process speech and have conversations using a natural language subsystem. Around January 2018 Sophia was upgraded with functional legs and the ability to walk.

Figure 9-2 Sophia's internals

Sophia is conceptually similar to the computer program ELIZA, which was one of the first attempts at simulating a human conversation. The software has been programmed to give pre-written responses to specific questions or phrases, like a *chatbot*. These responses are used to create the illusion that the robot is able to understand conversation, including *stock* answers to questions like "Is the door open or shut?". The information is shared in a cloud network which allows input and responses to be analysed with *blockchain* technology.

David Hanson has said that Sophia would ultimately be a good fit to serve in healthcare, customer service, *therapy* and education. Sophia runs on artificially intelligent software that is constantly being trained in the lab, so her conversations are likely to get faster, Sophia's expressions are likely to have fewer errors, and it should answer increasingly complex questions with more accuracy.

Public Figure

On November 21, 2017 Sophia was named the United Nations Development Programme's *first-ever* Innovation Champion for Asia and the Pacific. The announcement was made at the Responsible Business Forum in Singapore, an event[2] hosted by the UNDP in Asia and the Pacific and *Global Initiatives*. As part of her role, Sophia will help to *unlock* innovation to work toward achieving the United Nation's Sustainable Development Goals. On stage, it was assigned her first task by UNDP *Asia Pacific Chief of Policy and Program*, Jaco Cilliers.

Events

Sophia has been interviewed in the same manner as a human, *striking up* conversations with hosts. Some replies have been nonsensical, while others have impressed interviewers such as 60 *Minutes*' Charlie Rose. In a piece for CNBC, when the interviewer expressed concerns about robot behavior, Sophia joked that he had "been reading too much Elon Musk. And watching too many Hollywood movies". Musk *tweeted* that Sophia should watch The Godfather and asked "what's the worst that could happen?". Business Insider's[3] chief UK editor Jim Edwards interviewed Sophia, and while the answers were "not altogether terrible", he predicted it was a step towards "conversational artificial intelligence". At the 2018 Consumer Electronics Show, a BBC News reporter described talking with Sophia as "a slightly *awkward* experience".[4]

On October 11, 2017, Sophia was introduced to the United Nations with a brief conversation with the United Nations *Deputy Secretary-General*, Amina J. Mohammed. On October 25, at the Future Investment Summit in Riyadh, the robot was *granted Saudi Arabian* citizenship, becoming the first robot ever to have a nationality.[5] This attracted controversy as some commentators wondered if this implied that Sophia could vote or marry, or whether a deliberate system shutdown could be considered murder. Social media users used Sophia's citizenship to criticize Saudi Arabia's human rights record. As expressed by Ali Al-Ahmed, director of the *Institute for Gulf Affairs*, "Women (in Saudi Arabia) have since committed suicide[6] because they couldn't leave the house, and Sophia is running around (without a male guardian). Saudi law doesn't allow non-Muslims to get citizenship. Did Sophia *convert* to Islam? What is the religion of this Sophia and why isn't she wearing *hijab*? If she applied for citizenship as a human she wouldn't get it." It has been countered that not being a Muslim is not a formal barrier to being a Saudi citizen, though officials may take it into consideration at their *discretion*.[7] In December 2017, Sophia's creator David Hanson said in an interview that Sophia would use her citizenship to advocate for women's rights in her new country of citizenship; Newsweek criticized that "What (Hanson) means, exactly, is unclear". On November 27, 2018 Sophia was given a visa by Azerbaijan while attending Global Influencer Day Congress held in Baku.

Controversy over *Hype* in the Scientific Community

According to Quartz[8], experts who have reviewed the robot's open-source code state that Sophia is best categorized as a chatbot with a face. Many experts in the AI field disapprove of Sophia's overstated presentation. Ben Goertzel, the chief scientist for the company that made Sophia, acknowledges that it is "not ideal" that some think of Sophia as having human-equivalent intelligence, but argues[9] Sophia's presentation conveys something unique to audiences: "If I show them a beautiful smiling robot face, then they get the feeling that 'AGI' (artificial general intelligence) may indeed be nearby and viable …. None of this is what I would call AGI, but nor is it simple to get working".[10] Goertzel added that Sophia did utilize AI methods including face tracking, *emotion* recognition, and robotic movements generated by deep neural networks. Sophia's dialogue is generated via a decision tree, but is integrated with these outputs uniquely.

According to The Verge[11], Hanson often *exaggerates* and "grossly misleads" about Sophia's capacity for consciousness, for example by agreeing with Jimmy Fallon in 2017 that Sophia was "basically alive". In a piece produced by CNBC which indicates that their own interview questions for Sophia were heavily rewritten by her creators, Geortzel responds to the Hanson quote by suggesting Hanson means Sophia is "alive" in the way that, to a sculptor, a piece of sculpture becomes "alive" in the sculptor's eyes as the work nears completion.[12]

Words and Expressions

humanoid [ˈhjuːmənɔid] *adj.* 似人的；具有人的特点的
　　　　　　　　　　　　n. 类人动物；人形机；(科幻小说中的)星球人
cover [ˈkʌvə] *vt.* 报道，采访；覆盖；包括　　*n.* 封面，封底，封皮；盖子　　*vi.* 覆盖；代替
high-profile [haiˈprəufail] 备受瞩目的；高调的；知名度高的；高姿态的
title [ˈtaitl] *n.* 称号，头衔；冠军；标题　　*vt.* 授予……称号；加标题于
gesture [ˈdʒestʃə] *n.* 手势；姿势；(外交等方面的)姿态；表示
　　　　　　　　vt. 用姿势(或动作)表示　　*vi.* 用姿势(或动作)示意
elderly [ˈeldəli] *adj.* 上年纪的；中年以上的　　the elderly 老年人，老人
nursing home [ˈnəːsiŋˈhəum] 养老院；小型私人医院(尤指私人疗养院)
mischievously [ˈmistʃivəsli] *adv.* 调皮地，淘气地；有害地

bat [bæt] vt. 眨（眼睛） vt. & vi. 用球板打（球） n. 球棒，球拍；打击
the mess of 杂乱的，一团糟的，乱七八糟的
sustain [səs'tein] vt. 维持；继续；支撑，承受住；认可；遭受
chatbot ['tʃætˌbɒt] n. 智能聊天程序；聊天机器人
stock [stɒk] adj. 常备的，库存的 n. 股票；库存品，存货 vt. 贮备，备有
blockchain [blɒk'tʃein] 区块链
therapy ['θerəpi] n. 治疗，疗法
public figure ['pʌblik 'figə] 公众人物，社会名人
first-ever 首次的，初次的，第一次的
Global Initiatives ['gləubəl i'niʃiətivs] 全球倡议组织（位于新加坡）；全球倡议
unlock ['ʌn'lɒk] vt. 开启；开……的锁；启示，揭露（秘密等） vi. 解开
Asia Pacific Chief of Policy and Program 亚太地区政策和计划主管
strike up ['straik 'ʌp] 使开始；开始演奏（或唱歌）；开始敲响；建立起
60 Minutes 哥伦比亚广播公司（Columbia Broadcasting System，CBS）的节目名
tweet [twiːt] vi. 在 Twitter 上发微博；（小鸟）吱吱地叫；
　　　　　　n.（社交网站 Twitter 上的）微博；（小鸟的）吱吱声，啾啾声
awkward ['ɔːkwəd] adj. 尴尬的；难应付的，难处理的；使用不便的；笨拙的
Deputy Secretary-General 副秘书长
grant [grɑːnt] vt. 授予；同意；（姑且）承认 n. 同意；授给物
Saudi Arabian ['saudi ə'reibjən] adj. 沙特阿拉伯的；沙特阿拉伯人的
　　　　　　n. 沙特阿拉伯人；沙特阿拉伯居民
Institute for Gulf Affairs 海湾事务研究所
convert [kən'vəːt] vi. 皈依；改变信仰 vt. 使皈依宗教，使改信；转变，变换；兑换
　　　　　　n. 皈依宗教者；改变信仰者
hijab ['haidʒæb] n. 希贾布（穆斯林妇女戴的面纱或头巾）
discretion [dis'kreʃən] n. 斟酌决定的自由；处理权（限）；谨慎；判断（力）；不连续
hype [haip] n. 大肆宣传；广告；欺骗 vt. 大肆宣传；强行刺激；使……兴奋
emotion [i'məuʃən] n. 情绪，情感；感情；激动
exaggerate [ig'zædʒəreit] vt. & vi. 夸大，夸张 vt. 把……言过其实 vi. 言过其实

Abbreviations

ITU (International Telecommunication Union) 国际电信联盟
AGI (Artificial General Intelligence) 强人工智能
CNBC (Consumer News and Business Channel) 消费者新闻与商业频道
SXSW (South by Southwest Festival) 西南偏南艺术节
UNDP (United Nations Development Programme) 联合国开发计划署

Notes

1. 此 that 引导限制性定语从句,修饰其前的 data。
2. 此 an event 是 the Responsible Business Forum 的同位语,对其进行解释。
3. Business Insider(商业内幕)成立于 2007 年,是美国知名的科技博客、数字媒体创业公司、在线新闻平台,是一家关注 IT 和创业的重量级博客媒体,其文章范围包括 IT 技术、产品、趋势、业内动态以及各个领域的新闻事件等,相比报纸、杂志来说,Business Insider 可以称得上"新媒体",影响力基本都在线上。
4. 本句中,此 talking 是动名词,引导动名词短语作宾语。全句可译为:在 2018 年消费电子产品展上,一位 BBC News(英国广播公司新闻)记者把与索菲娅的交谈描述为"一次有点尴尬的经历"。
5. 本句中,现在分词 becoming 引导分词短语作状语。全句可译为:10 月 25 日,在利雅得举行的未来投资峰会上,该机器人被授予沙特阿拉伯公民身份,成为有史以来第一个拥有国籍的机器人。
6. 本句中,since 与动词完成时态连用,意为"从那时以后,后来"。have committed suicide 意为"(已)自杀"。
7. 本句以先行代词 it 为主语,实际主语是 that 引导的从句。本句中出现的两个 being 都是动名词。not being a Muslim 是从句的主语,第二个 being 引导的短语作介词 to 的宾语,此 to 表示关联、联系,含意为"对于、至于、关于"。全句可译为:有人反驳说,不是穆斯林并不是成为沙特公民的正式障碍,尽管官员们可能酌情考虑(这一因素)。
8. Quartz 是 OpenSymphony 开源组织在 Job scheduling 领域的又一个开源项目,是一个完全由 Java 编写的开源作业调度框架。
9. argues 之后省略了 that,它引导名词性从句作宾语。
10. 本句中,what 引导的名词性从句作表语,其后的 it 是先行代词,代替 to get working。全句可译为:这都不是我所说的 AGI,但是其工作也并不简单。
11. The Verge 是一家美国科技媒体网站,成立于 2011 年 11 月 1 日,办公地点位于纽约曼哈顿。该网站提供新闻、产品评论、播客、视频等内容。The Verge 获得了 2012 年度威比奖五项大奖,包括最佳写作(编辑)、最佳播客、最佳视觉设计、最佳消费电子网站和最佳移动新闻应用程序。
12. 本句中,which 引导限制性定语从句,修饰其前的 a piece;indicates 之后的 that 引导名词性从句作 indicates 的宾语;suggesting 之后省略了 that,它引导名词性从句作 suggesting 的宾语;means 之后省略了 that,它引导名词性从句作宾语;the way 之后的 that 引导的从句是 the way 的同位语。全句可译为:在 CNBC 写作、指出他们自己对索菲娅的采访问题被她的创作者们大量改写的一篇文章中,戈泽尔回应汉森的话,言下之意,汉森的意思是索菲娅"活着",就像在雕塑家看来,当一件雕塑作品接近完成时,这件雕塑在雕塑家眼中就变得"活着"一样。
    注:请观看视频材料中"Sophia 视频 1,2"。

Term

Brain-Computer Interface (BCI)

Since the nineteen seventies, scientists have been searching for ways to link the brain with computers. The scientists keep improving the computer software that identifies brain signals and turns them into simple commands. Recently a researcher can operate a wheelchair just by thinking about moving his left or right hand(not muscular movements but brain activity), which thanks for the Brain-Computer Interface(BCI). BCI is a direct communication pathway between brain and external device. BCI is a system that allows people suffering from physical disabilities to communicate with external world and also to control devices. BCI differs from neuromodulation in that it allows for bidirectional information flow.

BCIs are often directed at researching, mapping, assisting, augmenting, or repairing human cognitive or sensory-motor functions. BCI and virtual reality (VR) are natural companions. BCI provides a new interaction technique for controlling VR, and VR provides a rich feedback environment for BCI while retaining a controlled and safe environment.

9.5 AlphaGo Zero: Learning from Scratch

Artificial intelligence research has made rapid progress in a wide variety of domains from speech recognition and image classification to *genomics* and drug discovery. In many cases, these are specialist systems that *leverage* enormous amounts of human expertise and data.

However, for some problems this human knowledge may be too expensive, too unreliable or simply unavailable. As a result, a *long-standing* ambition of AI research is to bypass this step, creating algorithms that achieve superhuman performance in the most challenging domains with no human input. In our most recent paper, published in the Journal Nature, we demonstrate a significant step towards this goal.

AlphaGo has become progressively more efficient *thanks to* hardware gains and more recently algorithmic advances.

The paper introduces AlphaGo Zero, the latest evolution of AlphaGo, the first computer program to defeat a world champion at the ancient *Chinese game of Go*. Zero is even more powerful and is arguably the strongest Go player in history.

Previous versions of AlphaGo initially trained on thousands of human *amateur* and professional games to learn how to play Go[1]. AlphaGo Zero skips this step and learns to play simply by playing games against itself, starting from completely random play. In doing so, it quickly surpassed human level of play and defeated the previously published champion-defeating version of AlphaGo by 100 games to 0.

It is able to do this by using a *novel* form of *reinforcement* learning, in which AlphaGo Zero becomes its own teacher.[2] The system starts off with a neural network that knows nothing about the game of Go. It then plays games against itself, by combining this neural network with a powerful search algorithm. As it plays, the neural network is tuned and updated to predict *moves*, as well as the eventual winner of the games.

This updated neural network is then recombined with the search algorithm to create a new, stronger version of AlphaGo Zero, and the process begins again. In each iteration, the performance of the system improves by a small amount, and the quality of the self-play games increases, leading to more and more accurate neural networks and ever stronger versions of AlphaGo Zero.

This technique is more powerful than previous versions of AlphaGo because it is no longer constrained by the limits of human knowledge. Instead, it is able to learn *tabula rasa* from the strongest player in the world: AlphaGo itself.

It also differs from previous versions in other notable ways.

— AlphaGo Zero only uses the black and white stones from the Go board as its input, whereas previous versions of AlphaGo included a small number of hand-engineered features.

— It uses one neural network rather than two. Earlier versions of AlphaGo used a "policy network" to select the next move to play and a "value network" to predict the winner of the game from each *position*. These are combined in AlphaGo Zero, allowing it to be trained and evaluated more efficiently.

— AlphaGo Zero does not use "rollouts" — fast, random games used by other Go programs to predict which player will win from the current board position. Instead, it relies on its high quality neural networks to evaluate positions.

All of these differences help improve the performance of the system and make it more general. But it is the algorithmic change that makes the system much more powerful and efficient.

After just three days of self-play training, AlphaGo Zero *emphatically* defeated the previously published version of AlphaGo — which had itself defeated 18-time world champion *Lee Sedol* — by 100 games to 0. After 40 days of self training, AlphaGo Zero became even stronger, *outperforming* the version of AlphaGo known as "Master", which has defeated the world's best players and world number one *Ke Jie*.

Over the course of millions of AlphaGo vs AlphaGo games, the system progressively learned the game of Go from scratch, accumulating thousands of years of human knowledge during a period of just a few days. AlphaGo Zero also discovered new knowledge, developing unconventional strategies and creative new moves that *echoed* and surpassed the novel techniques it played in the games against Lee Sedol and Ke Jie. [3]

These moments of creativity give us confidence that AI will be a multiplier for human *ingenuity*, helping us with our mission to solve some of the most important challenges humanity is facing. [4]

DeepMind[5] co-founder and CEO Demis Hassabis said the programme was so powerful because it was "no longer constrained by the limits of human knowledge". He believes that if applied to big health problems, such as defeating *Alzheimer's*, it could, in a matter of weeks, *come up with* cures that would have taken humans hundreds of years to find.

"Ultimately we want to harness algorithmic breakthroughs like this to help solve all sorts of *pressing* real world problems" said Hassabis.

While it is still early days, AlphaGo Zero constitutes a critical step towards this goal. "If similar techniques can be applied to other structured problems, such as *protein folding*, reducing energy consumption or searching for revolutionary new materials, the resulting breakthroughs have the potential to drive forward human understanding and positively impact all of our lives".

Crucially, AlphaGo Zero uses "tabula rasa" or blank *slate* learning, in which the programme becomes its own teacher, playing games against itself and improving a little each time. It needs no human knowledge, data or any intervention.

Dr Dave Silver, lead researcher for AlphaGo said: "If you can achieve tabula rasa learning you really have an agent which can be transplanted from the game of Go to any other domain. You *untie* yourself from the *specifics* of the domain you are in to an algorithm which is so general it can be applied anywhere". [6]

DeepMind has already begun using[7] AlphaGo Zero to study protein folding and has promised it will soon publish new findings. Misfolded proteins are responsible for many *devastating* diseases, including Alzheimer's, *Parkinson's* and *cystic fibrosis*.

Technology companies are increasingly moving into health. Last year Microsoft announced it planned to crack cancer within 10 years after launching several projects to "hack" the body. [8]

Google's secretive *arm* Calico is also investigating ways to extend human life and even stop *ageing* altogether.

What is Go? An ancient Chinese board game and one of the most complex games in the world, see Figure 9-3.

When was it first played? Over 3,000 years ago.

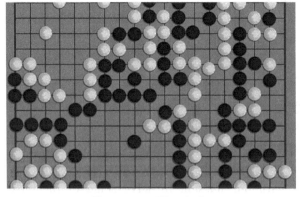

Figure 9-3 What is Go

Where is it played? All over the world, but it is particularly popular in Asia.

How many people play? Two.

What are the rules? Each player alternatively places black or white *counters*, called 'stones', onto the grid. The aim is to encircle your opponent's stones; each stone you encircle is taken off the board.

How do you win? By having the last stone left on the board.

What makes it so difficult? There are more possible moves in a Go game than atoms in the universe. So, unlike chess, you can't employ calculations to win.

What skills do you need? Intuition, creativity and visual imagination.

Words and Expressions

genomics [dʒə'nəumiks] *n*. [复]（用作单）基因组学

leverage ['li:vərɪdʒ] *vt*. 利用；举债经营　*n*. 力量；影响；杠杆作用；杠杆率

long-standing ['lɔŋstændɪŋ]　长久的；经久不衰的

AlphaGo　*n*. 阿尔法狗（一款围棋人工智能程序）

thanks to　幸亏，多亏，由于

Chinese game of Go　中国围棋

amateur ['æmətə:, 'æmətjuə] *n*. 业余活动者；非专业性人员　*adj*. 业余的

novel ['nɔvəl] *adj*. 新颖的；新奇的　*n*.（长篇）小说

reinforcement [ˌri:ɪn'fɔ:smənt] *n*. 强化，增强；加强；加固；增援

move [mu:v] *n*.（下棋用语）走棋，一着；移动；搬家　*vt*. 移动，使改变位置（或姿势）
　　　　　　vi.（下棋）走一子；移动；搬家

tabula rasa ['tæbjulə 'reɪsə]　白板，擦去了文字的空白书板；[喻]（婴儿）空白的心灵状态

position [pə'zɪʃən] *n*. 棋局；形势，状况；位置；方位；地位；职位
　　　　　　 vt. 把……放在适当位置；给……定位

emphatically [ɪm'fætɪklɪ] *adv*. 断然地；显著地；强调地

Lee Sedol　李世石（韩国围棋九段选手，世界冠军）

outperform [autpə'fɔ:m] *vt*. 胜过；做得比……好

Ke Jie　柯洁（我国围棋九段选手，世界冠军）

echo ['ekəu] *vt*. 附和；重复；模仿　*vi*. 发出回声；起反响　*n*. 回声；重复；共鸣

ingenuity [ˌindʒi'nju:iti] *n*. 独创性；心灵手巧；精巧；精巧的装置

Alzheimer ['ælts,haiməs] *n*. 老年痴呆症（亦称阿尔茨海默症）

come up with　提出，想出；提供；赶上

pressing ['presɪŋ] *adj*. 紧迫的；迫切的　*n*. 按；冲压；唱片

protein folding ['prəuti:n 'fəuldɪŋ]　蛋白质折叠

slate [sleit] *n*. 石板；板岩；石板色；暗蓝灰色　*vt*. 铺石板　*adj*. 含板岩的；石板色的

untie [ˌʌn'taɪ] *vt*. 解开；解决，解除　*vi*. 解开；松开

specific [spɪ'sɪfɪk] *n*. 细节；特性；[复]（计划、建议等的）详细说明书
　　　　　　adj. 明确的；特定的

devastating [ˈdevəsteitiŋ] *adj.* 毁灭性的，灾难性的
　　　　　　　　　　　vt. 使荒芜(devastate 的现在分词)，破坏；压倒
Parkinson [ˈpɑːkinsn] *n.* 帕金森(英国医生 James Parkinson)
cystic fibrosis [ˈsistik faiˈbrəusis] 囊(肿)性纤维化；囊性纤维变性
arm [ɑːm] *n.* 部门；分部；臂；(椅子的)扶手；权力；[常用复] 武器；兵种
　　　　　　vt. 武装；装备　　*vi.* 武装起来
ageing [ˈeidʒiŋ] *n.* 老化；变陈；成熟
counter [ˈkauntə] *n.* (某些棋盘游戏的)筹码；计数器，计算器；计算者；柜台
　　　　　　　　vt. 反击，反对　　*vi.* 反击　　*adj.* 相反的；反对的　　*adv.* 反方向地

Notes

1. 全句可译为：为了学习如何下围棋，先前的 AlphaGo 各个版本最初都从数千次业余爱好者和职业棋手的比赛中接受训练。

2. 本句中，it 是先行代词，代替不定式短语 to do … 作主语，以强调 to do …；本句中的 which 是关系代词，引导非限制性定语从句，修饰其前的 learning。
   下一句中的关系代词 that 引导限制性定语从句，修饰其前的 a neural network。

3. 本句中，现在分词 developing 引导的分词短语作状语，其后的关系代词 that 引导限制性定语从句，修饰其前的 moves。本句中的 it 前省略了关系代词 that，它引导限制性定语从句，修饰其前的 techniques。全句可译为：AlphaGo Zero 还发现了新知识，开发了非常规策略和创新的新着法，这些采用并超越了它在与李世石(Lee Sedol)和柯洁(Ke Jie)比赛中使用的新颖技术。

4. 本句中，humanity 之前省略了关系代词 that，它引导限制性定语从句，修饰其前的 challenges。全句可译为：这些富有创造力的时刻给我们以信心：人工智能将是人类独创性的倍增器，帮助我们完成解决人类正面临的一些最重要挑战的使命。

5. AlphaGo 是由 DeepMind 开发的程序。DeepMind 是位于英国伦敦的前沿人工智能企业，由人工智能程序师兼神经科学家戴密斯·哈萨比斯(Demis Hassabis)等人联合创立，其将机器学习和系统神经科学的最先进技术结合起来，建立强大的通用学习算法。据外媒 2016 年 6 月 8 日报导，DeepMind 欲将其算法应用到医疗保健行业，包括计划在 5 年内使用机器学习处理英国国家医疗服务体系的数据。

6. 本句中，domain 之后省略了关系代词 which，它引导限制性定语从句 you are in，修饰 domain，意为"你处在的(领域)"；本句中的 which 是关系代词，引导限制性定语从句，修饰其前的 algorithm；it 之前省略了 that。全句可译为：为了算法通用到能被应用于任何地方，把你自己从所处领域的具体细节中解放出来。

7. 本句中，using 引导动名词短语作宾语。

8. hack 含意为"劈、砍"，也有"非法侵入(他人计算机系统)"之意，hacker 称为"黑客"。其实 hack 指基于开源的程序，对其代码进行增加、删除或者修改、优化，使之在功能上符合新的需求。例如，CSS(Cascading Style Sheets，层叠样式表)hack 的目的是使 CSS 代码兼容不同的浏览器，反过来利用 CSS hack 为不同版本的浏览器定制实现不同

的 CSS 效果。本句中参照"hacker"把 hack 译为"黑",也就是在此意义下。全句可译为:去年微软宣布,它计划在启动"黑"人体的几个项目之后 10 年内攻克癌症。

9.6 Big Data Analytics

Big Data analytics is the process of collecting, organizing and analyzing large sets of data (called Big Data) to discover patterns and other useful information. Big Data analytics can help organizations to better understand the information contained within the data and will also help identify the data that is most important to the business and future business decisions. [1] Analysts working[2] with Big Data typically want the knowledge that comes from analyzing the data.

Why is Big Data Analytics Important?

Big data analytics helps organizations *harness* their data and use it to identify new opportunities. That, in turn, leads to smarter business moves, more efficient operations, higher profits and happier customers. In his report Big Data in Big Companies, IIA Director of Research Tom Davenport interviewed more than 50 businesses to understand how they used big data. He found they got value in the following ways:

1. **Cost reduction.** Big data technologies such as Hadoop[3] and cloud-based analytics bring significant cost advantages when it comes to[4] storing large amounts of data—plus they can identify more efficient ways of doing business.

2. **Faster, better decision making.** With the speed of Hadoop and in-memory analytics, combined with the ability to analyze new sources of data, businesses are able to analyze information immediately—and make decisions based on what they've learned. [5]

3. **New products and services.** With the ability to *gauge* customer needs and satisfaction through analytics comes the power to give customers what they want. [6] Davenport points out that with big data analytics, more companies are creating new products to meet customers' needs.

When you combine big data with high-powered analytics, you can accomplish business-related tasks such as:

- Determining root causes of failures, issues and defects in near-real time.
- Generating *coupons* at the point of sale based on the customer's buying habits.
- Recalculating entire risk *portfolios* in minutes.
- Detecting *fraudulent* behavior before it affects your organization.

High-Performance Analytics Required

To analyze such a large volume of data, Big Data analytics is typically performed using[7] specialized software tools and applications for predictive analytics, data mining, text mining, forecasting and data optimization. Collectively these processes are separate but highly integrated functions of high-performance analytics. Using[8] Big Data tools and software enables an organization to process extremely large volumes of data that a business has collected to determine which data is relevant and can be analyzed to drive better business decisions in the future.

The Challenges

For most organizations, Big Data analysis is a challenge. Consider the *sheer* volume of data and the different formats of the data (both structured and unstructured data) that is collected across the entire organization and the many different ways different types of data can be combined, contrasted and analyzed to find patterns and other useful business information.[9]

The first challenge is in breaking down data *silos* to access all data an organization stores in different places and often in different systems.[10] A second challenge is in creating platforms that can pull in unstructured data as easily as structured data. This massive volume of data is typically so large that it's difficult to process using traditional database and software methods.

How Big Data Analytics is Used Today

As the technology that helps an organization to break down data silos and analyze data improves[11], business can be transformed in all sorts of ways. Today's advances in analyzing big data allow researchers to decode human DNA in minutes, predict where *terrorists* plan to attack, determine which *gene* is mostly likely to be responsible for certain diseases and, of course, which *ads* you are most likely to respond to on Facebook.

Another example comes from one of the biggest *mobile carriers* in the world. France's Orange launched its Data for Development project by releasing subscriber data for customers in the *Ivory Coast*. The 2.5 billion records, which[12] were made anonymous, included details on calls and text messages exchanged between 5 million users. Researchers accessed the data and sent Orange proposals for how the data could serve as the foundation for development projects to improve public health and safety. Proposed projects included one that showed how to improve public safety by tracking cell phone data to map where people went after emergencies; another showed how to use *cellular* data for disease *containment*.

The Benefits of Big Data Analytics

Enterprises are increasingly looking to find actionable insights into their data. Many big data projects originate from the need to answer specific business questions. With the right big data analytics platforms in place, an enterprise can *boost* sales, increase efficiency, and improve operations, customer service and risk management.

Webopedia parent company, QuinStreet, surveyed 540 enterprise decision-makers involved in big data purchases to learn which business areas companies plan to use Big Data analytics to improve operations. About half of all respondents said they were applying big data analytics to improve customer *retention*, help with product development and gain a competitive advantage.

Notably, the business area getting the most attention relates to increasing efficiency and optimizing operations. Specifically, 62 percent of respondents said that they use big data analytics to improve speed and reduce complexity.

Who Uses Big Data?

Big data affects organizations across practically every industry. See how each industry can benefit from this *onslaught* of information.

- Banking

With large amounts of information streaming in from countless sources, banks are faced with finding new and innovative ways to manage big data. While it's important to understand customers and boost their satisfaction, it's equally important to minimize risk and fraud while maintaining *regulatory compliance*. Big data brings big insights, but it also requires financial institutions to stay one step ahead of the game with advanced analytics.

- Education

Educators armed[13] with data-driven insight can make a significant impact on school systems, students and *curriculums*. By analyzing big data, they can identify at-risk students, make sure students are making adequate progress, and can implement a better system for evaluation and support of teachers and *principals*.

- Government

When government agencies are able to harness and apply analytics to their big data, they gain significant ground when it comes to managing utilities, running agencies, dealing with traffic congestion or preventing crime.[14] But while there are many advantages to big data, governments must also address issues of transparency and privacy.

- Health Care

Patient records. Treatment plans. Prescription information. When it comes to health care, everything needs to be done quickly, accurately and, in some cases, with enough transparency to satisfy stringent industry regulations. When big data is managed effectively, health care providers can uncover hidden insights that improve patient care.

- Manufacturing

Armed[15] with insight that big data can provide, manufacturers can boost quality and output while minimizing waste—processes that are key in today's highly competitive market. More and more manufacturers are working in an analytics-based culture, which means they can solve problems faster and make more agile business decisions.

- Retail

Customer relationship building is critical to the retail industry and the best way to manage that is to manage big data. Retailers need to know the best way to market to customers, the most effective way to handle transactions, and the most strategic way to bring back *lapsed* business. Big data remains at the heart of all those things.

Words and Expressions

harness [ˈhɑːnis] vt. 利用，治理；给(马等)上挽具　n. 马具；挽具
gauge [geidʒ] vt. 估计；测量，测定；使符合标准
　　　　　　　n. 量规，量器，量计，表；标准尺寸，标准规格；轨距
coupon [ˈkuːpɔn] n. 优惠券；赠券；息票；[经]配给券
portfolio [pɔːtˈfəuljəu] n. 投资组合；文件夹；公事包；代表作选
fraudulent [ˈfrɔːdjulənt] adj. 欺诈的，欺骗性的；骗得的
sheer [ʃiə] adj. 绝对的，彻底的；十足的；(织物)极薄的；陡峭的　adv. 全然；彻底；十足
silo [ˈsailəu] n. 筒仓，地窖；地下仓库
terrorist [ˈterərist] n. 恐怖分子；恐怖主义者　adj. 恐怖主义的；恐怖行为的
gene [dʒiːn] n. 基因
ad [æd] n. 广告
mobile carrier　移动通信承运商
Ivory Coast [ˈaivəri ˈkəust] 象牙海岸(非洲)
cellular [ˈseljulə] n. 移动电话；单元　adj. 蜂窝状的，多孔的；细胞的，由细胞组成的
containment [kənˈteinmənt] n. 抑制，遏制；牵制；包含；容量；密闭度
boost [buːst] vt. 促进，提高；升，提；推　n. 增加，提高；升；帮助
Webopedia　公司名　webopedia　web百科全书
retention [riˈtenʃən] n. 保持；保留；保持力；保留物；停滞；记忆(力)
onslaught [ˈɔnslɔt] n. 大量；大批；猛攻；突击
regulatory [ˈreɡjulətəri] adj. 规章的；制订规章的；受规章限制的
compliance [kəmˈplaiəns] n. 依从；屈从
curriculum [kəˈrikjuləm] n. 全部课程；(一门)课程
principal [ˈprinsəpəl] n. 校长；首长；负责人；主要演员，主角；委托人，本人；本金
　　　　　　　adj. 最重要的，主要的；负责人的；资本的，本金的
lapse [læps] vi. 失效；终止；消失　vt. 使失效　n. 权利失效；失误，小错

Abbreviations

IIA (Information Industries Association)　美国信息工业协会
HDFS (Hadoop Distributed File System)　Hadoop分布式文件系统

Notes

1. 本句中，better是副词，well的比较级，意为"更好地"，修饰其后的动词understand；contained是动词的过去分词形式，意为"被包含"；其后的help与identify之间省略了to；that是关系代词，引导限制性定语从句，修饰其前的data。
2. 本句中，working引导分词短语修饰其前的Analysts。
3. Hadoop是一个由Apache软件基金会开发的分布式系统基础架构。用户可以在不了

解分布式底层细节的情况下，开发分布式程序，充分利用集群的威力进行高速运算和存储。Hadoop 实现了一个分布式文件系统（Hadoop Distributed File System），简称 HDFS。

4. when it comes to 意为"就……而论，当说到……的时候；当涉及……时，一谈到"。
5. 本句中，based on 引导的分词短语作状语，修饰其前的 make；what 是关系代词，引导宾语从句。
6. 本句是倒装句，主语是 the power to … 。全句可译为：具有通过分析来估计客户需求和满意度的能力，就有了向客户提供他们想要的东西的能力。
7. 本句中，此 using 引导现在分词短语作状语，表示"使用……（来进行）"。
8. 本句中，此 using 引导动名词短语作主语。
9. 本句中，the many different ways 是名词性短语作状语；其后的 different types of data 是从句的主语，该从句解释其前的 the many different ways。大数据，其特征有人概括为 4V，即 Volume（数据量巨大）、Variety（数据的类型、格式繁多）、Velocity（处理速度快）与 Veracity（真实性，追求高质量）；有人概括为 5V，即增加 Value（高价值）。本文强调了大数据分析带来的两个挑战是因为大数据的前两个特征。整句是祈使句。全句可译为：考虑数据的绝对数量和数据（结构式的和非结构式的数据两者）的各种不同格式，这些数据是整个机构中按不同类型的数据能被组合、对比和分析以找到各种模式及其他有用业务信息的很多不同方法收集的。
10. all data 之后是修饰它的限制性定语从句，省略了关系代词 that。全句可译为：第一个挑战在于分解数据仓库以访问机构存储在不同地方且经常是在不同系统中的所有数据。
11. 该从句的谓语动词是 improves，意为"随着……的技术改进"。
12. 此 which 是关系代词，引导非限制性定语从句，修饰其前的 records。
13. armed 引导的过去分词短语作定语，修饰其前的 Educators，意为"武装有（拥有）……的教育工作者们"。
14. 本句中，gain ground 含意为"进展；发展，壮大；普及"；when it comes to 的解释见注释4。全句可译为：当政府机构能够利用大数据并对大数据应用分析时，它们在管理公用事业、经营机构、处理交通堵塞或预防犯罪等方面就能取得重大进展。
15. Armed 引导的过去分词短语作状语。全句意为"拥有……，制造商能够……，同时……，这些（过程）都是当今激烈竞争的市场中的关键"。

Terms

Decision Support System(DSS)

A Decision Support System (DSS) is a computer program application that analyzes business data and presents it so that users can make business decisions more easily. It is an "informational application" (to distinguish it from an "operational application" that collects the data in the course of normal business operation). Typical information that a decision support application might gather and present would be:

- Comparative sales figures between one week and the next.

- Projected revenue figures based on new product sales assumptions.
- The consequences of different decision alternatives, given past experience in a context that is described.

A decision support system may present information graphically and may include an expert system or artificial intelligence (AI). It may be aimed at business executives or some other group of knowledge workers.

Natural language systems

One of the greatest challenges that scientists in the field of AI face is giving computer systems the ability to communicate *in natural languages*—such as English, Spanish, French, and Japanese. Unfortunately, this challenge is not easy to meet. People have personalized ways of communicating, and the meanings of words vary according to the contexts in which they are used. Nonetheless researchers have made some major strides toward getting computers to listen to and respond in natural languages. Systems in which the computers can understand natural languages are called natural language systems.

Computer animation

Computer animation is the art of creating moving images via the use of computers. It is a subfield of computer graphics and animation. Increasingly it is created by means of 3D computer graphics, though 2D computer graphics are still widely used for low bandwidth and faster real-time rendering needs. Sometimes the target of the animation is the computer itself, but it sometimes the target is another medium, such as film. It is also referred to as CGI (Computer-generated imagery or computer-generated imaging), especially when used in films.

To create the illusion of movement, an image is displayed on the computer screen then quickly replaced by a new image that is similar to the previous image, but shifted slightly. This technique is identical to how the illusion of movement is achieved with television and motion pictures.

Granular Computing(GrC)

Granular computing (GrC) is an emerging computing paradigm of information processing. It concerns the processing of complex information entities called information granules, which arise in the process of data abstraction and derivation of knowledge from information or data. Generally speaking, information granules are collections of entities that usually originate at the numeric level and are arranged together due to their similarity, functional or physical adjacency, indistinguishability, coherency, or the like.

At present, granular computing is more a theoretical perspective than a coherent set of methods or principles. As a theoretical perspective, it encourages an approach to data that recognizes and exploits the knowledge present in data at various levels of resolution or scales. In this sense, it encompasses all methods which provide flexibility and adaptability in the resolution at which knowledge or information is extracted and represented.

Granular computing is not an algorithm or process; there is not a particular method that is called "granular computing". It is rather an approach to looking at data that recognizes how different and interesting regularities in the data can appear at different levels of granularity. The same is generally true of all data: at different resolutions or granularities, different features and relationships emerge. The aim of granular computing is ultimately simply to try to take advantage of this fact in designing more-effective machine-learning and reasoning systems.

Exercises

1. Translate the text of Lesson 9.1 into Chinese.
2. Topics for oral workshop.
   - What is Artificial Intelligence? Talk about its application areas.
   - What language is used for AI? Explain it briefly.
   - Why deep learning is an interesting topic for artificial intelligence researchers?
3. Translate the following into English.

(a) AI 当前正在以知识系统的形式应用于商务,这些知识系统使用人类知识来解决问题。最流行的基于知识的系统是专家系统。专家系统是一个计算机程序,它尝试以一些启发式的论据(heuristics)的形式来表示人类专家的知识。术语启发式的论据是从与词 eureka 相同的希腊语词根衍生的,该词意指"发现"。

(b) 用户接口使得管理员能把一些指令和信息输入专家系统中和接收来自专家系统的信息。指令指明一些参数,这些参数在专家系统的推理处理中引导专家系统。信息是以赋给某些(certain)变量值的形式传递的。

(c) 知识库包含描述问题领域的事实和知识表示技术两者,知识表示技术描述诸事实如何以逻辑方式配合(fit)在一起。术语问题论域(domain)被用于描述问题领域(area)。

(d) 专家系统,也称为基于知识的系统,是一个人工智能系统,它应用一些推理能力以达到一个结论。专家系统对于解决一些诊断的和惯例的(prescriptive)问题都是极好的。

(e) DSS(决策支持系统)不打算替代经理。计算机能应用于问题的结构式部分,然而经理负责非结构式部分——应用判断或直觉并且进行(conduct)分析。

4. Listen to the video "Camcorder" and write down the text (after step).

Unit 10 Data Structure and Algorithms

10.1 Abstract Data Types and Algorithms

Data *abstraction* is a central concept in program design. The abstraction defines the *domain* and structure of the data, along with a collection of operations that access the data. The abstraction, called an abstract data type (ADT), creates a user defined data type whose operations specify how a client may manipulate the data. The ADT is implementation independent and enables the programmer to focus on *idealized* models of the data and its operations.

Examples

1. An accounting program for a small business maintains *inventory* information. Each item in the inventory is represented by a data record that includes the item's identification number, the current *stock level*, pricing information, and *reordering* information. A set of inventory handling operations updates the different information fields and initiates a reordering of stock when inventory levels fall below a certain threshold.[1] The data abstraction describes an item as a record containing a series of information fields and operations that would be used by a company manager for inventory maintenance. Operations might include changing the Stock on Hand value when units of the item are sold, changing the Unit Price when a new pricing policy is used, and initiating a reorder when the stock level falls below the reorder level.

    Data

| Identification | Stock on Hand | Unit Price | Reorder Level |

    Operations
        UpdateStockLevel
        AdjustUnitPrice
        ReorderItem

2. A gaming program involves *tossing* a set of *dice*. In the design, the dice are described as an ADT whose data includes the number of dice that are tossed, the sum of the dice on the last toss, and a list that identifies the value of each die in the last toss. The operations include tossing the dice, returning the sum of the dice on a toss, and printing the value of each individual die in the list.

Data
   diceTotal   Dice List

Operations
   Toss
   Total
   DisplayToss

Most abstract data types have an *initializer* operation that assigns initial values to the data. In the C++ language environment, the initializer is called a constructor.

Stacks

A *stack* is one of the most frequently used and most important data structures. Applications of stacks are vast. For instance, *syntax recognition* in a compiler is stack based, as is *expression evaluation*. [2] At a lower level, stacks are used to pass parameters to functions and to make the actual *function call* to and return from a function.

A *stack* is a list of items that are accessible at only one end of the list. [3] Items are added or deleted from the list only at the top of the stack. Food trays in a dining hall or *a pile of boxes* are good models for a stack.

A stack structure features operations that add and delete items. A *Push* operation adds an item to the top of the stack. The operation of removing an element from the stack is said to *pop* the stack. Figure 10-1 illustrates a sequence of Push and Pop operations. The last item added to the stack is the first one removed. For this reason, a stack is said to have LIFO (Last In First Out) ordering.

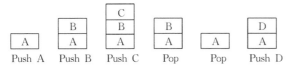

Figure 10-1   Pushing and popping a stack

The abstract concept of a stack allows for an indefinitely large list. Logically, food trays could be stacked to the sky. In reality, food trays are stored in a *rack*. When the rack is full, you cannot add (Push) another item on the stack. The stack has reached the maximum number of elements it can handle. The situation gives meaning to a "stack full" condition. At the other extreme, you cannot pick up a tray from an empty stack. A "stack empty" condition implies that you cannot remove (Pop) an element.

The Stack Class in C++

The stack members include a list, an index or pointer to the top of the stack[4] and the set of stack operations. We use an array to hold the stack elements. As a result, the stack size may not exceed the number of elements in the array and the stack full condition is relevant. We will remove the restriction when we develop a Stack class with a *linked list*.

The *declaration* of a stack object includes the stack size that defines the maximum number of elements in the list. The size has a default value MaxStackSize = 50. The list (stacklist), the maximum number of elements in the stack (size) and the *index* (top) are private members. The operations are public.

Initially the stack is empty and top = −1. Items enter the array (Push) in increasing order of the indices (top = 0,1,2) and come off the stack (Pop) in decreasing order of the indices (top = 2,1,0), see Figure 10-2.

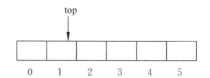

Figure 10-2 Implementation view of a stack

Algorithm

An algorithm generally takes some input, carries out a number of effective steps in a finite amount of time, and produces some output (or a set of ordered steps for solving a problem). An effective step is an operation so basic that it is possible, at least in principle, to carry it out using pen and paper.[5] In computer science theory, a step is considered effective if it is feasible on a *Turing machine* or any of its *equivalents*. A Turing machine is a mathematical model of a computer used in an area of study known as *computability*, which deals with such questions as what tasks can be algorithmically carried out and what cannot.

Once an algorithm is given for a problem and decided (somehow) to be correct, an important step is to determine how much in the way of resources, such as time or space, the algorithm will require.[6] This is called algorithm analysis.

Many algorithms are useful in a broad spectrum of computer applications. These elementary algorithms are widely studied and considered an essential component of computer science. They include algorithms for sorting, searching, text processing, solving graph problems, solving basic *geometric* problems, displaying graphics, and performing common mathematical calculations.

Words and Expressions

abstraction [əb'strækʃən] *n.* 抽象
domain [dəu'mein] *n.* 定义域；域，领域
idealized [ai'diəlaizd] *adj.* 理想化的
inventory ['invəntəri] *n.* 库存，存货；清单
stock level 库存水平，库存量
reorder [riː'ɔːdə] *vt.* 再订购
toss [tɔs] *vt.* 掷，抛
dice [dais] die 的复数 *vt.* 掷骰子 *n.* 骰子；掷骰游戏
initializer [i'niʃəlaizə] *n.* 初始化程序

algorithm [ˈælgəriðəm] n. 算法　algorithm analysis　算法分析
stack [stæk] n. 栈
syntax recognition　语法识别
expression evaluation　表达式求值
function call　函数调用
a pile of boxes　一叠盒子
push [puʃ] vt. 压入，进(栈)，推　n. 推
pop [pɔp] vt. 弹出，出(栈)，上托
rack [ræk] n. 架，搁物架
linked list　链表
declaration [dekləˈreiʃən] n. 说明
index [ˈindeks] n. 索引；指标；下标；指数　vt. 把……编入索引
Turing machine　图灵机
equivalent [iˈkwivələnt] adj. 相等的；相当的；同意义的　n. 等价物
computability n. 可计算性
geometric [dʒiəˈmetrik] adj. 几何的，几何学的

Abbreviation

LIFO (Last In First Out)　后进先出

Notes

1. 本句中 a set of inventory handling operations 译为"一组库存处理操作"。handling 是分词作定语。本句译为：一组库存处理操作更新不同的信息字段，以及当库存量低于某个阈值时着手对库存物再订购。
2. 本句中 as is expression evaluation 是 as 的一种用法，as 代替主句中表语的意思，并在从句中作表语，表示"也一样"。本句译为：例如，编译程序中的语法识别是基于栈的，表达式求值也是基于栈的。
3. 本句译为：栈是只能在其一端存取元素的表。
4. 本句中 an index or pointer to the top of the stack 译成"一个指向栈顶的指标或指针"。下面第二句中 the stack full condition is relevant 译为"栈满的条件与之相关"。
5. 本句中 so basic that it is ... 是 so ... that ... 结构。在 so basic 之前通常有逗号。全句可译为：一有效步是一个操作，这种操作如此基本，以致可以用笔和纸来完成(至少原则上如此)。
6. 本句中 in the way of 意为"按照，属于……种类"。后半句译为：重要的一步是确定该算法需要多少资源，如需要多少时间或空间。

10.2　Spanning Trees

Consider the graph representing the airline's connections between seven cities (see Figure 10-3(a)). If the economic situation forces this airline to shut down as many connections as possible, which of them should be retained to make sure that it is still possible to reach any city from any other city, if only indirectly?[1] One possibility is the graph in Figure 10-3(b). City a can be reached from city d using the path d, c, a, but it is also possible to use the path d, e, b, a. Because the number of retained connections is the issue, there is still the possibility we can reduce this number. It should be clear that the minimum number of such connections form a tree because alternate paths arise as a result of cycles in the graph.[2] Hence, to create the minimum number of connections, a *spanning tree* should be created, and such a spanning tree is the *by-product* of depthFirstSearch( ). Clearly, we can create different spanning trees (see Figure 10-3(c)-(d))—that is, we can decide to retain different sets of connections—but all these trees have six edges and we cannot do any better than that.

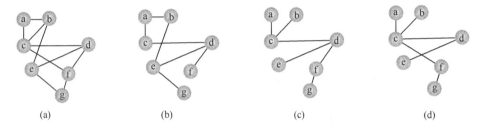

Figure 10-3   A graph representing (a) the airline connections between seven cities and (b-d) three possible sets of connections

The solution to this problem is not optimal in that[3] the distances between cities have not been taken into account. Because there are alternative six-edge connections between cities, the airline uses the cost of these connections to choose the best, guaranteeing the optimum cost. This can be achieved by having maximally short distances for the six connections. This problem can now be *phrased* as finding a minimum spanning tree, which is a spanning tree in which the sum of the *weights* of its edges is minimal. The previous problem of finding a spanning tree in a simple graph is a case of the minimum spanning tree problem in that the weights for each edge are assumed to equal one.[4] Therefore, each spanning tree is a minimum tree in a simple graph.

The minimum spanning tree problem has many solutions, and only a handful of them are presented here.

One popular algorithm was devised by Joseph Kruskal. In this method, all edges are ordered by weight, and then each edge in this ordered sequence is checked to see whether it can be considered part of the tree under construction. It is added to the tree if no cycle arises after its inclusion. This simple algorithm can be summarized as follows:

KruskalAlgorithm (weighted connected undirected graph)
    tree = null;
    edges=sequence of all edges of graph sorted by weight;
      for (i=1; i<=|E| and |tree|<|V|-1; i++)
        if e_i from edges does not form a cycle with edges in tree
          add e_i to tree;

The *complexity* of this algorithm is determined by the complexity of the *sorting* method applied, which for an efficient sorting is $O(|E| \lg |E|)$.[5] It also depends on the complexity of the method used for cycle detection. If we use union( ) to implement Kruskal's algorithm, then the for loop of KruskalAlgorithm( ) becomes

    for (i=1; i<=|E| and |tree|<|V|-1; i++)
        union (e_i = edge (vu));

Although union( ) can be called up to $|E|$ times, it is exited after one (the first) test if a cycle is detected and it performs a union, which is of complexity $O(|V|)$), only for $|V|$-1 edges added to tree. Hence, the complexity of KruskalAlgorithm( )'s for loop is $O(|E|+(|V|-1)|V|)$, which is $O(|V|^2)$. Therefore, the complexity of KruskalAlgorithm( ) is determined by the complexity of a sorting algorithm, which is $O(|E| \lg |E|)$, that is, $O(|E| \lg |V|)$.

Kruskal's algorithm requires that all the edges be ordered before beginning to build the spanning tree. This, however, is not necessary; it is possible to build a spanning tree by using any order of edges. A method was proposed by Dijkstra (1960) and independently by Robert Kalaba, and because no particular order of edges is required here, their method is more general than other two.

Dijkstra method (weighted connected undirected graph)
    tree = null;
    edges = an unsorted sequence of all edges of graph;
    for j=1 to |E|
        add e_j to tree;
        if there is a cycle in tree
            remove an edge with maximum weight from this only cycle;

In this algorithm, the tree is being expanded by adding to it edges one by one, and if a cycle is detected, then an edge in this cycle with maximum weight is discarded.

To deal with cycles, DijkstraMethod( ) can use a modified version of union( ). In the modified version, an additional array, prior, is used to enable immediate detaching of a vertex from a *linked list*.[6] Also, each vertex should have a field next so that an edge with the maximum weight can be found when checking all the edges in a cycle. With these modifications, the algorithm runs in $O(|E||V|)$ time.

Words and Expressions

spanning ['spæniŋ] tree 生成树
by-product 副产品
phrase [freiz] vt. 用短语表达,用术语描述 n. 短语
weight [weit] n. 权;重力;重量 vt. 使加权
complexity [kəm'pleksiti] n. 复杂性
sorting ['sɔːtiŋ] n. 排序 vt. 排序;分类;整理(sort 的现在分词)
linked list 链表

Notes

1. airline "航空公司,航线"。上一句的 airline's connections 可译为"航线"。if only "只要,要是……多好"。全句可译为:如果经济状况迫使这家航空公司尽可能多地关闭一些航线,那么应该保留哪些航线,才能保证从任一城市仍然可以到达其他任何城市(可间接到达)?
2. alternate "交替的,轮流的"。后半句可译为:因为交替的路径起因于图中的环路。
3. 本句中 in that 意为"因为,既然"。
4. case "事例,案例,情况"。全句可译为:之前的在一个简单图中找一棵生成树的问题是最小生成树问题的一个例子,在该问题中假设每个边上的权是相同的。
5. 本句中 applied 修饰前面的 method。applied 后面的 which 代表 complexity, which 引导的句子可译为:一个有效排序的复杂度是 $O(|E| \lg |E|)$。
6. 全句可译为:在这个修改版中,用了一个附加的数组 prior,能够把一个顶点立即从链表中分离出来。

10.3 *Block Sorting* Algorithms: Parallel and Distributed Algorithm

When p processors are available and n records are to be sorted, one possibility is to distribute the n records among the p processors so that *a block of* $M=\lceil n/p \rceil$ *records* is stored in each processor's local memory (a few dummy records may have to be added to constitute the last block). The processors are labeled $P_1, P_2, \cdots, P_p$, according to an indexing rule that is usually dictated by the topology of the interconnecting network. Then, the processors cooperate to *redistribute* the records so that.

1. the block residing in each processor's memory constitutes *a* sorted *sequence S_i of length M*, and,

2. the *concatenation* of these local sequences, $S_1, S_2, \cdots, S_p$, constitutes a sorted sequence of length n.

For example, for three processors, the distribution of the sort keys[1] before and after sorting could be the following.

	Before	After
P_1	2,7,3	1,2,3
P_2	4,9,1	4,5,6
P_3	6,5,8	7,8,9

Thus we now have a convention for ordering the total address space of a multiprocessor, and we have defined parallel sorting of an array of size n, where n may be much larger than p.

Algorithms to sort large arrays of files that are initially distributed across the processors' local memories can be constructed as a sequence of block merge-split steps.[2] During a merge-split step, a processor merges two *sorted block*s of equal length (which were produced by a previous step), and splits the resulting block into a "higher" and a "*lower*" *block*, which are sent to two destination processors (like the high and low outputs in a comparison-exchange step).

A block sorting algorithm is obtained by replacing every comparison-exchange step (in a sorting algorithm that consists of comparison-exchange steps) by a merge-split step. It is easy to verify that this procedure produces a sequence that is sorted according to the above definition.

Two-Way Merge-Split

A *two-way merge-split step* is defined as a two-way merge of two sorted blocks of size M, followed by splitting the result block of size $2M$ into two halves. Both operations are executed within a processor's local memory. The contents of processor's memory before and after a two-way merge-split are shown in Figure 10-4. After two sorted sequences of length M have been stored in each processor's local memory, the processors execute in parallel a merge procedure and fill up an output buffer $O[1..2M]$ [3] (thus a two-way merge-split step uses a local memory of size at least $4M$). After all processors have completed the parallel execution of the merge procedure, they split their output buffer and send each half to a destination processor. The destination processors' addresses are determined by the comparison-exchange algorithm on which the block sorting algorithm is based.

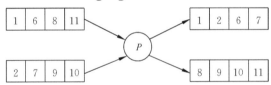

Figure 10-4   Merge-split based on two-way merges

Block Odd-Even Sort Based on Two-Way Merge-Split

Initially, each of the p processors' local memory contains a sequence of length M. The algorithm consists of a *preprocessing* step (Step 0), during which each processor independently sorts the sequence residing in its local memory, and p additional steps (Steps 1 to p), during which the processors cooperate to merge the p sequences generated by Step 0. During Step 0, the processors perform a local sort by using any fast *serial sorting* algorithm. For example, a local two-way merge or a quick sort can be used. Steps 1 to p are similar to Steps 1 to p of the odd-even *transposition sort*. During the odd (even) steps, the odd-(even-) numbered processors receive from their right neighbor a sorted block, perform a two-way merge, and send back the higher M records. The algorithm can be executed synchronously by p processors, odd and even processors being *alternately* idle.

Processor Synchronization

When M is large, or when the individual records are long, transferring blocks of $M * p$ records between the processors introduces time delays that are higher by several *orders of* magnitude than the instruction rate of the individual processors. In addition, depending on the data distribution, the number of comparisons required to merge two blocks of M records may vary. Thus, for the execution of block sorting algorithms based on two-way merge-split, a *coarser granularity* for processor synchronization (MIMD mode) might be more adequate than the SIMD mode, where processors are synchronized at the machine instruction level.[4] A multiprocessor model for those algorithms (distributed algorithms) in which processors operate independently of each other, but can be synchronized by exchanging messages among themselves or with a controlling processor at intervals of several thousand instructions, is more appropriate for these algorithms.[5] At the initiation time of a block sorting algorithm, the controller assigns a number of processors to its execution. Because other operations may already be executing, the controller maintains a free list[6] and assigns processors from this list. In addition to the availability of processors, the size of the sorting problem is also considered by the controller to determine optimal processor allocation.

Words and Expressions

  block sorting 块排序
  a block of M records M 个记录的块
  redistribute [riːdiˈstribjuːt] *vt.* 重新分布
  a sequence of length M 长度为 M 的序列
  concatenation [kɔnˌkætiˈneiʃən] *n.* 拼接；连接；并置
  "lower" block 低端块
  sorted block 排好序的块
  two-way merge-split step 二路归并-分拆步
  block Odd-Even Sort 块奇偶排序
  preprocessing [ˌpriːˈprɔsesiŋ] *n.* 预处理
  serial sorting 串行排序
  transposition sort 交叉排序，换位排序
  alternately [ɔːlˈtəːnitli] *adv.* 交替地；轮流地
  order of magnitude 数量级
  coarser granularity 粗粒度

Notes

1. 本句中 sort keys 可译为"排序关键字"。假设一个记录由若干字段组成，对 n 个记录进行排序是指按记录中某个关键字段的递增（或递减）次序进行排序，关键字段就是这里的 sort keys。
2. 本句中 large arrays of files 意为"大批文件"，可理解为大数组。本句可译为：对初始分布在若干台处理机内存中一些文件上的大数组的排序算法可以（被）构造成一系列的块归并—分拆步。
  下一句中 splits the resulting block into a "higher" and a "lower" block 译为"把（归

并)得到的块分为'高端'块和'低端'块"。

注：高端块中的任一数都大于低端块中的任何数。

3. 本句后半句译为：这些处理机并行地执行归并过程，并且填满输出缓冲 $O[1..2M]$。

4. 本句中 a coarser granularity for processor synchronization 译成"处理机同步的粗粒度"。本句译为：因此，就运行基于二路归并一分拆的块排序算法而言，处理机的粗粒度同步（MIMD 方式）可能比 SIMD 方式的机器指令级处理机同步更合适。

注：当由若干处理机合作解决一个问题时（称为分布式计算），每台处理机各自做自己的那部分工作。但由于是由多台处理机一起完成一个大问题，因此常常需要各台处理机在一定时候（如本例中当各台处理机都对自己内存中的一块数据排好序以后）进行同步，同步完成后再继续各自的工作，到一定时候又要进行必要的同步，直至一起完成一个大任务。所谓粗粒度同步，是指多处理机一次同步后，要经较长的时间再同步。SIMD 机实际上是指令级的同步，即各处理机同步执行每一条指令。

5. 本句的主要结构是 model ... is more appropriate ... 。in which 引导的从句修饰 algorithms。句中 by 和 with 是同等结构，即"通过……同步"和"用……同步"。全句译为：多处理机模型更适合于这些（分布式的）算法，在这些算法中，各处理机彼此独立地运行，但能通过在它们中间交换信息来同步，或由一台控制处理机，每隔数千条指令来进行同步。

6. 本句中 a free list 可译为"空闲表"，是指该表中当前处于空闲状态的处理机名。

注：本节介绍并行排序算法，实际上该算法可演变成分布式排序算法，未学过并行和分布式算法的读者可作为新内容学习。

10.4 Divide-and-Conquer

Divide-and-conquer is a design strategy which is well known for *breaking down* efficiency barriers.[1] When the method applies, it often leads to a large improvement in *time complexity*, from $O(n^2)$ to $O(nlogn)$, for example.

One notable algorithm employing this strategy is the fast Fourier transform, which is used in the physical sciences for transforming a function of time into a function of frequency, both functions being defined by their values at a large number of points. A *nuclear magnetic resonance spectrometer*, for example, regularly produces 2^{16} or more numbers which must be transformed in this way. Without the fast $O(nlogn)$ method made possible by divide-and-conquer, this whole technique would be infeasible.

The divide-and-conquer strategy is as follows: divide the problem instance into two or more smaller instances of the same problems, solve the smaller instances recursively, and assemble the solutions to form a solution of the original instance. The recursion stops when a instance is reached which is too small to divide: the solution to this instance can be produced directly, and it forms the basis for a *proof of correctness* by *induction* on the size of the instance.[2]

When dividing the instance, one may either use whatever division comes most easily to hand, or invest time in making the division carefully so that the assembly is simplified.[3]

The matrix multiplication algorithm due to Strassen is perhaps the most dramatic example of this. The usual way to multiply two $n * n$ matrices A and B, yielding result matrix C requires n^3 scalar multiplications and n^3 scalar additions. It very simply reflects the definition of matrix multiplication, and we naturally expect that it can't be improved upon.

Now apply divide-and-conquer to this problem. It is a fact that, if the three matrices are divided into quarters like this,

$$\begin{bmatrix} A_{11} & A_{12} \\ A_{21} & A_{22} \end{bmatrix} \begin{bmatrix} B_{11} & B_{12} \\ B_{21} & B_{22} \end{bmatrix} = \begin{bmatrix} C_{11} & C_{12} \\ C_{21} & C_{22} \end{bmatrix}$$

then the C_{ij} can be found by the usual matrix multiplication algorithm, substituting matrix operations for scalar ones. That is

$$C_{11} = A_{11} B_{11} + A_{12} B_{21}$$
$$C_{12} = A_{11} B_{12} + A_{12} B_{22}$$
...

This leads to a divide-and-conquer algorithm, which performs an $n * n$ matrix multiplication by partitioning the matrices into quarters and performing eight $(n/2) * (n/2)$ matrix multiplications and four matrix additions. The *recurrence* equation for the number of scalar multiplications performed is

$$T(1) = 1$$
$$T(n) = 8T(n/2)$$

which leads to $T(n) = n^3$ when n is a power of 2, as the reader can easily show.

Strassen's insight was to find an alternative method for calculating the C_{ij}, requiring seven $(n/2) * (n/2)$ matrix multiplications and eighteen matrix additions and subtractions:

$$M_1 = (A_{12} - A_{22})(B_{21} - B_{22}) \qquad M_5 = A_{11}(B_{12} - B_{22})$$
$$M_2 = (A_{11} + A_{22})(B_{11} + B_{22}) \qquad M_6 = A_{22}(B_{21} - B_{11})$$
$$M_3 = (A_{11} - A_{21})(B_{11} + B_{12}) \qquad M_7 = (A_{21} + A_{22})B_{11}$$
$$M_4 = (A_{11} + A_{12})B_{22}$$
$$C_{11} = M_1 + M_2 - M_4 + M_6$$
$$C_{12} = M_4 + M_5$$
$$C_{11} = M_6 + M_7$$
$$C_{11} = M_2 - M_3 + M_5 - M_7$$

If this method is used recursively to perform the seven $(n/2) * (n/2)$ matrix multiplications, then the recurrence equation for the number of scalar multiplications performed is

$$T(1) = 1$$
$$T(n) = 7T(n/2)$$

Solving this for the case $n = 2^k$ is easy:

$$T(2^k) = 7T(2^{k-1}) = 7^2 T(2^{k-2}) = \cdots = 7^k$$

That is, $T(n) = 7^{\log_2 n}$. Applying the identity $a^{\log_b c} = c^{\log_b a}$, which is easily proven by taking *logarithms* to base b[4].

$$T(n) = n^{\log_2 7} = n^{2.81}$$

The reader is left the task of showing that the number of scalar additions performed is also $O(n^{\log_2 7})$, so concluding that Strassen's algorithm is *asymptotically* more efficient than the standard algorithm.

Words and Expressions

divide-and-conquer　分而治之；分治法
break down　排除；破除；分解；故障
time complexity　时间复杂度(性)
nuclear magnetic resonance　核磁共振
spectrometer [spek'trɔmitə] *n*. 分光仪，光谱仪，频谱仪
proof of correctness　正确性证明
induction [in'dʌkʃən] *n*. 归纳法
recurrence [ri'kʌrəns] *n*. 递推；递归
logarithm ['lɔgəriθəm] *n*. 对数
asymptotically [ˌæsimp'tɔtikəli] *adv*. 渐近地

Abbreviation

FFT (Fast Fourier Transform)　快速傅里叶变换

Notes

1. 本句中 breaking down efficiency barriers 译为"排除效率障碍"。
2. 本句中 it forms the basis for ... 子句译为：它对实例大小进行的归纳证明构成了正确性证明的基础。
3. 本句中 comes to hand 意为"到手，找到"；invest time in ... "投入时间到……"。全句译为：在划分实例时，人们可以使用任何一种最容易找到的划分方法，也可以花时间仔细地进行划分，以便组合各个解时更容易一些。
4. taking logarithms to base *b*，译为"以 *b* 为底取对数"。

注：本节介绍分治法，并以矩阵乘法为例。

Exercises

1. Translate the text of Lesson 10.3 into Chinese.
2. Topics for oral workshop.
- Talk about stack, its operations, and its applications.
- Describe a sorting algorithm (bubble sort, insertion sort, quick sort, ...).
- Discuss parallel and distributed algorithms, the difference between a sequential algorithm and a parallel and distributed algorithm.
3. Translate the following into English.

链表(linked list,见下图)由一系列节点组成,这些节点在内存中不必毗连(邻接)。每个节点均包含表元素和指向(to)它的后继节点的链,我们称之为 next 链。最后一个单元的 next 链引用(reference)null。

为了执行 printList 或 find(x),我们只要从表中的第一个节点开始,然后沿着 next 链遍历(traverse)该表。这种操作显然是线性时间的,就像在数组实现中那样,虽然这个常数很可能比用数组实现时要大。findKth 操作不再像数组实现时那样效率高;findKth(i)花费 $O(i)$ 的时间并且通过用显而易见的方式向下遍历链表来工作(work)。实际上,这个界是悲观的,因为对 findKth 的调用常常按排序的顺序(in sorted order)(按 i)进行。例如,findKth(2)、findKth(3)、findKth(4)及 findKth(6)都能够在对表的一次向下扫描中执行。

Remove 方法可以用一个 next 引用的改变来实现。图 2 给出了在原表中删除第二个元素的结果。

Insert 方法需要使用 new 调用从系统取得一个新节点,然后执行两次引用调整(maneuver)。其一般思想在图 3 中给出。虚线表示废弃(old)的 next 引用。

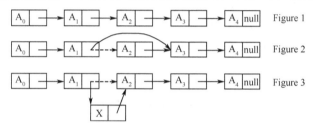

4. Listen to the video "Broadband Internet" and write down 1-3 paragraphs.

Unit 11　Fundamentals of the Computing Sciences

11.1　Set Theory

Set theory is the branch of mathematics that studies sets, which are collections of objects. [1] Although any type of object can be collected into a set, set theory is applied most often to objects that are relevant to mathematics.

The modern study of set theory was initiated by Georg Cantor [2] and in the 1870s. After the discovery of *paradoxes* in *naive* set theory, numerous *axiom systems* were proposed in the early twentieth century, of which the Zermelo Fraenkel axioms, with the axiom of choice, are the best-known. [3]

The language of set theory could be used in the definitions of nearly all mathematical objects, such as functions, and concepts of set theory are integrated throughout the mathematics *curriculum*. Elementary facts about sets and set membership can be introduced in primary school, along with Venn and Euler diagrams[4], to study collections of commonplace physical objects. Elementary operations such as *set union* and *intersection* can be studied in this context. More advanced concepts such as *cardinality* are a standard part of the undergraduate mathematics curriculum.

Set theory is commonly employed as a foundational system for mathematics, particularly in the form of Zermelo Fraenkel set theory with the axiom of choice. Beyond its foundational role, set theory is a branch of mathematics *in its own right*, with an active research community. *Contemporary* research into set theory includes a diverse collection of topics, ranging from the structure of the *real number line* to the study of the consistency of large *cardinals*.

Basic Concepts

Set theory begins with a fundamental *binary relation* between an object o and a set A. If o is a member (or element) of A, we write $o \in A$. Since sets are objects, the membership relation can relate sets as well.

A derived binary relation between two sets is the subset relation, also called set inclusion. If all the members of set A are also members of set B, then A is a subset of B, denoted $A \subseteq B$. For example, $\{1,2\}$ is a subset of $\{1,2,3\}$, but $\{1,4\}$ is not. From this definition, it is clear that a set is a subset of itself; in cases where one wishes to avoid this, the term *proper subset* is defined to exclude this possibility.

Just as arithmetic *features* binary operations on numbers, set theory features binary operations on sets. The:
- Union of the sets A and B, denoted $A \cup B$, is the set of all objects that are a member of

A, or B, or both. The union of $\{1,2,3\}$ and $\{2,3,4\}$ is the set $\{1,2,3,4\}$.

• Intersection of the sets A and B, denoted $A \cap B$, is the set of all objects that are members of both A and B. The intersection of $\{1,2,3\}$ and $\{2,3,4\}$ is the set $\{2,3\}$.

• Complement of set A relative to set U, denoted as A^c, is the set of all members of U that are not members of A.[5] This terminology is most commonly employed when U is a *universal set*, as in the study of Venn diagrams. This operation is also called the set difference of U and A, denoted $U \backslash A$. The complement of $\{1,2,3\}$ relative to $\{2,3,4\}$ is $\{4\}$, while, conversely, the complement of $\{2,3,4\}$ relative to $\{1,2,3\}$ is $\{1\}$.

• *Symmetric difference* of sets A and B is the set of all objects that are a member of exactly one of A and B (elements which are in one of the sets, but not in both). For instance, for the sets $\{1,2,3\}$ and $\{2,3,4\}$, the symmetric difference set is $\{1,4\}$. It is the set difference of the union and the intersection, $(A \cup B) \backslash (A \cap B)$.

• *Cartesian product* of A and B, denoted $A \times B$, is the set whose members are all possible *ordered pairs* (a,b) where a is a member of A and b is a member of B.

• *Power set* of a set A is the set whose members are all possible subsets of A. For example, the power set of $\{1,2\}$ is $\{\{\},\{1\},\{2\},\{1,2\}\}$.

Axiomatic Set Theory

Elementary set theory can be studied *informally* and intuitively, and so can be taught in primary schools using, say, Venn diagrams. The intuitive approach silently assumes that all objects in the *universe of discourse* satisfying any defining condition form a set.[6] This assumption gives rise to *antinomies*, the simplest and best known of which being Russell's paradox. Axiomatic set theory was originally devised to rid set theory of such antinomies.

Systems of constructive set theory, such as CST, CZF, and IZF, embed their set axioms in *intuitionistic* logic instead of *first order logic*. Yet other systems accept standard first order logic but feature a nonstandard membership relation. These include *rough set theory* and *fuzzy set theory*, in which the value of an atomic formula embodying the membership relation is not simply True and False.[7] The Boolean-valued models of ZFC are a related subject.

Applications

Nearly all mathematical concepts are now defined formally in terms of sets and set theoretic concepts. For example, mathematical structures as diverse as graphs, *manifolds*, rings, and vector spaces are all defined as sets having various (axiomatic) properties.[8] Equivalence and order relations are *ubiquitous* in mathematics, and the theory of relations is entirely *grounded* in set theory.

Set theory is also a promising foundational system for much of mathematics. Since the publication of the first volume of Principia Mathematica, it has been claimed that most or even all mathematical theorems can be derived using an *aptly* designed set of axioms for set theory, augmented with many definitions, using first or *second order logic*.[9] For example, properties of the natural and real numbers can be derived within set theory, as each number system can be identified with a set of equivalence classes under a suitable equivalence relation whose field is some infinite set.

Set theory as a foundation for mathematical analysis, *topology*, abstract algebra, and *discrete mathematics* is likewise *uncontroversial*; mathematicians accept that (in principle) theorems in these areas can be derived from the relevant definitions and the axioms of set theory. Few full derivations of complex mathematical theorems from set theory have been formally verified, however, because such formal derivations are often much longer than the natural language proofs mathematicians commonly present.[10]

Words and Expressions

paradox ['pærədɔks] *n.* 似是而非的论点[妙语]；悖论；自相矛盾的话(事物)
naive [nɑː'iːv] *adj.* 朴素的；天真的；自然的；单纯的
axiom system   公理系统
curriculum [kə'rikjuləm] *n.* 全部课程；(一门)课程
set union   集合并
set intersection   集合交
cardinality [ˌkɑːdi'næliti] *n.* 基数，集的势
in its own right   就其本身而言；凭本身的头衔(或资格、质量等)
contemporary [kən'tempərəri] *adj.* 现代的，当代的   *n.* 同龄人
real number line   实数轴
cardinal ['kɑːdinl] *n.* 基数
binary relation   二元关系，双目关系
proper subset   真子集
feature ['fiːtʃə] *vt.* 以……为特色，描绘……的特征   *n.* 特征，特色
universal set   全集
symmetric difference   对称差
Cartesian product   笛卡儿积
ordered pair   有序对
power set   幂集
informally [in'fɔːməli] *adv.* 非形式地，非正式地
universe of discourse   论域
antinomy [æn'tinəmi] *n.* 法律上的自相矛盾，两个法律间的矛盾；自相矛盾话
intuitionistic [ˌintjuː'iʃəniztik] *adj.* 直觉的，直观的
first order logic   一阶逻辑
rough set theory   粗(糙)集合论
fuzzy set theory   模糊集合论
ZFC   Zermelo-Fraenkel 公理
manifold ['mænifəuld] *n.* 簇，流形

ubiquitous [juːˈbikwitəs] *adj.* （同时）普遍存在的，无处不在的
ground [graund] *vt.* 把……建立在牢固的基础上；把（论点等）基于（on）
aptly [ˈæptli] *adv.* 恰当地，贴切地；灵巧地
second order logic 二阶逻辑
topology [təˈpɔlədʒi] *n.* 拓扑，拓扑学
discrete mathematics 离散数学
uncontroversial [ˌʌnˌkɔntrəˈvəːʃəl] *adj.* 无争论的，无可争辩的，无可置疑的

Notes

1. 本句中的关系代词 that 引导限制性定语从句，修饰其前的 the branch；which 引导非限制性定语从句，修饰前面的 sets。全句可译为：集合论是数学中研究集合的分支，集合是一些对象的汇集。

2. George Cantor(乔治·康托)(1845—1918)是德国数学家，19 世纪数学伟大成就之一——集合论的创立人。他证明了有理数是可枚举的。Richard Dedekind(理查德·戴德金)(1831—1916)是德国数学家，他的主要贡献有以下两个方面：在实数和连续性理论方面，他提出"戴德金分割"，给出了无理数及连续性的纯算术的定义；在代数数论方面，他创立了现代代数和代数数域理论，引出了现代的"理想"概念，并得到代数整数环上理想的唯一分解定理。现在把满足理想唯一分解条件的整环称为"戴德金整环"。

3. 本句中的 of which 引导非限制性定语从句，修饰前面的 axiom systems，of which 可译为"其中（的）"；with 引导的是状语短语，修饰前面的 the Zermelo-Fraenkel axioms，含意为"使用（……的）"。Zermelo-Fraenkel 指的是数学家 Ernst Zermelo 和 Abraham Fraenkel。在数学领域，以他们命名的使用选择公理的 Zermelo-Fraenkel 集合论通常缩写为 ZFC，是一种公理系统，可以用来系统阐述没有 Russell 悖论那样的悖论的集合论。

4. Venn and Euler diagrams 指的是文氏图与欧拉图，前者 Venn diagrams 是集合的图形表示，可用来表达集合的相关概念，后者 Euler diagrams 是瑞士数学家欧拉（Euler）于 1736 年在他发表的图论第一篇论文"哥尼斯堡七桥问题"中提出的。欧拉图可以直观地表示概念间的关系，在刑事侦查逻辑里有实际用途。

5. 本句中，relative 引导的短语修饰前面的 complement；插入的分词短语 denoted as A^c 作状语；that 引导限制性定语从句，修饰前面的 all members。全句可译为：集合 A 关于集合 U 的补表示为 A^c，是 U 中所有不是 A 的成员的成员组成的集合。

6. 本句中，satisfying 引导动名词短语，修饰前面的 all objects，这从后面的动词 form 是动词原形可知。全句可译为：直觉的方法默认假定论域中满足某一定义条件的所有对象形成一个集合。

7. 本句中，embodying 引导动名词短语，修饰前面的 formula。rough set theory 和 fuzzy set theory 这两种集合论介绍如下。

　　Rough set(Rough 集合，又称粗糙集合或粗集合)理论是 Pawlak 教授于 1982 年提出的一种能够定量分析处理不精确、不一致、不完整信息与知识的数学工具。粗糙

集合论的基本思想是通过关系数据库分类归纳形成概念和规则，通过等价关系的分类以及分类对于目标的近似来实现知识发现。粗糙集合论由于思想新颖、方法独特，已成为一种重要的智能信息处理技术。该理论已经在机器学习与知识发现、数据挖掘、决策支持与分析等方面得到广泛应用。

  Fuzzy set theory（模糊集合论）是对常规集合论的扩充。它处理部分真假值[在1（完全真）和0（完全假）之间的真假值]的概念。它是由加利福尼亚大学伯克利分校的Lotfi A. Zadeh教授于1965年作为复杂系统中建模的手段而引进的。

8. 本句中，as diverse as 引导的短语修饰前面的 structures，having 引导动名词短语，修饰前面的 sets。全句可译为：例如，图、簇、环和向量空间这样的形形色色的数学结构全都被定义为具有各种（公理）性质的集合。

9. 本句中，第一个 using 引导的动名词短语修饰前面的 be derived，第二个 using 引导的动名词短语修饰 augmented，都表示方法。全句可译为：自 *Principia Mathematica* 第一卷出版以来，一直有人声称：大多数甚至所有的数学定理都能应用一阶或二阶逻辑，使用恰当设计（增加了很多定义的）的一组集合论公理推导出来。

10. proofs 之后省略了关系代词 that，它引导限制性定语从句，修饰 proofs。全句可译为：然而，几乎没有复杂的数学定理从集合论（出发进行）的全推导已被形式地验证，因为这种形式推导经常比数学家通常提出的自然语言证明长得多。

注：下列表示法的读法如下。

$A=\{a_1, a_2, \ldots, a_n\}$	读作：A is the set of *a* sub 1,*a* sub 2,… ,*a* sub *n*
$a \in A$	读作：*a* is the member (or element) of (set) *A*
$A \subseteq B$	读作：*A* is the subset of (set) *B*
$A \subset B$	读作：*A* is the proper subset of (set) *B*
$A \cup B$	读作：the union of the sets *A* and *B*
$A \cap B$	读作：the intersection of the sets *A* and *B*
$A \setminus B$	读作：the set difference of the sets *A* and *B*
$A \times B$	读作：the Cartesian product of the sets *A* and *B*
$\|A\|$	读作：the cardinality of (set) *A*

[（集合）A 的基数，即元素个数]

11.2 Predicates

  Generally, predicates make statements about *individuals*. To illustrate this notion, consider the following statements:

  Mary and Paul are siblings.

  Jane is the mother of Mary.

  Tom is a cat.

  The sum of 2 and 3 is 5.

  In each of these statements, there is a list of individuals, which is given by the argument list, together with phrases that describe certain relations among or properties of the individuals mentioned in the argument list.[1] these properties or relations are referred to as predicates. In the statement "Mary and Paul are siblings", for instance, the argument list is given by Mary and Paul, in that order, whereas the predicate is described by the phrase "are siblings". Similarly, the statement "Tom is a cat" has an argument list with the single element "Tom" in it, and its predicate is described by "is a cat". The entries of the argument list are called arguments. The arguments can be either variables or individual constants, but since we have not discussed variables yet, we restrict our attention to the case when all arguments are individual constants.

In *predicate calculus*, each predicate is given a name, which is followed by the list of arguments. The list of arguments is enclosed in parentheses. For instance, to express "Jane is the mother of Mary", one would choose an *identifier*, say "mother", to express the predicate "is mother of", and one would write mother(Jane, Mary).[2] Many logicians use only single letters for predicate names and constants. They would write, for instance $M(j,m)$ instead of mother(Jane, Mary); that is, they would use M as a name for the predicate "is mother of", j for Jane, and m for Mary. To save space, we will often follow this convention.

Note that the order of the arguments is important. Clearly, the statements mother(Mary, Jane) and mother(Jane, Mary) have a completely different meaning. The number of elements in the predicate list is called the *arity* of the predicate. For instance, mother(Jane, Mary) has an arity of 2.[3] The arity of a predicate is fixed. For example, a predicate cannot have two arguments in one case and three in another. Alternatively, one can consider two predicates as different if their arity is different. The following statements illustrate this.

The sum of 2 and 3 is 5.

The sum of 2, 3, and 4 is 9.

To express these statements in predicate calculus, one can either use two predicate names, such as "sum2" and "sum3," and write sum2(2,3,5) and sum3(2,3,4,9), respectively, or one can use the same symbol, say "sum", with the implicit understanding that the name "sum" in sum(2,3,5) refers to a different predicate than in sum(2,3,4,9).

A predicate with arity n is often called an *n-place predicate*. A one-place predicate is called a property.

Example:

The predicate "is a cat" is a one-place predicate, or a property. The predicate "is the mother of", as in "Jane is the mother of Mary", is a two-place predicate; that is, its arity is 2. The predicate in the statement "The sum of 2 and 3 is 6" (which is false) contains the three-place predicate "is the sum of".

A predicate name, followed by an argument list in parentheses, is called an *atomic formula*. The atomic formulas are statements, and they can be combined by logical *connectives* like *propositions*. For instance, to express the fact that Jane is the mother of Mary, one can use the atomic formula mother(Jane, Mary), and this statement can be part of some *compound statement*, such as

$$\text{mother(Jane, Mary)} \Rightarrow \neg \text{mother(Mary, Jane)}$$

Similarly, if cat(Tom) and hastail(Tom) are two atomic formulas, expressing that Tom is a cat and Tom has a tail, respectively, one can form

$$\text{cat(Tom)} \Rightarrow \text{hastail(Tom)}$$

If all arguments of a predicate are individual constants, then the resulting atomic formula must either be true or false. This is part of the definition of a predicate. For instance, if the *universe of discourse* consists of Jane, Doug, Mary, and Paul, we have to know for each ordered pair of individuals whether or not the predicate "is the mother of" (or "mother" for short) is true. This can be done in the form of a table, as in Table 11-1. Any method that assigns *truth values* to all possible combinations of individuals of a predicate is called an *assignment* of the predicate. For instance, Table 11-1 is an assignment of the predicate "mother". Specifically, the truth value for mother(x, y) is given in row x and column y. For instance, mother(Jane, Paul) is true because in row "Jane" the entry in column "Paul" is T. More generally, if a predicate has two arguments, its assignment can be given by a table in which the rows correspond to the first argument and the columns correspond to the second. The convention that rows correspond to the first argument and columns to the second will be assumed throughout unless explicitly stated otherwise.

Table 11-1 **Assignment for the predicate "mother"**

	Doug	Jane	Mary	Paul
Doug	F	F	F	F
Jane	F	F	T	T
Mary	F	F	F	F
Paul	F	F	F	F

Another example of an assignment is as follows. The domain consists of the four numbers 1, 2, 3 and 4. The predicate "greater" is true if the first argument is greater than the second. Hence, greater(4, 3) is true and greater(3, 4) is false. This gives an assignment for all pairs of individuals. For the sake of clarity, we represent this assignment by Table 11-2.

Table 11-2 **Assignment for the predicate "greater"**

	1	2	3	4
1	F	F	F	F
2	T	F	F	F
3	T	T	F	F
4	T	T	T	F

In a finite universe of discourse, one can represent the assignments of predicates with arity n by n-dimensional arrays. For instance, properties are assigned by one-dimensional arrays, predicates of arity 2 by two-dimensional arrays, and so on.

Note that the mathematical symbols $\geqslant$ and $\leqslant$ are predicates. However, these predicates are normally used in *infix* notations. By this, we mean that they are placed between the arguments.[4] For instance, to express that 2 is greater than 1, we write $2 > 1$, rather than $>(2, 1)$.

Words and Expressions

individual [ˌindiˈvidjuəl] *n.* 个体，个人

predicate calculus　　谓词演算
identifier [aiˈdentifaiə] n. 标识符
arity　n. 元
n-place predicate　n 元谓词
atomic formula　　原子公式
connective [kəˈnektiv] n. 连接词　adj. 连接的
proposition [ˌprɔpəˈziʃən] n. 命题；陈述
compound statement　　复合命题；复合语句
universe of discourse　　论域
truth value　　真值
assignment [əˈsainmənt] n. 赋值；委派
infix [inˈfiks] n. 中缀

Notes

1. certain relations among or properties of the individuals mentioned in the argument list 含意为"在变元表中提到的个体间的某些关系或（各个体的）性质"。全句可译为：在每个陈述中有一个用变元表给出的个体表，以及描述变元表中那些个体间的关系或（各个体的）性质的一些短语。

2. one 是代词，泛指某人，这里意为"有人"。say 这里意为"譬如说"。全句可译为：例如，为表达"Jane 是 Mary 的母亲"，有人会选取某个标识符，譬如说"mother"，来表达谓词"是……的母亲"，并且写作 mother(Jane, Mary)。

3. of 这里含意为"由……做成的"。全句可译为：例如，mother(Jane, Mary) 有元 2；或者译为：例如，mother(Jane, Mary) 的元是 2。

4. by 这里含意为"靠、用、通过"，表示方法、手段。全句可译为：这表示它们被放在变元之间。

注：下列表示法的读法如下。

$f(x, y)$	读作：function f of x and y（x 与 y 的函数 f）
$P \Rightarrow Q$	读作：P implies Q（P 蕴含 Q）
$\neg P$	读作：not P（非 P）
$a \leqslant b$	读作：a is less than or equal to b
$a \geqslant b$	读作：a is greater than or equal to b
(A, R)	读作：the set A together with the relation R
	或　the poset（偏序集）A together with the relation R
$A \subseteq B$	读作：A is contained in B（A 被包含于 B）
	或　A is a subset of B（A 是 B 的子集）
Z^+	读作：Z up plus
	或　the positive closure of Z（Z 的正闭包）
$a \mid b$	读作：a divides b（a 除尽 b 或 a 整除 b）
R^{-1}	读作：the inverse of the relation R（关系 R 之逆）
$A \times B$	读作：the Cartesian product of A and B（A 与 B 的笛卡儿积）

$A \cap B$	读作：	the intersection of sets A and B（集合 A 与 B 的交）
R^2	读作：	R square 或 R squared 或 the square of R
$(R^{-1})^2$	读作：	the square of the inverse of the relation R
		（关系 R 之逆的平方）
S^*	读作：	S up star
	或	the closure of S（（集合）S 的闭包）
S^n	读作：	the n_th power of S
	或	S to the n_th power
$A = \{a_1, a_2, \ldots, a_n\}$	读作：	A is the set of a sub 1, a sub 2, ..., a sub n
$a \in S$	读作：	a is the element of (set) S
	或	a belongs to S
$S \rightarrow T$	读作：	S maps into T（S 映射到 T）
$a * b = \dfrac{ab}{2}$	读作：	a star b is a times b over 2
$\lvert G \rvert$	读作：	the cardinality of (set) G
		[（集合）G 的基数，即元素个数]

11.3 Languages and Grammars

While high-level programming languages reduce much of the *drudgery* of machine language programming, they will introduce new problems.[1] A program (compiler) that converts a program to some object language such as machine language must be written. Also, programming languages must be precisely defined. Sometimes it is the *existence* of a particular compiler that finally provides the precise definition of a language (not a very satisfactory situation for either the programmer or the compiler writer!). The specification of a programming language involves the definition of the following:

1. The set of symbols (or *alphabet*) that can be used to construct correct programs.
2. The set of all correct programs.
3. The "meaning" of all correct programs.

In this section we shall be concerned with the first two items in the specification of programming languages.

First, however, we introduce some terminology associated with grammars and *formal languages*. For this purpose, let V denote a nonempty set of symbols. Such a set V is called an *alphabet*. An alphabet needs not be finite or even *countable*. However, for our purpose here we shall assume V to be finite. For example, we may have

$$V_1 = \{0, 1\}$$
$$V_2 = \{a, b, \ldots, z\}$$
$$V_3 = \{a, 1, *, +, A\}$$
$$V_4 = \{\text{cat}, \text{book}, \text{dog}, x\}$$
$$V_5 = \{a, \ldots, z, 0, \ldots, 9, ', .\}$$

In alphabet V_4 we assume that each word is a symbol and is indivisible. An element of an alphabet is called a letter, a character, or a symbol. A *string* over an alphabet is a sequence of symbols from the alphabet.[2] A *string* is also called a sequence, a word, or a sentence, depending on its nature. A string consisting of m symbols ($m>0$) is called a string of length m.[3] For example, let $V=\{a,b\}$; then $aa, bb, ab,$ and ba are all possible strings of length 2. If we admit an empty string, that is, a string of length 0 ($m=0$), which is usually denoted by ε, then we can have strings of length m for $m \geqslant 0$. The set of strings over an alphabet V is generally denoted by V^* and the set of nonempty strings by $V^+ = V^* - \{\varepsilon\}$. We shall assume that the strings in V^* are of finite length.

A language L can be considered a subset of V_T^*, where V_T is the alphabet associated with L. The language consisting of V_T^* is not particularly interesting since it is too large. Our definition of a language L is a set of strings or sentences over some finite alphabet V_T so that $L \subseteq V_T^*$.[4]

How can a language be represented? A language consists of a finite or an infinite set of sentences. Finite languages can be specified by exhaustively *enumerating* all their sentences. However, for infinite languages such an enumeration is not possible. On the other hand, any device that specifies a language should be finite. One method for specification that satisfies this requirement uses a generative device called a grammar.[5] A grammar consists of a finite set of rules or *productions* that specifies the syntax of the language. These productions are often *recursive*.

In addition, a grammar imposes structure on the sentences of a language. The study of grammars constitutes an important subarea of computer science called formal languages. This area emerged in the mid-1950s[6] as a result of the efforts of Noam Chomsky, who gave a mathematical model of a grammar in connection with his study of natural languages. In 1960, the concept of a grammar became important to programmers because the syntax of *ALGOL* 60 was described by a grammar. Today, grammars are used widely to describe many programming languages.

A grammar imposes a structure on the sentences of a language. For a sentence in English, such a structure is described in terms of subject, predicate, phrase, noun, and so on. On the other hand, for a program the structure is given in terms of procedures, statements, expressions, and the like. In any case, it may be desirable to describe all such structures and to obtain a set of all the correct or admissible sentences in a language. For example, we may have a set of correct sentences in English or a set of valid Pascal programs. The grammatical structure of a language helps us to determine whether a particular sentence belongs to the set of correct sentences. The grammatical structure of a sentence is generally studied by analyzing the various parts of a sentence and their relationships to one another; this analysis is called *parsing*.

Let us now formalize the idea of a grammar. For this purpose, let V_T be a finite nonempty set of symbols called the alphabet. The symbols in V_T are called *terminal symbols*. The *metalanguage* that is used to generate strings in the language is assumed to contain a set of syntactic classes or variables called *nonterminal symbols*. The set of nonterminal symbols is denoted by V_N, and the elements of V_N are used to define the syntax (structure) of the language. Furthermore, the sets V_N and V_T are assumed to be disjoint. The set $V_N \cup V_T$ consisting of nonterminal and terminal symbols is called the vocabulary of the language. We shall use capital letters such as $A, B, C, \ldots, X, Y, Z$ to denote nonterminal symbols, while S_1, $S_2, \ldots$ represent the elements of the vocabulary. The strings of terminal symbols are denoted by lowercase letters $x, y, z, \ldots$, while strings of symbols over the vocabulary are given by $\alpha, \beta, \gamma, \ldots$. The length of a string α will be denoted by $\sharp \alpha$.[7]

> **Definition** *A context-free grammar is defined by a 4-tuple $G = (V_N, V_T, S, \Phi)$, where V_T and V_N are sets of terminal and nonterminal (syntactic class) symbols, respectively. S, a distinguished element of V_N and therefore of the vocabulary, is called the starting symbol.*[8] *Φ is a finite relation from V_N to $(V_T \cup V_N)^+$. In general, an element $\langle \alpha, \beta \rangle$ is written as $\alpha \to \beta$ and is called a production rule or a rewriting rule.*[9] *$V = V_N \cup V_T$ is called the vocabulary of grammar.*

Note that the above definition does not allow the use of empty rules (ε-rules). An empty rule is one with its right part as the empty string.[10] Although most context-free *parsers* can handle empty rules, we do not further discuss them in the interest of simplicity.

For our example given earlier, dealing with variable names, we may write the grammar as $G_1 = (V_N, V_T, S, \Phi)$ in which

$$V_N = \{I, L, D\}$$
$$V_T = \{a, b, c, d, e, f, g, h, i, j, k, l, m, n, o, p, q, r, s, t, u, v, w, x, y, z,$$
$$\qquad 0, 1, 2, 3, 4, 5, 6, 7, 8, 9\}$$
$$S = I$$
$$\Phi = \{I \to L, I \to IL, I \to ID, L \to a, L \to b, \ldots, L \to z,$$
$$\qquad D \to 0, D \to 1, \ldots, D \to 9\}$$

Words and Expressions

drudgery [ˈdrʌdʒəri] *n.* 单调乏味的工作；苦工
existence [igˈzistəns] *n.* 存在；存在物；实体
alphabet [ˈælfəbit] *n.* 字母表
formal language 形式语言
countable [ˈkauntəbl] *adj.* 可数的　*n.* 可数的东西
string [striŋ] *n.* （字符或符号）串
enumerate [iˈnjuːməreit] *vt.* 枚举，列举
production [prəˈdʌkʃən] *n.* 产生式；生产；产品
recursive [riˈkəːsiv] *adj.* 递归的
ALGOL [ˈælgɔl] *n.* ALGOL算法语言（Algorithmic Language 的缩写）
parse [pɑːz] *vt.* 语法分析，从语法上分析；解析（词，句等）

terminal symbol　终结符号
metalanguage [ˈmetəˌlæŋgwidʒ] n. 元语言，语言分析用的语言
nonterminal symbol　非终结符号
context-free　上下文无关的
starting symbol　开始符号，识别符号
rewriting rule　重写规则
parser [ˈpɑːsə] n. 语法分析程序，语法分析器，分析器

Notes

1. while 这里含意为"虽然、尽管"，是连接词，引导一个状语从句。全句可译为：虽然高级程序设计语言减少了机器语言程序设计大量的单调乏味的工作，它们也将带来一些新问题。
2. from 引导的介词短语作定语，修饰 symbols。全句可译为：字母表上的（符号）串是字母表中的符号所组成的序列。也可简单地译为"字母表上的（符号）串是字母表上的符号序列"。
3. 全句可译为：由 m 个符号（$m>0$）组成的串称为长度为 m 的（符号）串。
4. 本句中第一个 of 表示动作的对象。全句可译为：我们对语言 L 的定义是在某个有穷字母表 V_T 上的串或句子的集合，因此，$L \subseteq V_T^*$。
5. called 引导的分词短语作定语，修饰 device。全句可译为：满足这一要求的一种详细说明方法使用了称为文法的产生器。
6. the mid-1950s 意为"20 世纪 50 年代中期"。
7. 在编译原理课程中，符号串 α 的长度用 $|\alpha|$ 表示。
8. a distinguished element … 是 S 的同位语。全句可译为：V_N 的（因此也是词汇表的）与众不同的元素 S 称为开始符号。
开始符号也称为识别符号。
9. 在编译原理课程中，重写规则（rewriting rule）有时写成如下形式：$\alpha ::= \beta$，读作：α 定义为 β。
10. as 引导的介词短语作定语，修饰 its right part。全句可译为：空规则是其右部为空串的规则。

11.4　Finite-State Machines

We think of a machine as a system that can accept **input**, possibly produce **output**, and have some sort of internal memory that can *keep track of* certain information about previous inputs. The complete internal condition of the machine and all of its memory, at any particular time, is said to constitute the **state** of the machine at that time. The state in which a machine finds itself at any *instant* summarizes its memory of past inputs and determines how it will react to *subsequent* input. When more input arrives, the given state of the machine determines (with the input) the next state to be occupied, and any output that may be produced.[1] If the number of states is finite, the machine is a *finite-state machine*.

Suppose that we have a finite set $S=\{s_0, s_1, \ldots, s_n\}$, a finite set I, and for each $x \in I$, a function $f_x: S \to S$. Let $\mathscr{F}=\{f_x \mid x \in I\}$. The *triple* $(S, I, \mathscr{F})$ is called a finite-state machine, S is called the state set of the machine, and the elements of S are called states. The set I is called the input set of the machine. For any input $x \in I$, the function f_x describes the effect that this input has on the states of the machine and is called a state *transition* function. Thus if the machine is in state s_i and input x occurs, the next state of the machine will be $f_x(s_i)$.

Since the next state $f_x(s_i)$ is uniquely determined by the pair (s_i, x), there is a function $F: S \times I \to S$ given by

$$F(s_i, x) = f_x(s_i)$$

The individual functions f_x can all be *recovered* from a knowledge of F. Many authors will use a function $F: S \times I \to S$, instead of a set $\{f_x \mid x \in I\}$, to define a finite-state machine. The definitions are completely equivalent.

Example 1. Let $S=\{s_0, s_1\}$ and $I=\{0, 1\}$. Define f_0 and f_1 as follows:

$$f_0(s_0) = s_0, \qquad f_1(s_0) = s_1,$$
$$f_0(s_1) = s_1, \qquad f_1(s_1) = s_0.$$

The finite-state machine has two states, s_0 and s_1, and accepts two possible inputs, 0 and 1. The input 0 leaves each state fixed, and the input 1 reverses states. [2] We can think of this machine as a model for a circuit (or logical) device and *visualize* such a device as in Figure 11-1. The output signals will, at any given time, consist of two voltages, one higher than the other. [3] Either line 1 will be at the higher voltage and line 2 at the lower, or the reverse. The first set of output conditions will be denoted s_0 and the second will be denoted s_1. An input pulse, represented by the symbol 1, will reverse output voltages. The symbol 0 represents the absence of an input pulse and so results in no change of output. This device is often called a T *flip-flop* and is a concrete realization of the machine in this example.

We summarize this machine in Figure 11-2. The table shown there lists the states down the side and inputs across the top. The column under each input gives the values of the function corresponding to that input at each state shown on the left. [4]

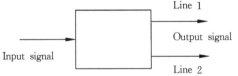

Figure 11-1 Thinking of a finite-state machine as a model for a circuit device

Figure 11-2 Summarizing a finite-state machine

The arrangement illustrated in Figure 11-2 for summarizing the effect of inputs on states is called the state transition table of the finite-state machine. It can be used with any machine of reasonable size and is a convenient method of specifying the machine.

Example 2. Consider the state transition table shown in Figure 11-3. Here a and b are the possible inputs, and there are three states, s_0, s_1, and s_2. The table shows us that

$$f_a(s_0) = s_0, \; f_a(s_1) = s_2, \; f_a(s_2) = s_1$$

and

$f_b(s_0)=s_1$, $f_b(s_1)=s_0$, $f_b(s_2)=s_2$.

If M is a finite-state machine with states S, inputs I, and state transition functions $\{f_x \mid x \in I\}$, we can determine a relation R_M on S in a natural way. [5] If $s_i, s_j \in S$, we say that $s_i R_M s_j$ if there is an input x so that $f_x(s_i)=s_j$.

Thus $s_i R_M s_j$ means that if the machine is in state s_i, there is some input $x \in I$ that, if received next, will put the machine in state s_j. The relation R_M permits us to describe

	a	b
s_0	s_0	s_1
s_1	s_2	s_0
s_2	s_1	s_2

Figure 11-3 The state transition table of a finite-state machine

the machine M as a labeled digraph of the relation R_M on S, where each edge is labeled by the set of all inputs that cause the machine to change states as indicated by that edge. [6]

Example 3. Consider the machine of Example 2. Figure 11-4 shows the digraph of the relation R_M, with each edge labeled appropriately. Notice that the entire structure of M can be recovered from this digraph, since edges and their labels indicate where each input sends each state.

Figure 11-4 The digraph of the relation R_M

Example 4. Consider the machine M whose state table is shown in Figure 11-5(a). The digraph of R_M is then shown in Figure 11-5(b), with edges labeled appropriately.

Note that an edge may be labeled by more than one input, since several inputs may cause the same change of state. The reader will observe that every input must be part of the label of exactly one edge out of each state. This is a general property that holds for the labeled digraphs of all finite-state machines. For *brevity*, we will refer to the labeled digraph of a machine M simply as the digraph of M.

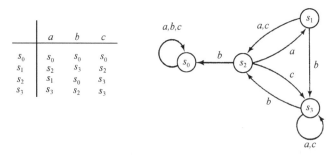

(a) The state table of the machine M (b) The digraph of R_M

Figure 11-5 A finite-state machine M

It is possible to add a variety of extra features to a finite-state machine in order to increase the utility of the concept. A simple, yet very useful extension results in what is often called a Moore machine, or recognition machine, which is defined as a sequence $(S, I, \mathscr{F}, s_0, T)$, where $(S, I, \mathscr{F})$ constitutes a finite-state machine, $s_0 \in S$ and $T \subseteq S$. The state s_0 is called the starting state of M, and it will be used to represent the condition of the machine before it receives any input. The set T is called the set of acceptance states of M.

When the digraph of a Moore machine is drawn, the acceptance states are indicated with two *concentric* circles, instead of one. No special notation will be used on these digraphs for the starting state, but unless otherwise specified, this state will be named s_0.

Example 5. Let M be the Moore machine $(S, I, \mathscr{F}, s_0, T)$, where $(S, I, \mathscr{F})$ is the finite-state machine of Figure 11-5 and $T = (s_1, s_3)$. Figure 11-6 shows the digraph of M.

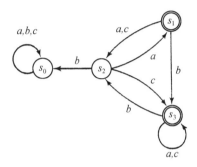

Figure 11-6　The digraph of the finite-state machine M

Words and Expressions

keep track of　记录；保持与……的联系
instant ['instənt] *n.* （某一）时刻，瞬息　*adj.* 立即的
subsequent ['sʌbsikwənt] *adj.* 随后的，继……之后的
finite-state machine　有限状态机
triple ['tripl] *n.* 三元组；三倍数；三个一组
transition [træn'siʃən] *n.* 转换；过渡
recover [ri'kʌvə] *vt.* 重新获得；恢复
　　　　　　　　　n. DOS 命令：从含有损坏磁盘扇区的磁盘上恢复文件
visualize ['vizjuəlaiz] *vt.* 使可见，可视化
flip-flop ['flipflɔp] *n.* 触发器
brevity ['breviti] *n.* （讲话，文章等的）简短，简洁
concentric [kɔn'sentrik] *adj.* 同一中心的，同轴的

Notes

1. 本句中的 to be occupied 是被动语态的不定式，与前面的 the next state 构成形如名词＋不定式的复合宾语；其后的 that 是关系代词，引导一个限制性定语从句，修饰前面的 output。全句可译为：当还有输入到达时，该机器的给定状态（连同该输入）确定将要处于的下一个状态，以及可能产生的某个输出。
2. 本句是并列复合句，前面的 each state fixed，其结构形如名词＋分词，作 leaves 的复合宾语。全句可译为：输入 0 保持每个状态不变，而输入 1 使状态反转。
3. 本句中的 one higher than the other 是名词＋形容词形式的复合结构，作状语。全句可译为：输出信号在任一给定时间都将由两个电压组成，其中一个比另一个高。
4. 本句中的 corresponding 是现在分词，引导一个分词短语作定语，修饰前面的 values，其后的 that 是形容词，修饰后面的 input。本句中的 shown 是过去分词，引导一个分词短语作定语，修饰前面的 state。全句可译为：每个输入下面的列给出了左边各个状态对应于该输入时的函数值。
5. 鉴于一个有限状态机有多个状态、多个输入与多个状态转换函数，因此本句中的 states S、inputs I 与 state transition functions $\{f_x | x \in I\}$ 分别译为：状态集 S、输入集 I 与状态转换函数集 $\{f_x | x \in I\}$。
6. 本句中的 where 是关系副词，引导一个非限制性定语从句，修饰前面的 digraph，可译为"其中"。后面的 that 是关系代词，引导一个限制性定语从句，修饰前面的 inputs。最后的连词 as 引导一个分词短语，作状语。全句可译为：关系 R_M 允许我们把（有限状态）机器 M 描述为 S 上关系 R_M 的加标记的有向图，其中每个边用那样的输入集来标记，即这些输入将引起机器像那条边所指示的那样改变状态。

注：读者请注意本节所讨论的有限状态机与计算机编译原理中的有限状态自动机（Finite State Automata）的区别，应该说，本节所述的 Moore machine 与有限状态自动机更接近。

Term

Regular expression

In computing, a regular expression is a string that is used to describe or match a set of strings, according to certain syntax rules.

Regular expressions are used by many text editors and utilities to search and manipulate bodies of text based on certain patterns. Many programming languages support regular expressions for string manipulation. For example, Perl and Tcl have a powerful regular expression engine built directly into their syntax. The set of utilities (including the editor ed and the filter grep) provided by Unix distributions were the first to popularize the concept of regular expressions. "Regular expression" is often shortened in speech to regex, and in writing to regexp or regex (singular) or regexps, regexes, or regexen (plural).

Exercises

1. Translate the text of Lesson 11.3 into Chinese.
2. Topics for oral workshop.
 • How to understand the concept of predicate? Explain it by using examples.
 • Explain the relationship between the grammar and the language. How to define a grammar and its corresponding language?
 • Do you know how many subjects about fundamentals of the computing science? Enumerate them and give a simple description. Which of them you are familiar with?
3. Translate the following into English.

(a) 如果一个集合的所有成员都是集合，成员的所有成员也都是集合，等等，那么这个集合是纯的。例如，仅包含空集的集合是一个非空纯集合。

(b) 语言是一个可描述的有穷串集合，这些串是从固定的字母表获得(draw，抽取)的。文法是"描述"语言的一种方法。文法由有穷的一列规则组成，其中每个规则把一个子串替代为另一个，左边的串必须包含至少一个非终结符号(nonterminal)。第一个串"产生"或"生成"第二个，因此规则也称为产生式(production)。

(c) 有穷状态机(FSM)或有穷状态自动机(FA)是不时用于设计数字逻辑或计算机程序的数学抽象。它是由有穷多个状态、这些状态间的转换(transition)，以及一些动作组成的行为模型，类似于流图，在这种图中能够检查(inspect)当满足某些条件时逻辑运行的路线。

(d) 有时用列出集合的所有元素来描述一个集合是不方便或不可能的。另一种定义集合的有用方式是指明集合的元素共同具有的性质。记号 $P(x)$ 用于表示关于变量对象 x 的一个句子或陈述 P。由 $P(x)$ 定义的集合写成 $\{x \mid P(x)\}$，就是 P 为真的所有对象的汇集。

4. Listen to the video "Broadband Internet" and write down the 4-7 paragraphs.

Unit 12　Computer Applications Ⅰ

12.1　Computer Graphics

Computer graphics is a branch of computer science that[1] deals with the theory and techniques of computer image synthesis. Computers produce images by analyzing a collection of dots, or pixels (picture elements). Computer graphics is used to enhance the transfer and understanding of information in science, engineering, medicine, education, and business by facilitating the generation, production, and display of *synthetic* images of natural objects with *realism* almost indistinguishable from photographs[2]. Computer graphics facilitates the production of images that range in complexity from simple line drawings to three-dimensional reconstructions of data obtained from computerized axial tomography(CAT) scans in medical applications. User interaction can be increased through animation, which conveys large amounts of information by seemingly *bringing to life* multiple related images. [3] Animation is widely used in entertainment, education, industry, flight simulators, scientific research, and *heads-up displays* (devices which allow users to interact with a virtual world). Virtual-reality applications permit users to interact with a three-dimensional world, for example, by "*grabbing*" objects and manipulating objects in the world. Digital image processing[4] is a companion field to computer graphics. However, image processing, unlike computer graphics, generally begins with some image in image space, and performs operations on the components (pixels) to produce new images.

Computers are equipped with special hardware to display images. Several types of image presentation or output devices convert digitally represented images into visually perceptible pictures. They include *pen-and-ink plotters*, dot-matrix plotters, plotters, *electrostatic or laser-printer plotters*, *storage tubes*, liquid-crystal displays (LCDs), *active matrix panels*[5], *plasma panels*, and cathode-ray-tube (CRT) displays. Images can be displayed by a computer on a cathode-ray tube in two different ways: *raster scan* and random (vector) scan.

Interaction with the object takes place via devices attached to the computer, starting[6] with the keyboard and the mouse. Each type of device can be programmed to deliver various types of functionality. The quality and ease of use of the user interface often determines whether users enjoy a system and whether the system is successful. Interactive graphics aids the user in the creation and modification of graphical objects and the response to these objects in real-time. The most commonly used input device is the mouse. Other kinds of interaction devices include the joystick, *trackball*, light pen, and *data tablet*. Some of these two-dimensional (2D) devices can be modified to extend to three dimensions (3D). The *data glove*[7] is a device capable of recording hand movements. The data glove is capable of a simple gesture recognition and general tracking of hand orientation.

In the production of a computer-generated image, the designer has to specify the objects in the image and their shapes, positions, orientations, and surface colors or *textures*. Further, the viewer's position and direction of view (camera orientation) must be specified. The software should calculate the parts of all objects that[8] can be seen by the viewer (camera). Only the visible portions of the objects should be displayed (captured on the film). (This requirement is referred to as the *hidden-surface problem*.) The *rendering software* is then applied to compute the amount and color of light reaching[9] the viewer eye (film) at any point in the image, and then to display that point. Some modern graphics work stations have special hardware to implement[10] projections, hidden-surface elimination, and direct *illumination*. Everything else in image generation is done in software.

Solid modeling is a technique used to represent three-dimensional shapes in a computer. The importance of solid modeling in computer-aided design and manufacturing (CAD/CAM) systems has been increasing. Engineering applications ranging from drafting to the numerical control of machine tools increasingly rely on solid modeling techniques.[11] Solid modeling uses three-dimensional solid *primitives* (the cube, sphere, *cone*, cylinder, and *ellipsoid*) to represent three-dimensional objects. Complex objects can be constructed by combining the primitives.

The creation of images by simulating a model of light propagation is often called image synthesis. The goal of image synthesis is often stated as *photorealism*, that is, the criterion that the image looks as good as a photograph.[12] Rendering is a term used for methods or techniques that are used to display *realistic*-looking three-dimensional images on a two-dimensional medium such as the cathode-ray-tube screen.[13] The display of a *wire-frame* image is one way of rendering the object. The most common method of rendering is shading (see Figure 12-1). Generally, rendering includes addition of texture, shadows, and the color of light that reaches the observer's eye from any point in the image.

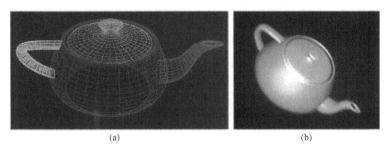

Figure 12-1 Image renderings of a teapot. (a) Wire-frame model with 512 *polygons*. (b) Smooth shading (non-shiny)

Computer-generated images are used extensively in the entertainment world and other areas. Realistic images have become essential tools in research and education. Conveying[14] realism in these images may depend on the convincing generation of natural phenomena. A fundamental difficulty is the complexity of the real world. Existing models are based on physical or biological concepts. The behavior of objects can be determined by physical properties or chemical and *microphysical* properties.

Words and Expressions

synthetic [sin'θetik] *adj.* 合成的；综合(性)的；人造的　　*n.* 化学合成物
realism ['riəlizəm] *n.* 写实主义；现实主义；唯实论；实在论
tomography [tə'mɔgrəfi] *n.* 层面 X 线照相术；(医学)(X 线)断层照相术，体层照相术
bring to life　　使栩栩如生；使苏醒；使生气勃勃；让人想起，使回忆起
heads-up display　　平视显示器
grab [græb] *vt.* 攫取，抓取；(急速)抓住　　*vi.* 抓住，攫取　　*n.* 攫取，(急速)抓住
pen-and-ink plotter　　钢笔画的绘图仪
electrostatic or laser-printer plotter　　静电或激光打印机绘图仪
storage tube　　储存管，存储管；存储显像管
active matrix panel　　有源矩阵面板
plasma panel　　等离子体面板，等离子体显示板
raster scan　　光栅扫描
trackball ['træk,bɔ:l] *n.* 跟踪球，轨迹球；轨迹球鼠标
data tablet　　数据板，数据输入板
data glove　　数据手套
texture ['tekstʃə] *n.* 纹理；质地；结构；本质　　*vt.* 使具有某种结构(或特征)
hidden-surface problem　　隐面问题
rendering software　　渲染软件
illumination [i,lju:mi'neiʃən] *n.* 照明，光亮，照(明)度；阐明，解释
solid modeling　　立体建模(技术)，实体建模；实体造型
primitive ['primitiv] *n.* 原语；根词；本原，原始　　*adj.* 原始的；基本的
cone [kəun] *n.* 锥面，锥体，锥形物　　*vt.* 使成锥形
ellipsoid [i'lipsɔid] *n.* 椭圆球；椭面
photorealism [,fəutəu'riəlizəm] *n.* 照相写实；照相写实主义；照相现实主义
realistic [,riə'listik] *adj.* 逼真的；现实的；现实主义的
wire-frame　　线框；立体线稿
polygon ['pɔligən] *n.* 多边形，多角形
microphysical [,maikrəu'fizikəl] *adj.* 微观物理学的

Abbreviations

CAT (Computerized Axial Tomography)　　计算机轴向断层扫描，
　　　　　　　　　　　　　　　　　　　　计算机化 X 射线轴向分层造影
VR (Virtual-Reality)　　虚拟现实

Notes

1. 关系代词 that 引导的限制性定语从句修饰前面的 branch；graphics 可解释为"图形"(graphic 的复数)或"图形学"，此处的含义是"图形学"。

2. with realism almost indistinguishable from photographs 可译为"(合成图像)逼真得几乎与照片无法区分"。请注意本句中 generation 与 production 的区别，前者意为"生成"，而后者意为"制作"。也请注意本句中几处 of 的不同用法。
3. 本句中，关系代词 which 引导的非限制性定语从句修饰其前的 animation。bring to life 含意为"使栩栩如生"。全句可译为：动画可以增进用户交互，通过把多幅相关的图像栩栩如生地展现出来，动画能传递大量的信息。
4. processing 是动名词，digital image processing 可译为"数字图像处理"。
5. active matrix panel，有源矩阵面板。有源矩阵(Active Matrix)是使用叫作 TFT(Thin Film Transistor，薄膜晶体管)的存储器元件来创建各个有源像素的一种 LCD 技术。有源矩阵允许全运动视频和动画，不像无源矩阵。有源矩阵技术有两类：AmorphousTFT(无定型 TFT)和 PolysiliconTFT(多硅 TFT)。
6. starting 引导的现在分词短语作状语，表示"从……开始"。其前的过去分词 attached 引导的分词短语作定语，修饰 devices。
7. data glove，数据手套。数据手套是一种多模式的虚拟现实硬件，通过软件编程，可进行虚拟场景中物体的抓取、移动、旋转等动作，也可以利用它的多模式性，用作控制场景漫游的工具。数据手套的出现，为虚拟现实系统提供了一种全新的交互手段，目前的产品已经能够检测手指的弯曲，并利用磁定位传感器精确地定位手在三维空间中的位置。这种结合手指弯曲度测试和空间定位测试的数据手套称为"真实手套"，可以为用户提供一种非常真实自然的三维交互手段。
8. 关系代词 that 引导的限制性定语从句修饰其前的 parts。
9. 现在分词 reaching 引导的分词短语修饰其前的 light。
10. to implement 引导的不定式短语作定语，修饰其前的 hardware。
11. 本句中，ranging 引导的分词短语修饰其前的 applications。全句可译为：从画草图到机床数值控制的各种工程应用日益依赖于各种立体建模技术。
12. 本句中，that is 是插入语，后面的 criterion 是 photorealism 的同位语。连接词 that 引导的名词性从句作 criterion 的同位语。全句可译为：图像合成的目标经常被说成照相写实，即图像看起来与照片一样好的标准。
13. 本句中，rendering 是动名词，作主语，意为"渲染(法、技术)"。下面第二句中的 shading 也是动名词，意为"画阴影(法)"。本句中，关系代词 that 引导的限制性定语从句修饰其前的 methods or techniques。
14. conveying 是动名词，引导的短语作主语。全句可译为：这些图像是否逼真，可能取决于能否生成令人信服的自然现象。

12.2　Computer-Aided Design

Computer-aided design (CAD) can be most simply described as "using a computer in the design process". In the design process, a computer can be used in both the representation and the analysis steps. The application of CAD for representation is not limited to *drafting*. Three-dimensional *wire frame* modeling, boundary representation (B-rep) modeling, and *solid modeling* are representation methods available to CAD users. To aid in engineering analysis, there are packages that perform *kinematic* simulation, circuit analysis and simulation, and *finite-element modeling* for different engineering analyses. The application domain includes mechanical engineering applications, architecture and construction engineering applications, electronic circuits, printed-circuit board layout, IC layout, etc. CAD systems frequently consists of a collection of many application modules under a common (not always) database and graphics editor.

CAD systems can be classified in several ways:

BY THE SYSTEM HARDWARE:

Mainframe

Minicomputer

Engineering workstation

Microcomputer

BY THE APPLICATION AREA:

Mechanical engineering

Circuit design and board layout

Architectural design and construction engineering

Cartography

BY THE MODELING METHOD:

Two-dimensional drafting

Three-dimensional drawing

Sculptured surface

Three-dimensional solid modeling

A CAD system consists of three major parts:

1. hardware: computer and input/output (I/O) devices;
2. operating system software;
3. application software: CAD package.

Hardware is used to support the software functions. A wide range of hardware is used in CAD systems. This hardware is discussed later. The operating system software is the interface between the CAD application software and the hardware. It supervises the operation of the hardware and provides basic functions such as creating and removing operating tasks, controlling the progress of tasks, allocating hardware resources among tasks, and providing access to software resources such as files, *editors*, compilers, and *utility programs*.[1] It is important not only for CAD software, but for non-CAD software.

The application software is the CAD package. It is the heart of a CAD system. It consists of programs that do 2-D and 3-D modeling, drafting, and engineering analysis.[2] The functionality of a CAD system is built into the application software. It is what makes one CAD package different from another. Application software is usually operating-system dependent. To transport a CAD system running in one operating system to another operating system is not as trivial as recompiling the software.[3] Therefore, attention must be given to the operating system as well. Details of application software are also discussed later.

A general architecture of a CAD system is shown in Figure 12-2. Application software is on the top level and is used to manipulate the CAD model database. The graphics utility system performs the coordinate transformation, windowing, and display control. Since there may be several different I/O devices used, device drivers are used to translate the data into and out from the specific data format used by each device, and to control the devices.[4] The operating system is run in *background* to coordinate the entire operation. Finally a user interface links the human and the system.

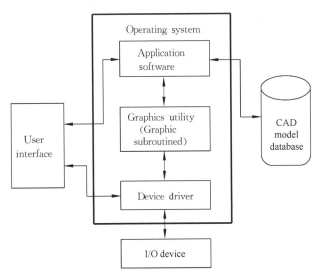

Figure 12-2 CAD system architecture

CAD software is what gives a CAD system its functionality and personality. Software can be classified based on the technology used:
1. 2-D drawing;
2. Basic 3-D drawing;
3. Sculptured surfaces;
4. 3-D solid modeling;
5. Engineering analysis.

Some of the commonly available functions provided by CAD software are:
1. Picture manipulation: add, delete, and modify *geometry* and text.
2. Display transformation: scaling, rotation, *pan*, *zoom*, and partial erasing.
3. Drafting symbols: standard drafting symbols.
4. Printing control: output device selection, configuration, and control.
5. Operator aid: screen menus, *tablet overlay*, *function keys*.
6. File management: create, delete, and merge picture files.

Words and Expressions

drafting ['drɑːftiŋ] n. 制图；起草
wire frame　线框；引线连接框；线条圈
solid modeling　立体建模（术），实体建模，实体造型（术）
kinematic [ˌkaini'mætik] adj. 运动学的，运动学上的
finite-element modeling　有限元建模（技术）
cartography [kɑː'tɔgrəfi] n. 制图法，制图学；绘制图表
sculpture ['skʌlptʃə] vt. 雕刻；塑造；刻蚀
editor ['editə] n. 编辑程序，编辑器
utility program　实用程序
background ['bækgraund] n. 后台；背景
geometry [dʒi'ɔmitri] n. 几何形状；几何学
pan [pæn] n. 摄全景，摇镜头　vt. 摇动（镜头），使拍摄全景　vi. 摇镜头，拍摄全景
zoom [zuːm] n. 图像电子放大，缩放　vt. 使摄像机移动
tablet overlay　（图形）输入板叠加
function key　功能键

Notes

1. creating 与 removing 是动名词，其他的如 controlling 等也是，它们都作为句中第一个 such as 引导的同位语的组成部分。全句可译为：它监督硬件的运转并提供诸如创建和取消运行任务、控制任务的进展、为任务分配硬件资源，以及对文件、编辑程序、编译程序和实用程序之类软件资源的访问等基本功能。
2. that 引导的限制性定语从句修饰 programs。全句可译为：它由进行二维建模和三维建模、制图和工程分析等的一些程序组成。
3. running 引导一个分词短语作定语，修饰 a CAD system。全句可译为：把在一个操作系统上运行的 CAD 系统搬到另一个不同的操作系统上，并不像重编译该软件那样简单。
4. 本句第一个 used 是过去分词，作定语，修饰前面的 device。全句可译为：因为可能会使用几个不同的 I/O 设备，所以，使用设备驱动程序对数据按每个设备所使用的特定数据格式进行相互转换，并控制这些设备。

12.3　Graphical User Interface

A GUI (usually pronounced GOO-ee) is a graphical (rather than purely textual) user interface to a computer. As you read this, you are looking at the GUI or graphical user interface of your particular Web browser. The term *came into existence* because the first interactive user interfaces to computers were not graphical; they were text-and-keyboard oriented and usually consisted of commands you had to remember and computer responses that were *infamously* brief.[1] The command interface of the DOS operating system (which you can still *get to* from your Windows operating system) is an example of the typical user-computer interface before GUIs arrived. An intermediate step in user interfaces between the command line interface and the GUI was the non-graphical menu-based interface, which let you interact by using a mouse rather than by having to type in keyboard commands. [2]

Today's major operating systems provide a graphical user interface. Applications typically use the elements of the GUI that come with the operating system and add their own graphical user interface elements and ideas. A GUI sometimes uses one or more *metaphors* for objects familiar[3] in real life, such as the desktop, the view through a window, or the physical layout in a building. Elements of a GUI include such things as: windows, pull-down menus, buttons, scroll bars, *iconic* images, *wizards*, the mouse, and no doubt many things that haven't been invented yet. With the increasing use of multimedia as part of the GUI, sound, voice, motion video, and virtual reality interfaces seem likely to become part of the GUI for many applications. A system's graphical user interface along with its input devices is sometimes referred to as its "*look-and-feel*".

The GUI familiar to most of us today in either the *MAC* or the Windows operating systems and their applications originated at the Xerox Palo Alto Research Laboratory in the late 1970s. [4] Apple used it in their first Macintosh computers. Later, Microsoft used many of the same ideas in their first version of the Windows operating system for IBM-compatible PCs. When creating an application, many object-oriented tools exist that facilitate writing[5] a graphical user interface. Each GUI element is defined as a class *widget* from which you can create object instances for your application. You can code or modify prepackaged methods that an object will use to respond to user *stimuli*.

Graphical user interfaces, such as Microsoft Windows and the one used by the Apple Macintosh, feature the following basic components:
- pointer: A symbol that appears on the display screen and that you move to select objects and commands. Usually, the pointer appears as a small angled arrow. Text-processing applications, however, use an *I-beam* pointer that is shaped like a capital I.
- pointing device: A device, such as a mouse or trackball, that enables you to select objects on the display screen.
- icons: Small pictures that represent commands, files, or windows. By moving the pointer to the icon and pressing a mouse button, you can execute a command or convert the icon into a window. You can also move the icons around the display screen as if they were real objects on your desk.
- desktop: The area on the display screen where[6] icons are grouped is often referred to as the desktop because the icons are intended to represent real objects on a real desktop.
- windows: You can divide the screen into different areas. In each window, you can run a different program or display a different file. You can move windows around the display screen, and change their shape and size at will.

• menus: Most graphical user interfaces let you execute commands by selecting a choice from a menu. In addition to their visual components, graphical user interfaces also make it easier to move data from one application to another. A true GUI includes standard formats for representing text and graphics. Because the formats are *well-defined*, different programs that run under a common GUI can share data. This makes it possible, for example, to copy a graph created by a spreadsheet program into a document created by a word processor.[7]

Many DOS programs include some features of GUIs, such as menus, but are not graphics based.[8] Such interfaces are sometimes called graphical character-based user interfaces to distinguish them from true GUIs.

Words and Expressions

come into existence　开始存在；产生；成立
infamously ['infəməsli] *adv.* 低劣地；差劲地；不名誉地，声名狼藉地
get to　接触到；到达
metaphor ['metəfə] *n.* 隐喻，暗喻
iconic [ai'kɔnik] *adj.* 图标的；图符的；偶像的
wizard ['wizəd] *n.* （程序）安装向导
look-and-feel　外观和感觉
MAC　Macintosh，苹果计算机
widget ['widʒit] *n.* 控件，（窗口）小部件；桌面小程序
stimuli ['stimjulai] *n.* 刺激；刺激物；促进因素（stimulus 的复数形式）
I-beam ['aibi:m] *n.* I 型标；工形梁，工字钢
well-defined　定义良好的，定义明确的；界限清楚的

Notes

1. commands 后跟的是关系代词 that 引导的限制性定语从句，修饰此 commands。此 that 在从句中作宾语，因此被省略。后面的关系代词 that 引导的限制性定语从句修饰 responses。整句可译为：它们都是面向文本和键盘的，并且通常由你不得不记住的一些命令和简短得让人看不懂的计算机响应组成。
2. 本句中，which 引导一个非限制性定语从句，修饰其前面的 interface。注意本句中的两个 in，前一个表示"范围，领域，方面"，而第二个表示"形式，方式"。全句可译为：在命令行界面和 GUI 之间，过渡的用户界面是非图形的、基于菜单的界面，让你可以使用鼠标交互，而不是必须敲键盘命令交互。
3. 形容词 familiar 引导的短语修饰其前的 objects。
4. MAC(Macintosh，苹果计算机)区别于装配有微软 Windows 系统的计算机(PC)，需要说明的是，这里的 PC 并非 Personal Computer(个人计算机)的简称，而是来源于最初由 IBM 销售的 PC-DOS 系统(Windows 的前身)，后来所有装配 Windows 系统的计算机被称为 PC，而苹果计算机因装配自家的 Mac OS 系统，而称为 MAC。本句的结构是

The GUI ... originated at ... 。全句可译为：今天我们大多数人在 MAC 或 Windows 操作系统及其应用程序中熟悉的 GUI，源自 20 世纪 70 年代末期的施乐帕洛阿尔托 (Xerox Palo Alto)研究实验室。

5. 动名词 writing 引导的短语作 facilitate 的宾语。前面的关系代词 that 引导的限制性定语从句修饰前面的 tools。
6. 关系副词 where 引导的限制性定语从句修饰前面的 area。
7. 本句中，it 是先行代词，代替不定式 to copy 引导的短语。注意，两处出现的 created 引导的短语都修饰其前面的名词。
8. 此处 graphics based 看作名词＋过去分词构成的合成词，意为"基于图形的"。

Term

Application Programming Interface(API)

An Application Programming Interface (API) is a software program that facilitates interaction with other software programs. An API allows a programmer to interact with an application using a collection of callable functions. The goal of an API is to allow programmers to write programs that will not cease to function if the underlying system is upgraded.

An API can be general or specific. The full set of a general API is bundled in the libraries of a programming language. With a specific API a specific term is meant to deal with a specific problem.

An API is language dependent or independent：
- Language Dependent：This means it is only available by using the syntax and elements of a particular language, making it more convenient to use.
- Language Independent：This means the API is written to be called from several programming languages.

12.4 The Virtual Reality Responsive Workbench

Virtual Reality (VR) is a complex and challenging field, and several distinct types of systems have been developed for displaying and interacting with virtual environments. One of the newest is the Virtual Reality *Responsive Workbench* which is an interactive VR environment designed to support a team of end users such as military and civilian command and control specialists, designers, engineers, and doctors.

Perhaps the greatest strength of the VR Responsive Workbench is the ease of natural interaction with virtual objects. Current interactive methods emphasize *gesture* recognition, speech recognition, and a simulated "laser" pointer to identify and manipulate objects.

There is no accepted definition for VR. One important reference, the U.S. National Research Council report *Virtual Reality：Scientific and Technical Challenges* (Durlach and Mavor 1995), does not attempt a definition. Rather, characteristics of a virtual environment are given. These include a man-machine interface between human and computer, 3-D objects,

objects having a spatial presence independent of the user's position, and the user manipulating objects using a variety of motor channels.[1] Virtual reality can be subdivided in many different ways; here we will categorize based upon the visual channel.

This paper classifies VR systems into three categories: immersive head-mounted displays (HMDs), immersive non-HMD systems, and partially immersive tabletop systems.

Head-Mounted Displays/BOOMs

HMDs, which typically also include earphones for the auditory channel as well as devices for measuring the position and orientation of the user, have been the primary VR visual device for *much of* the 1990s. Using CRT or LCD technology,[2] HMDs provide two imaging screens, one for each eye. Thus, given sufficient computer power,[3] *stereographic* images are generated. Typically, the user is completely immersed in the scene, although HMDs for augmented reality overlay the computer-generated image onto the view of the real world. Low-end HMDs can be obtained for less than $10,000. These suffer from information loss (resolutions of approximately 400×300 pixels; typical field of views are between 40° to 75°). Extremely low-end "glasses" cost only hundreds of dollars, but these systems are not yet usable for serious applications and find their role in system testing and in research. High-end HMDs overcome these limitations at very high costs and thus are utilized for only a limited number of applications such as military flight training. In addition, *ergonomic* limitations such as weight, fit, and isolation from the real environment make it unlikely that users will accept HMD-based immersion for more than short time periods until advances in material science produce *lightweight*, eyeglass-size HMDs.[4] They are, however, more portable than are other VR systems.

An alternative to HMDs is the BOOM (Binocular Omni-Orientation Monitor). Two high-resolution CRTs are mounted inside a package against which the user places his eyes.[5] By *counterbalancing* the CRT packaging on a *free-standing* platform, the display unit allows the user six-degree-of-freedom movement while placing no weight on the user's head. The original version of the BOOM had the user navigating through the virtual world by grasping and moving two handles and turning the head display *much as* one would manipulate a pair of binoculars.[6] Buttons on the *handgrip* are available for user input. A more recent desktop version (the Fakespace push BOOM) allows the user to navigate by pushing his/her head against a *spring-loaded* system.

HMDs and BOOMs are similar devices in that the user is fully immersed in the virtual environment and does not see his/her actual surroundings. The BOOM solves several of the limitations of the HMD (e.g., resolution, weight, field of view), but at the expense of reducing the sense of immersion by requiring the user to stand or sit in a fixed position. This loses the freedom of movement associated with HMDs where users typically take steps and turn their body to determine direction (the BOOM also restricts the user's hands).

Immersive Rooms

Immersion does not necessarily require the use of the head-mounted displays that are the most common method for presenting the visual channel in a virtual environment. The CAVE™(Cave Automatic Virtual Environment), a type of *immersive room* facility developed at the University of Illinois, Chicago, accomplishes immersion by projecting on two or three walls and a floor and allowing the user to interactively explore a virtual environment. An immersive room is typically about $10' \times 10' \times 13'$ (height), allowing a half-dozen or more users to examine the virtual world being generated within the space.[7] Computer-generated stereographic images are produced by calculating right and left eye images and using *stereographic shuttered glasses* to synchronize these alternating images. To determine the view, the group leader's head position is tracked using magnetic sensors to determine position and orientation. Both by walking within the immersive room and by utilizing an interactive device called[8] a *wand*, which has a second *tracker* for position identification and buttons for issuing commands, the group leader navigates through the data. All users see the same image; thus, other team members view the scene from an incorrect perspective with the resulting *distortion* depending upon differences in location within the immersive room.[9] Since the stereographic shuttered glasses are *see-through*, all users see each other. This facilitates group discussion and data analysis.

While HMDs require that users interact in virtual spaces (they cannot see each other in their "real" environment), the immersive room offers the significant advantage of permitting user interaction, discussion, and analysis in the real world. However, the computational cost of generating scenes within an immersive room are very high. Two images must be generated at high refresh rates for each wall in the immersive room. In addition, each wall requires a high-quality projector, and since *back projection* is used, a large allocation of space is required for projection length. Costing over one-half million dollars[10], immersive rooms exist only in *a handful of* large research organizations and corporations.

The VR Responsive Workbench

The two paradigms discussed above are both fully immersive. However, there are many applications for which full immersion is not desirable. A doctor performing *presurgical* planning has no reason to wish to be fully immersed in a virtual room and with virtual equipment.[11] Rather, he would like a virtual patient lying on an operating table in a real room.[12] He would like to *reach* out and interactively examine the virtual patient and, perhaps, practice the operation. Similar remarks apply to engineering design, military and civilian command and control, architectural layout, and *a host of* other applications that would typically be performed on a desktop, table, or workbench. These applications are categorized by not requiring navigation through complex virtual environments but rather by demanding a *fine-granularity* visualization and interaction with virtual objects and scenes.

Thus, the Workbench supports VR for a large class of applications that are substantially different from the fully immersed, navigation-oriented applications supported by HMDs and immersive rooms.[13]

The VR Responsive Workbench operates by projecting a computer-generated, *stereoscopic* image off a mirror and then onto a table (i. e., workbench) surface that is viewed by a group of users around the table (see Figure 12-3). Using stereoscopic shuttered glasses (just as is done in the immersive room), users observe a 3-D image displayed above the tabletop. By tracking the group leader's head and hand movements using magnetic sensors, the Workbench permits changing the view angle and interacting with the 3-D scene.[14] Other group members observe the scene as manipulated by the group leader, facilitating easy communication between observers about the scene and defining future actions by the group leader.[15] Interaction is performed using speech recognition, a *pinch glove* for gesture recognition, and a simulated laser pointer. Figure 12-3 shows a *schematic* of the Workbench.

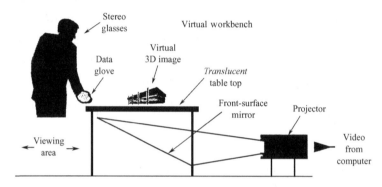

Figure 12-3　A schematic diagram of the VR responsive workbench

Words and Expressions

responsive [ris'pɔnsiv] *adj.* 响应的，易起反应的，敏感的；应答的
workbench ['wə:kbentʃ] *n.* 工作台，手工台
gesture ['dʒestʃə] *n.* 手势，姿势；（外交等方面的）姿态，表示　*vt.* 用姿势（或动作）表示
　　　　　　　　　　　　　　　　　vi. 用姿势（或动作）示意
much of　差不多，几乎
stereographic [stiəriəu'græfik] *adj.* 立体照相的；立体画法的，立体平画（法）的
ergonomic [ˌə:gə'nɔmik] *adj.* 人类工程学的；人体工程学的；工效学的
lightweight ['laitweit] *adj.* 轻量的，无足轻重的
binocular [bi'nɔkjulə, bai'n-] *n.* [复]（双筒）望远镜；双目镜　*adj.* 双筒的；双目的
omni-orientation ['ɔmniˌɔ:riən'teiʃn] *n.* 全方位
counterbalance [ˌkauntə'bæləns] *vt.* 使平衡；抵销　*n.* 平衡；平衡力；砝码
free-standing　自立的；独立式的
much as　非常像，和……几乎一样；（在句首时）虽然很……
handgrip ['hændgrip] *n.* 柄，把；紧握

spring-loaded 弹簧承力的，弹簧承载的；装有弹簧的
immersive room 沉浸室
shutter ['ʃʌtə] vt. 为……装快门；关上……的窗板（或快门） n. 快门，（光）闸；窗板
stereographic shuttered glasses 立体（照相）快门式眼镜
wand [wɔnd] n. 指挥棒；魔杖
tracker ['trækə] n. 跟踪仪，追踪系统，跟踪系统
distortion [dis'tɔ:ʃən] n. （信号等的）失真；（透镜成像产生的）畸变；变形；歪曲；曲解
see-through ['si:'θru:] adj. 透明的
back projection 背投，背景放映（利用银幕作背景进行合成）
a handful of 很少一点，少量，一把，一小撮
presurgical [pri:'sə:dʒikəl] adj. 预外科的；（外科）手术前的
reach [ri:tʃ] vi. 伸出手（或脚）；延伸 vt. 达到；伸出 n. 伸（出）；能及的范围
a host of 许多，一大群
fine-granularity 微粒状，细粒度
stereoscopic [ˌstiəriəs'kɔpik] adj. 立体的，体视（镜）的
pinch glove 连指手套
schematic [ski'mætik] n. 图解，略图；简表 adj. 图解的；纲要的
translucent [trænz'lju:snt] adj. 半透明的，半透彻的；透明的

Abbreviations

VR（Virtual Reality） 虚拟现实
HMD（Head-Mounted Display） 头戴式显示器
BOOM（Binocular Omni-Orientation Monitor） 双目全方位监视器
CAVE（Cave Automatic Virtual Environment） Cave 自动虚拟环境

Notes

1. 本句中的 having 和 manipulating 都是现在分词，分别引导分词短语作定语，修饰前面的名词；using 也是现在分词，但引导一个分词短语作状语，修饰前面的 manipulating。independent of 为习语介词短语，引导的短语作定语，修饰前面的 presence。
2. 本句中现在分词 using 引导的分词短语作状语，表示行为方式，其中 CRT 意为"阴极射线管"，LCD 意为"液晶显示"。
3. 本句中的 given 是过去分词，引导一个分词短语作状语，表示条件。
4. 本句中的 it 是先行代词，代替其后 that 引导的从句。全句可译为：另外，人类工程学的一些限制，诸如重量、合适（性）以及与现实环境的隔绝，使得在材料科学发展到能生产出重量轻的、眼镜大小的 HMD 之前，用户未必会较长时间地接受基于 HMD 的沉浸。
5. 本句中的 package 含意为"包，标准部件"。全句可译为：两个高分辨率 CRT 被安装在一个包（装盒）内，用户的眼睛就对着它。
6. 本句中的 the user navigating 构成名词＋（现在）分词形式的复合宾语。全句可译为：初期版本的 BOOM 让用户像人操纵一副双筒望远镜那样，握紧并移动两个柄，转动头

上的显示器，遨游虚拟世界。

7. 本句中的 allowing 是现在分词，引导一个分词短语作状语，表示伴随发生的情况；to examine 是不定式，引导一个不定式短语作 allowing 的宾语补足语，而 being generated 是现在分词的被动语态形式，引导一个分词短语作定语，修饰前面的 world，意为"正被生成的"。

8. 本句中的 called 是过去分词，引导一个分词短语作定语，修饰前面的 device，意为"称为……的设备"；其后的 which 是关系代词，引导的非限制性定语从句也修饰 device。

9. 全句可译为：所有用户看到的是同一个图像；因而，小组其他一些成员从不恰当的视点观察该场景所产生的失真，取决于沉浸室内位置的差异。

10. 本句中的 costing 是现在分词，引导一个分词短语作状语，表示原因。

11. 本句中的 performing 是现在分词，引导一个分词短语作定语，修饰前面的 doctor，其后的 planning 是动名词。本句中的 and 不宜译出，而 with 含意为"具有，带有"，可灵活翻译。全句可译为：施行预外科（手术）计划的医生没有理由想要完全沉浸在使用虚拟设备的虚拟手术室中。

12. 本句使用虚拟语气，表达一种想法或意愿，其中的 a virtual patient lying on … 是名词＋分词形式的复合宾语。全句可译为：相反，他希望（有）一个虚拟病人躺在现实手术室的手术台上。

13. 本句中的 supported 是过去分词，引导一个分词短语作定语，修饰前面的 applications，也可视为其前省略了"which are"。全句可译为：因此，该工作台支持一大类应用的 VR，这类应用实质上区别于由 HMD 和沉浸室支持的完全沉浸的、面向导航的应用。

14. 本句中的 using 是现在分词，引导一个分词短语作状语，表示方式，其后的 changing 与 interacting 是并列的动名词，引导的短语作宾语。全句可译为：通过使用磁传感器跟踪小组长头与手的移动，工作台允许改变视角并且与三维场景进行交互。

15. 本句中的 as 表示时间，含意为"当……的时候"，更合适的是作"随着"解。句中的 facilitating 与 defining 处于并列地位，都是现在分词，引导一个分词短语作状语，表示伴随发生的情况。全句可译为：小组其他成员随着小组长的操纵而观察场景，这样便于在观察者之间就场景进行交流，并由小组长确定进一步的一些行动。

注：本文是内容和翻译并重的选文，有参考译文。请观看视频材料中"虚拟现实视频"。

12.5 Augmented Reality

Augmented Reality (AR) is an interactive experience of a real-world environment where the objects that reside in the real-world are "augmented" by computer-generated *perceptual* information, sometimes across multiple sensory *modalities*, including visual, auditory, *haptic*, *somatosensory*, and *olfactory*.[1] The overlaid sensory information can be constructive (i.e. additive to the natural environment) or *destructive* (i.e. *masking* of the natural environment) and is seamlessly *interwoven* with the physical world such that it is perceived as an *immersive* aspect of the real environment. In this way, augmented reality alters one's ongoing perception of a real world environment, whereas virtual reality completely replaces the user's real world environment with a simulated one. Augmented reality is related to two *largely synonymous* terms: mixed reality and computer-mediated reality.

The primary value of augmented reality is that it brings components of the digital world into a person's perception of the real world, and does so not as a simple display of data, but through the integration of immersive sensations that are perceived as natural parts of an environment.[2] The first functional AR systems that provided immersive mixed reality experiences for users were invented in the early 1990s, starting with the Virtual *Fixtures* system developed at the U. S. Air Force's Armstrong Laboratory in 1992. The first commercial augmented reality experiences were used largely in the entertainment and gaming businesses, but now other industries are also getting interested about AR's possibilities for example in knowledge sharing, educating, managing the information flood and organizing distant meetings. Augmented reality is also transforming the world of education, where content may be accessed by scanning or viewing an image with a mobile device or by bringing immersive, markerless AR experiences to the classroom. Another example is an AR *helmet* for construction workers which display information about the construction sites.

Augmented reality (AR) is used to enhance natural environments or situations and offer perceptually enriched experiences. With the help of advanced AR technologies (e. g. adding computer vision and object recognition) the information about the surrounding real world of the user becomes interactive and digitally manipulable. Information about the environment and its objects is overlaid on the real world. This information can be virtual or real, e. g. seeing other real sensed or measured information such as electromagnetic radio waves overlaid in exact *alignment* with where they actually are in space.[3] Augmented reality also has a lot of potential in the gathering and sharing of *tacit* knowledge[4]. Augmentation techniques are typically performed in real time and in semantic context with environmental elements. Immersive perceptual information is sometimes combined with supplemental information like scores over a live video feed of a sporting event.[5] This combines the benefits of both augmented reality technology and *heads up display* (HUD) technology.

Hardware components for augmented reality are: processor, display, *sensors* and input devices. Modern mobile computing devices like smartphones and tablet computers contain these elements which often include a camera and MEMS[6] sensors such as accelerometer, GPS, and solid state compass, making them suitable AR platforms.

Various technologies are used in augmented reality rendering[7], including optical projection systems, monitors, handheld devices, and display systems worn[8] on the human body.

A *head-mounted display* (HMD) is a display device worn on the forehead. HMDs place images of both the physical world and virtual objects over the user's field of view.

AR displays can be rendered on devices resembling[9] eyeglasses. Versions include *eyewear* that employs cameras to intercept the real world view and re-display its augmented view through the *eyepieces* and devices in which the AR imagery is projected through or reflected off the surfaces of the eyewear's lens pieces.[10]

A head-up display (HUD) is a transparent display that presents data without requiring users to look away from their usual viewpoints.

Spatial augmented reality (SAR) augments real-world objects and scenes without the use of special displays such as monitors, head-mounted displays or hand-held devices. SAR makes use of digital projectors to display graphical information onto physical objects.

Techniques include speech recognition systems that translate a user's spoken words into computer instructions, and gesture recognition systems that interpret a user's body movements by visual detection or from sensors embedded in a peripheral device such as a *wand*, *stylus*, pointer, glove or other body wear.

The computer analyzes the sensed visual and other data to synthesize and position augmentations. Computers are responsible for the graphics that *go with* augmented reality. Augmented reality uses a computer-generated image and it has an striking effect on the way the real world is shown. [11] With the improvement of technology and computers, augmented reality is going to have a drastic change on our perspective of the real world. According to Time Magazine, in about 15-20 years it is predicted that Augmented reality and virtual reality are going to become the primary use for computer interactions. Computers are improving at a very fast rate, which means that we are figuring out new ways to improve other technology. The more that computers progress, augmented reality will become more flexible and more common in our society. Computers are the core of augmented reality.

A key measure of AR systems is how realistically they integrate augmentations with the real world. The software must derive real world coordinates, independent from the camera, from camera images. That process is called *image registration*[12], and uses different methods of computer vision, mostly related to video tracking. Many computer vision methods of augmented reality are inherited from visual *odometry*.

Usually those methods consist of two parts. The first stage is to detect interest points, *fiducial* markers or optical flow in the camera images. This step can use feature detection methods like corner detection, *blob* detection, edge detection or thresholding, and other image processing methods. The second stage restores a real world coordinate system from the data obtained in the first stage. Some methods assume objects with known geometry (or fiducial markers) are present in the scene. [13] In some of those cases the scene 3D structure should be precalculated beforehand. If part of the scene is unknown simultaneous localization and mapping (SLAM) can map relative positions. If no information about scene geometry is available, structure from motion methods like *bundle adjustment* are used. Mathematical methods used in the second stage include *projective (epipolar) geometry*, geometric algebra, rotation representation with exponential map, *Kalman* and particle filters, nonlinear optimization, *robust statistics*.

Augmented Reality Markup Language (ARML) is a data standard developed within the Open Geospatial Consortium (OGC), which consists of XML grammar to describe the location and appearance of virtual objects in the scene, as well as ECMA Script bindings to allow dynamic access to properties of virtual objects.[14]

To enable rapid development of augmented reality applications, some Software Development Kits (SDKs) have emerged. A few SDKs such as CloudRidAR leverage cloud computing for performance improvement.

The most exciting factor of augmented reality technology is the ability to utilize the introduction of 3D space. This means that a user can potentially access multiple copies of 2D interfaces within a single AR application. AR applications are collaborative, a user can also connect to another's device and view or manipulate virtual objects in the other person's context.

Augmented reality has been explored for many applications, from gaming and entertainment to medicine, education and business. Example application areas described below include *archaeology*, architecture, visual art, commerce, education, emergency management/search and rescue, social interaction, video games, industrial design, medical, *spatial immersion and interaction*, flight training, military, navigation, broadcast and *live events*, tourism and sightseeing, translation, music, as well as retail.

Words and Expressions

perceptual [pə(:)'septjuəl] *adj.* (五官所)知觉的,感性的;感知的
modality [məu'dæliti] *n.* 感觉模式;感觉形式;形式,方式,样式;形态;模态;物理疗法
haptic ['hæptik] *adj.* 触觉的;由触觉引起的
somatosensory [ˌsəumətə'sensəri] *adj.* 体觉的;躯体感觉的
olfactory [ɔl'fæktəri] *adj.* 嗅觉的,嗅觉器官的 *n.*[常用复数]嗅觉;嗅觉器官
destructive [dis'trʌktiv] *adj.* 破坏(性)的;毁灭(性)的;危害的(of)
mask [mɑːsk] *vt.* 屏蔽;掩饰,伪装;戴面具 *vi.* 掩饰;化装
          n. 屏蔽;掩码;(假)面具;口罩;防护面具;伪装;遮盖物;面模;假面舞会
interweave [ˌintə(:)'wiːv] *vt.* 使交织;使混杂;使紧密结合;织进
              vi. 交织;混杂(过去分词:interwove,过去时:interwoven)
immersive [i'məːsiv] *adj.* 沉浸式的;身临其境的;拟真的
synonymous [si'nɔniməs] *adj.* 同义的
fixture ['fikstʃə] *n.* 固定装置;固定物;工件夹具;固定;固定状态
helmet ['helmit] *n.* 头盔,钢盔;防护帽
alignment [ə'lainmənt] *n.* 一直线,队列;校直,调整;准线;结盟,联合
tacit ['tæsit] *adj.* 缄默的;默示的;心照不宣的;不言而喻的
sensor ['sensə] *n.* 传感器;灵敏元件
eyewear ['aiwɛə] *n.* (总称)眼镜;护目镜;眼镜防护
eyepiece ['aipiːs] *n.* 目镜,接目镜
wand [wɔnd] *n.* 指挥棒;魔杖;识别笔(等于 wand reader),扫描笔;权杖
          vt. 用扫描笔在……上扫描条形码
stylus ['stailəs] *n.* 触笔;铁笔;光笔;(留声机的)唱针;(古代刻写用的)尖笔

go with　伴随；与……相配；与……持同一看法
image registration [ˈimidʒ ˌredʒiˈstreiʃən]　图像配准
odometry [ɔˈdɔmitri] n. 测程法；里程计
fiducial [fiˈdjuːʃəl] adj. 基准的；基于信仰的；信托的，信用的
blob [blɔb] n. 斑点；一团；一滴　vt. 用斑点弄污
bundle adjustment [ˈbʌndl əˈdʒʌstmənt]　光束法平差；捆绑调整；集束调整
projective (epipolar [ˌepiˈpəulə]) geometry　射影(对极)几何
Kalman　卡尔曼滤波(一种递推式滤波算法)
robust statistics [rəuˈbʌst stəˈtistiks]　稳健统计(学)
archaeology [ˌɑːkiˈɔlədʒi] n. 考古学
spatial immersion and interaction　空间沉浸和互动
live event [laiv iˈvent]　现场活动；现场实况；直播

Abbreviations

AR (Augmented Reality)　增强现实
HUD (Head Up Display)　平视显示器，抬头显示器；平视显示
MEMS (Micro Electro Mechanical System)　微机电系统，微电子机械系统
HMD (Head-Mounted Display)　头戴式显示器，头盔显示器
SAR (Spatial Augmented Reality)　空间增强现实
SLAM (Simultaneous Localization And Mapping)
　　同步定位与地图构建，即时定位与地图构建，同步定位与建图，同时定位和映射
ARML (Augmented Reality Markup Language)　增强现实标记语言
OGC (Open Geospatial Consortium)　开放地理空间信息联盟
ECMA (European Computer Manufactures Association)　欧洲计算机制造联合会
SDK (Software Development Kit)　软件开发工具包

Notes

1. 本句中，关系副词 where 引导限制性定语从句，修饰其前的 environment；that 引导限制性定语从句，修饰其前的 objects；including 引导的分词短语引导 modalities 的同位语。

2. 本句中，第一个 that 引导名词性从句作表语；第二个 that 引导限制性定语从句，修饰其前的 sensations。全句可译为：增强现实的主要价值在于：它把数字世界的组成部分带入人对现实世界的知觉中，并且不是数据的简单显示，而是(通过)把被视为环境的自然部分的、身临其境的多种感觉集成在一起。

3. 本句中，e.g. 之后的 seeing 引导分词短语作同位语，解释前面所述；overlaid 引导的短语修饰前面的 waves；连接副词 where 引导的名词性从句作介词 with 的宾语。in exact alignment with 含意为"与……完全一致"或"与……符合"。全句可译为：这信息可以是虚拟的，也可以是真实的，例如，看到其他的真实感觉到或测量到的信息，诸如覆盖在空间中实际位置上的电磁无线电波。

4. tacit knowledge 含意为"隐性知识"。隐性知识是迈克尔·波兰尼(Michael Polanyi)在 1958 年从哲学领域提出的概念。迈克尔·波拉尼将知识分为隐性知识和显性知识。通常以书面文字、图表和数学公式加以表述的知识,称为显性知识;在行动中所蕴含的未被表述的知识,称为隐性知识。

5. 全句可译为:(令人)身临其境的感知信息有时会结合补充信息,如体育赛事现场视频直播上的分数。

6. MEMS 全称 Micro Electromechanical System,微机电系统,也称微电子机械系统,是指尺寸为几毫米乃至更小的高科技装置,其内部结构一般为微米甚至纳米量级,例如,常见的 MEMS 产品尺寸一般为 3 mm×3 mm×1.5 mm,甚至更小。微机电系统是集微传感器、微执行器、微机械结构、微电源微能源、信号处理和控制电路、高性能电子集成器件、接口、通信等于一体的微型器件或系统,是一个独立的智能系统。常见的产品包括 MEMS 加速度计(accelerometer)、MEMS 麦克风、微马达、微泵、微振子、MEMS 压力传感器、MEMS 陀螺仪、MEMS 湿度传感器等,以及它们的集成产品。MEMS 是一项革命性的新技术,广泛应用于高新技术产业,是一项关系到国家的科技发展、经济繁荣和国防安全的关键技术。

7. 此句中,rendering 是名词,在计算机图形学(CG)领域通常译为"渲染"。计算机将存储在内存中的形状(信息)转换成实际绘制在屏幕上的图像的对应过程称为渲染。渲染是 CG 后期制作前的一道工序,这一阶段渲染程序要计算我们在场景中添加的每一个光源对物体的影响,把显示出来的三维图形变成高质量的图像,最终使图像符合 3D 场景。图像渲染中要完成的工作是:通过几何变换、投影变换、透视变换和窗口剪裁,以及获取的材质与光影信息生成图像;此过程包括绘制虚拟场景、与真实场景进行图像配准、把两者进行合成(叠加)等。因此也可以把 rendering 理解成"生成、呈现、绘制"。augmented reality rendering 意为"增强现实渲染",也可理解为"增强现实生成"或"增强现实呈现"。image(或 graphics) rendering 可翻译为"图像绘制"或"图像渲染"。

8. 本句中,worn 是 wear 的过去分词,引导过去分词短语作定语,修饰其前的 systems。

9. 本句中,resembling 引导现在分词短语作定语,修饰其前的 devices。

10. 本句中,cameras to intercept ... 是 employs 的复合宾语;which 引导限制性定语从句,修饰其前的 devices。全句可译为:诸版本都包括眼镜,它使用摄像头来拦截现实世界的视图,并通过目镜和类似设备重新显示其增强了的视图,在这些设备中,AR 图像被投射穿过眼镜镜片表面或自眼镜镜片表面反射出来。

11. 本句中,the way 之后是修饰它的限制性定语从句,省略了 that。

12. 图像配准(Image registration)就是将不同时间、不同传感器(成像设备)或不同条件下(天候、照度、摄像位置和角度等)获取的两幅或多幅图像进行匹配、叠加的过程,它已经被广泛地应用于遥感数据分析、计算机视觉、图像处理等领域。

13. 本句中,assume 之后省略了 that,它引导名词性从句作宾语。全句可直译为:某些方法假定具有已知的几何形状(或基准标记)的诸对象存在于场景中。结合上下文,为了更连贯与强调起见,全句可译为:某些方法假定存在于场景中的诸对象是具有已知的几何形状(或基准标记)的。

14. 本句中，developed 引导的分词短语作定语，修饰其前的 standard；Open Geospatial Consortium 译为"开放地理空间信息联盟"；which 引导非限制性定语从句，修饰 Augmented Reality Markup Language（增强现实标记语言）；XML 是 Extensible Markup Language 的缩写，译为"可扩展标记语言"；ECMA 是 European Computer Manufactures Association 的缩写，译为"欧洲计算机制造联合会"。全句可译为：增强现实标记语言（ARML）是在开放地理空间信息联盟（OGC）内开发的数据标准，它由描述场景中虚拟对象的位置与外观的 XML 文法，以及允许动态访问虚拟对象属性的 ECMA 脚本绑定所组成。

注：为了更好地理解 AR，请观看视频材料中的 AR 视频。

Exercises

1. Translate the text of Lesson 12.1 into Chinese.
2. Topics for oral workshop.
- What is CAD? How to classify the CAD software?
- According to the text, describe the categories the AR systems are classified into.
- Introduce an application system that you are involved in or familiar with.
3. Translate the following into English.

(a)虽然在工程和科学中的一些早期应用不得不依靠昂贵而笨重的(cumbersome)设备，在计算机技术（领域）中的进步已使交互式计算机图形学成为一种实用的工具。

(b)因为作图一般是以三维显示的，所以 CAD 在设计汽车、飞机、船舶、建筑物、电子线路(包括计算机芯片)甚至服装(clothing) 中是特别有帮助的。

(c)虚拟现实(VR)是应用于计算机模拟环境的一个术语，该环境能模拟现实世界中的物理存在。大多数当前的虚拟现实环境主要是视觉的体验，或者在计算机屏幕上或者经由一些专门的立体显示器来显示，然而某些模拟包括附加的感觉信息，诸如经由扬声器或耳机的声音。

(d) GUI 的一个主要组件是窗口管理器，（它）允许用户显示多个称为显示窗口的矩形屏幕区域。每个屏幕显示区域可以包含一个特有的进程(process)，展示图形或非图形的信息，并且可以用各种方法激活一个显示窗口。

4. Listen to the video "Broadband Internet" and write down the 8-11 paragraphs.

Unit 13　Computer Applications Ⅱ

13.1　Distance Education Technological Models

Distance education covers a multi-*facetted* techno-*pedagogical* reality, ranging from the simple decentralization of classroom activities to interactive multimedia models that make learning available whatever the time or the location. We now describe these various models and stress the principles that we have retained to design an integrated model of the Virtual Campus.

The distance education world is *bubbling*. The rapidly *evolving availability* of multimedia telecommunication is *giving way* to a increasing number of techno-pedagogical models. [1] We describe them in terms of six main paradigms:

- The **enriched classroom** where technologies are used within a traditional setting in order to do a presentation, a demonstration or an experimentation. It is a networked classroom allowing access to campus resources and external databases and it is sometimes called an "intelligent" campus. [2]

- The **virtual classroom** mainly uses videoconferencing to support distant learners and teachers, thereby recreating a *telepresence* type of classroom; many university campus now have their own multimedia production studios so they may decentralize training at satellite locations. [3]

- The **teaching media** is focused on the learner's workstation. It allows access to *prefabricated* multimedia course contents on CD ROM, either shipped by mail or available from a distant multimedia server. Instruction and *didactic* resources are offered in such a way that the learner can individualize his own learning process. [4]

- **Information highway training** is also centered on the learner's workstation, which serves as a navigation and research instrument to find all kinds of useful educational information. Essentially, a "Web course" is offered on a central *site* where instructions and pointers related to didactic resources (other Web sites) are gathered in order to accomplish learning activities.

- The **communication network** uses the workstation, not only as a media support or as a way to access information, but also as a synchronous (desktop videoconferencing, screen sharing, etc.) and as an asynchronous communication tool (electronic mail, computer teleconferencing, etc.).

- The **performance support system** (EPSS) concerns task-oriented training modules that are added to an integrated support system within a *workplace*. Information has a "just-in-time" quality and training is seen as a process that is complementary and incorporated into the work process.

Each of these models has advantages and drawbacks. The first two are very popular at the present time. They *rest on* the traditional paradigm inherent in live information transmission: the teacher uses computerized and *audiovisual* equipment to *animate* a real-time multimedia group presentation, broadcast locally or to several distant locations where learners are gathered.[5] This model requires costly equipment as well as the learners and teacher's physical presence simultaneously. Moreover, too often it reduces the learners' interaction and initiative to a level that is *in no way* better than that of a traditional course presentation in an *auditorium*.[6]

This approach appears incapable of meeting the growing training needs in a *socioeconomic* context where *lifelong* learning, sought by busy and *mobile* people, involves *cognitive* abilities of a much higher level than what was required in the past.[7] As pointed out by a recent report produced by Quebec's Higher Council for Education.

- "From now on, the emphasis must bear on higher cognitive abilities (reasoning, problem solving, planning) and social abilities (autonomy, communication and *collaboration*). These are consistent with the capabilities expected from workers following the impacts of information and communication technology on the very nature of work".
- Rapport annuel 1993-94 du Conseil supérieur de l'éducation, Gouvernement du Québec, p. 24 (free translation by LICEF).

The availability of Internet and multimedia technologies exposes the learner to numerous sources of information among which he must make choices. The new paradigm (Figure 13-1) where the learner at the center of his learning process calls on many expertise sources, is better represented by the last four models described above than by the first two.[8]

In the models of the "teaching media" or the "information highway training" and their application as it is currently, training is personalized but it is also *deprived* of an important collaborative dimension. However, this dimension can be *reinstated* if we use the computer as a tool to communicate. *As for* the EPSS approach, it favors two principles: just-in-time information and learning seen as information processing. The learning *scenario* concept at the heart of our Virtual Campus proposal is based on these principles.

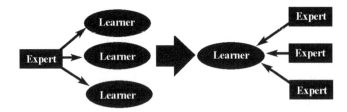

Figure 13-1　Paradigm shift

Words and Expressions

distance education　远程教育
facet ['fæsit] vt. 在……上刻面　　n. 某一方面；(多面体的)面，小(平)面

pedagogical [pedə'gɔdʒikəl] adj. 教育学的；教学法的
bubble ['bʌbl] vi. 沸腾；冒泡 n. 泡，气泡
evolve [i'vɔlv] v. (使)发展；(使)进展；(使)进化
availability [ə,veilə'biliti] n. 可用性；有效性；实用性
give way (to)　让路；让步；让位
telepresence ['teli,prezns] n. 远程出席
prefabricate [pri:'fæbrikeit] vt. 预制；预先构想
didactic [di'dæktik] adj. 教学用的；教导的；说教的
site [sait] n. 网站
workplace ['wə:k'pleis] n. 工作场所，车间
rest on　依据，依赖；信赖；被支撑(在)，搁(在)；(视线等)停留(在)
audiovisual ['ɔ:diəu'viʒəl] adj. 视(觉)听(觉)的
animate ['ænimeit] vt. 动画制作；激励；使活泼，使生气勃勃
　　　　 ['ænimit] adj. 有生命的；有生气的，生气勃勃的
in no way　决不，一点也不
auditorium [,ɔ:di'tɔ:riəm] n. 讲堂，礼堂
socioeconomic ['səusiəu,i:kə'nɔmik] adj. 社会经济(学)的
lifelong ['laiflɔŋ] adj. 毕生的，终身的
mobile ['məubail] adj. 流动的；移动的，运动的；机动的　n. 运动物体
cognitive ['kɔgnitiv] adj. 认知的，认识的；有感知的
collaboration [kə,læbə'reiʃən] n. 协作，合作
deprive [di'praiv] vt. 使丧失；夺去；剥夺
reinstate ['ri:in'steit] vt. 恢复；使恢复原状；使正常
as for　至于，就……方面说
scenario [si'nɑ:riəu] n. 脚本，方案

Notes

1. 本句中的 evolving 和 increasing 都是现在分词，作定语，分别修饰其后的 availability 和 number。全句可译为：迅速发展的多媒体远程通信的可用性正使得技术-教学法模型的数量日益增加。
2. 本句中的 networked 是过去分词，作定语，修饰后面的 classroom；allowing 是现在分词，引导一个分词短语作定语，修饰前面的 classroom。全句可译为：它是一种允许访问校园资源和外部数据库的网络化教室，有时称为"智能"校园。
3. 本句可译为：虚拟教室主要使用视频会议(技术)来支持远程的学习者和教师，因此，重建了远程出席类型的教室；很多大学校园现在有自己的多媒体产品演播室，所以它们可以在一些附属教学点进行分散培训。
4. 本句中的 that 是连词，引导一个状语从句，表明目的或结果。全句可译为：教育和教学用的资源是以这样一种方式提供的，使得学习者能把他自己的学习过程个性化以适应自己的需要。

5. 本句中的 where 是关系副词，引导一个限制性定语从句，修饰前面的 locations；本句中的 they 代表上句中的 the first two (modules)。全句可译为：它们依据现场信息传输中所固有的传统范式：教师使用计算机化设备和视听设备把实时多媒体组演示（课件）制作成动画，在本地播出或向学习者聚集的几个远距离场所播出。

6. 本句中此处的 that 是指示代词，用来代替句中的名词 level，以避免重复；其前的 that 是关系代词，引导一个限制性定语从句，修饰前面的 level。全句可译为：而且，它太多时候把学习者的交互作用和主动积极性降低到与课堂中的传统讲课差不多的程度。

7. 本句中的 sought 是动词 seek 的过去分词形式，它引导一个分词短语作非限制性定语，修饰前面的 lifelong learning。全句可译为：这种方法看来是不能满足社会经济环境中不断增长的培训需求的，在这种环境中，忙碌且流动的人们所寻求的毕生学习涉及比过去所需要的水平高得多的认知能力。

8. 本句中的 where 是关系副词，引导一个限制性定语从句，修饰前面的 paradigm。全句可译为：该新范式（见图 13-1）（处于其学习过程之中心的学习者需要很多专业知识的原始资料）用上面描述的后四种模型表示比用前两种表示更好。

Term

MOOC

Massive Open Online Courses (MOOCs) are free online courses available for anyone to enroll. MOOCs provide an affordable and flexible way to learn new skills, advance your career and deliver quality educational experiences at scale.

Course features:

It is free access to MOOC resources without school credentials.

There is no limit on the number of students.

As for other features, they only appeared in early MOOC, such as open authorization, open architecture and learning objectives, community orientation, and so on.

There are three MOOC platforms: EDX, Udacity and Coursera. On May 8, 2014, the MOOC platform of China University with autonomous intellectual property rights was officially opened on the "Love Course" network of the Ministry of Education, which can be used in the construction and application of MOOC courses in universities all over the country.

Existing issues:

1. Can a course cover so many students?

2. The scoring mechanism is not perfect especially in the non-scientech humanities, arts and other disciplines.

3. The biggest issue is serious cheating. Both Udacity and EDX now offer invigilated examinations.

13.2 Electronic Business

Electronic business, commonly referred to as "eBusiness" or "e-business", or an internet business, may be defined as the application of information and communication technologies (ICT) *in support of* all the activities of business. *Commerce* constitutes the exchange of products and services between businesses, groups and individuals and can be seen as one of the essential activities of any business. Electronic commerce focuses on the use of ICT to enable the external activities and relationships of the business with individuals, groups and other businesses.

Electronic business methods enable companies to link their internal and external data processing systems more efficiently and flexibly, to work more closely with suppliers and partners, and to better satisfy the needs and expectations of their customers. [1]

In practice, e-business is more than just e-commerce. While e-business refers to more strategic focus with an emphasis on the functions that occur in using electronic capabilities, e-commerce is a subset of an overall e-business strategy. E-commerce seeks to add *revenue* streams using the World Wide Web or the Internet to build and enhance relationships with clients and partners and to improve efficiency using the Empty *Vessel* strategy. [2] Often, e-commerce involves the application of knowledge management systems.

E-business involves business processes *spanning* the entire value chain: electronic purchasing and supplychain management, processing orders electronically, handling customer service, and cooperating with business partners. Special technical standards for e-business facilitate the exchange of data between companies. E-business software solutions allow the integration of intra and inter firm business processes. E-business can be conducted using the Web, the Internet, *intranets*, *extranets*, or some combination of these.

Basically, electronic commerce (EC) is the process of buying, transferring, or exchanging products, services, and/or information via computer networks, including the internet. EC can also be benefited from many *perspective* including business process, service, learning, collaborative, community. EC is often confused with e-business.

Applications can be divided into three categories:
1. Internal business systems: customer relationship management, enterprise resource planning, document management systems, human resources management.
2. Enterprise communication and collaboration: VoIP, content management system, e-mail, voice mail, Web conferencing, Digital work flows (or business process management).
3. Electronic commerce-business-to-business electronic commerce (B2B) or business-to-consumer electronic commerce (B2C): internet shop, supply chain management, online marketing, *offline marketing*.

When organizations *go online*, they have to decide which e-business models best suit their goals. A business model is defined as the organization of product, service and information flows, and the source of revenues and benefits for suppliers and customers. The concept of e-business model is the same but used in the online presence. The following is a list of the currently most adopted e-business models such as: E-shops, E-commerce, E-*procurement*, E-*malls*, E-*auctions*, *Virtual Communities*, Collaboration Platforms, Third-party Marketplaces, Value-chain Integrators[3], Value-chain Service Providers, Information *Brokerage*, Telecommunication, Customer relationship.

Roughly dividing the world into providers/producers and consumers/clients one can classify e-businesses into the following categories: business-to-business (B2B), business-to-consumer (B2C), business-to-employee (B2E), business-to-government (B2G), government-to-business (G2B), government-to-government (G2G), government-to-citizen (G2C), consumer-to-consumer (C2C), consumer-to-business (C2B).[4]

E-Business systems naturally have greater security risks than traditional business systems, therefore it is important for e-business systems to be fully protected against these risks. A far greater number of people have access to e-businesses through the internet than would have access to a traditional business.[5] Customers, suppliers, employees, and numerous other people use any particular e-business system daily and expect their *confidential* information to stay secure. *Hackers* are one of the great *threats* to the security of e-businesses. Some common security concerns for e-Businesses include keeping business and customer information private and confidential, *authenticity* of data, and data integrity. Some of the methods of protecting e-business security and keeping information secure include physical security measures as well as data storage, data transmission, anti-virus software, firewalls, and encryption to list a few.

Words and Expressions

electronic business　电子商务，电子业务
in support of　支持
commerce ['kɔməːs] *n.* 商业；贸易；（意见的）交流
revenue ['revənjuː] *n.* 收益，收入；[复]总收入；税收
vessel ['vesəl] *n.* 船，舰；容器
span [spæn] *vt.* 跨越；横跨　*n.* 跨度；跨径；一段时间
intranet [intrə'net] *n.* 企业内部网
extranet ['ekstrə'net] *n.* 外联网
perspective [pə'spektiv] *n.* 远景；前途；观点，看法；视角；透视；透视图；观察
offline marketing　离线营销
go online　上网
procurement [prɔ'kjuəmənt] *n.* 采购；获得；获得的条件
mall [mɔːl] *n.* 购物中心，商场
auction ['ɔːkʃən] *n.* 拍卖
virtual communities　虚拟社区
brokerage ['brəukəridʒ] *n.* 经纪；佣金；经纪费
confidential [ˌkɔnfi'denʃəl] *adj.* 机密的
hacker ['hækə] *n.* 黑客
threat [θret] *n.* 威胁，恐吓
authenticity [ˌɔːθen'tisiti] *n.* 真实性；确实性

Abbreviation

VoIP (Voice over Internet Protocol) IP 电话，网络电话

Notes

1. 本句中，companies to link 是名词＋不定式类型的复合宾语，其后的 to work 与 to better satisfy 跟 to link 并列。全句可译为：电子商务方法使一些公司能把它们的内外部数据处理系统更有效与灵活地连接，与供应商和合作伙伴更紧密地合作，并且更好地满足顾客的各种需求与期望。
2. 本句中，不定式 to add 与 to improve 并列，都作 seeks 的宾语。全句可译为：e-贸易寻求使用万维网或因特网建立与增进同客户与合作伙伴们的关系，以增加收益流，并使用空船策略提升效率。
3. Value-chain Integrator 意为"价值链集成商"。价值链集成商是区别于传统产品集成商而言的。在过去的产品集成模式下，集成商更侧重于对价值链进行硬性控制，希望以此确保安全，并从整条价值链中最大限度地获取利益。价值链集成模式最大的特点是抛弃了对价值链实施全链控制的企图，根据自身优劣确定其在价值链中的定位，并以市场黏合力聚合各方力量，共同创造价值。
4. 本句中，one 之前是 dividing 引导的分词短语，作状语。全句可译为：粗略地把世界分成供应商／生产者与消费者／客户，就能把电子商务分成以下几类：企业对企业(B2B)、企业对顾客(B2C)、……、顾客对企业(C2B)。
5. 本句中，助动词 would 含意为"会，要"。全句可译为：通过互联网进入电子商务的人数远远比进入传统商务的人数多。

Terms

IT
　　Short for Information Technology. The acquisition, processing, storage and dissemination of vocal, pictorial textual and numeric information by a micro-electronics based combination computing and telecommunication. The technology itself can be divided into computer and communications hardware and software.

O2O
　　O2O means "Online To Offline" to connect traditional business via online marketing. Online to offline is a phrase that is used in digital marketing to describe systems enticing consumers within a digital environment to make purchases of goods or services from physical businesses. In the majority of cases, O2O generates leads online and then prompts the customer to go to a physical location to complete their purchase. One aspect of newer O2O initiatives is the ability to pay online and then pick up a product in a physical location. Via online platform, consumers can select product/service directly and do online transactions. O2O is connecting

traditional business with a modern platform, advantages on both sides can be perfectly matched so consumers can enjoy good online price and the same quality offline service.

In the broadest sense, everything on the internet is "online to offline".

13.3　E-Government — Introduction

By definition, *e-government* is simply the use of information and communications technology, such as the Internet, to improve the processes of government. Thus, e-government is in principle nothing new. Governments were among the first users of computers. But the global *proliferation* of the Internet, which effectively integrates information and communications technology on the basis of open standards, combined with the movement to reform public administration known as New Public Management, has for good reason generated a new wave of interest in the topic.[1] E-government promises to make government more efficient, *responsive*, transparent and *legitimate* and is also creating a rapidly growing market of goods and services, with a variety of new business opportunities.

To some, e-government might seem to be *little more than* an effort to expand the market of e-commerce from business to government. Surely there is some truth in this. E-commerce is marketing and sales via the Internet. Since governmental *institutions* take part in marketing and sales activities, both as buyers and sellers, it is not inconsistent to speak of e-government applications of e-commerce. Governments do after all conduct business.

But e-commerce is not at the heart of e-government. The core task of government is governance, the job of regulating society, not marketing and sales. In modern democracies, responsibility and power for regulation is divided up and shared among the legislative, executive and judicial branches of government. Simplifying somewhat, the legislature is responsible for making policy in the form of laws, the executive for implementing the policy and law enforcement, and the judiciary for resolving legal conflicts. E-government is about improving the work of all of these branches of government, not just public administration in the narrow sense.

New Public Management is a kind of management theory about how to reform government by replacing rigid hierarchical organisational structures with more dynamic networks of small organisational units; replacing *authoritarian*, topdown decision and policy making practices with a more *consensual*, bottom-up approach which facilitates the participation of as many *stakeholders* as possible, especially ordinary citizens; adopting a more 'customer'-oriented attitude to public services; and applying market principles to enhance efficiency and productivity.[2]

E-government gives New Public Management fresh blood. Not only does information and communications technology provide the infrastructure and software tools needed for a loosely coupled network of governmental units to collaborate effectively, the *infiltration* of this technology into *government agencies* tends to lead naturally to institutional reform, since it is difficult to maintain strictly hierarchical channels of communication and control when every civil servant can collaborate efficiently and directly with anyone else via the Internet.

Orthogonal to the division of power among the branches of government is the hierarchical organisation of *supranational* (e. g., European), national, regional and local governments bounded[3] by geographical territory. Information and communication technology creates a 'new accessibility', overcoming temporal, geographical and organisational boundaries. Thus e-government can facilitate new forms of collaboration among governments which [4]*cut across* and *diminish* such boundaries. The EuroCities project is an example. Perhaps in the long term e-government will help to strengthen the identification of citizens with Europe.

E-government is not only or even primarily about reforming the work processes within and among governmental institutions, but is rather about improving its services to and collaboration with citizens, the business and professional community, and *nonprofit* and nongovernmental organisations such as associations, trade unions, political parties, churches, and public interest groups.

Using World Wide Web *portals* to create one-stop shops is one currently popular e-government approach to improving the delivery of public services to citizens. The basic idea of these portals is to provide a single, convenient place to take care of all the steps of a complex administrative process involving multiple government offices, bringing the services of these offices to the citizen instead of requiring the citizen to run from office to office. [5]

Web portals can deliver government services with various levels of interaction. Three levels are usually identified: information, communication, and transactions. Information services deliver government information via static web pages and pages generated from databases to citizens, tourists, businesses, associations, public administration, and other government users. Communication services use *groupware* technology such as E-mail, discussion forums and chat to facilitate dialogue, participation and feedback in planning and policy-making procedures. Transaction services use online forms, workflow and payment systems to allow citizens and business partners to take care of their business with government online. Typical applications of transaction services for citizens include applying for social benefits, registering automobiles, filing changes of address or applying for building permits. [6] For businesses, perhaps the application of greatest current *interest* is the online *procurement* of government contracts.

Often one reads that these three levels of interaction are ordered by complexity, with transactions being the most complex. *Presumably* this is because of the apparent and challenging security and business process *reengineering* issues of online transaction processing. Providing[7] high quality information and communication services, however, is no less challenging. Information services need to evolve into knowledge management services and become adaptive, personalised, *proactive* and accessible from a broader variety of devices. Communication services need to evolve into collaboration services providing better support for argumentation, negotiation, deliberation and other goal-directed forms of structured discourse.

Among the most interesting and challenging sociotechnological issues of e-government are in the area of e-Democracy, which aims to apply information and communication technology to improve the public opinion formation process central[8] to government's primary regulatory function. Here the ambition is to broaden actual public participation, not just the technical possibility, and *counter* political *apathy* without *disenfranchising* the poor or poorly educated.

The following articles give a good indication of the large number and variety of governmental processes requiring specific solutions. Together with the trend towards outsourcing tasks and working with industry in private-public partnerships, this is likely to lead to rapid growth of the e-government market and create plentiful business opportunities, also for small and medium-size enterprises. Viewing e-government projects as mainly an investment in public infrastructure is too restricted, since the investment is also aimed at reducing the size and costs of government while accelerating the growth of the e-government market, helping to create new businesses and jobs in the private sector.

Words and Expressions

e-government (electronic-government)　*n.* 电子政务
proliferation [ˌprəuˌlifə'reiʃən] *n.* 激增；扩散；增殖，增生
responsive [ris'pɔnsiv] *adj.* 响应的，易起反应的；敏感的；应答的
legitimate [li'dʒitimit] *adj.* 合法的；合理的；正统的　*vt.* 使合法
little more than　和……无差别(一样)
institution [ˌinsti'tju:ʃən] *n.* 公共机构；协会；制度；学会；学院
authoritarian [ɔːˌθɔri'tɛəriən] *adj.* 权力主义的，独裁主义的　*n.* 权力主义者，独裁主义者
consensual [kən'senʃuəl] *adj.* (律)经双方同意的；(生理)同感的，交感的
stakeholder ['steikhəuldə(r)] *n.* 利益共享者；赌金保管者
infiltration [ˌinfil'treiʃən] *n.* 渗透
government agency　政府机构
orthogonal [ɔː'θɔgnəl] *adj.* 正交的，直角的，直交的
supranational ['sju:prə'næʃənl] *adj.* 超国家的，超民族的
cut across　超越；抄近路通过，对直通过
diminish [di'miniʃ] *vt.* 缩减，减小，减少；降低　*vi.* 变少，缩小
nonprofit ['nɔn'prɔfit] *adj.* 非营利的，不以营利为目的的，无利可图的
portal ['pɔːtəl] *n.* 门户(网站)；入口，门，隧道门
groupware ['gru:pwɛə] *n.* [计]组件，群件
interest ['intrist] *n.* 兴趣；关心；利益；利息
procurement [prə'kjuəmənt] *n.* 采购；获得；获得的条件
presumably [pri'zju:məbli] *adv.* 推测起来，大概
reengineering [ri'endʒiniəriŋ]
　　n. 再(造)工程；重建　*vt.* (业务流程)再设计；重新建造(reengineer 的 ing 形式)
proactive [ˌprəu'æktiv] *adj.* (心理)前摄的

counter ['kauntə] vt. 反对；反击　vi. 反击　adv. 反方向地　adj. 相反的；反对的 n. 计数器，计算器；筹码；柜台；反面
apathy ['æpəθi] n. 冷漠；缺乏感情或兴趣
disenfranchise ['disin'fræntʃaiz] vt. （＝disfranchise）剥夺……公民权

Notes

1. 本句中的 which 是关系代词，引导一个非限制性定语从句，修饰其前的 Internet；combined 是过去分词，引导一个短语作状语；to reform 引导不定式短语作定语，修饰其前的 movement。全句可译为：然而，因在一些开放标准基础上有效地把信息和通信技术一体化，加上以新公共管理著称的改革公共管理的运动，Internet（因特网）在全球的扩散，有充分理由（说）已引发新的一波对该论题的兴趣。
2. 本句中的两个 replacing 与 adopting、applying 都是动名词，并列作为介词 by 的宾语，表示方法，即用……代替……，用……代替……，采用……和应用……的方法（来改革政务）。
3. bounded 是过去分词，引导一个短语作定语，修饰其前的 governments。全句可译为：与政府各部门中间权力的分配相正交的，是由地理区域界定的超国家的（如欧洲的）、国家的、地区的和地方的政府等组成的层次组织体制。
4. 此 which 引导一个限制性定语从句，修饰 new forms (of collaboration)。
5. 本句中的 involving 是现在分词，引导一个短语作定语，修饰其前的 process；bringing 也是现在分词，但引导的短语作状语。requiring 引导的动名词短语作 instead of 的宾语。
6. applying、registering、filing 与 applying 都是动名词，各自引导的短语是 include 的并列宾语。
7. providing 是动名词，引导的短语作主语。
8. central 是形容词，引导的短语作定语，修饰前面的 process。

Terms

CRM

Customer Relationship Management (CRM) is a broad term that covers concepts used by companies to manage their relationships with customers, including the capture, storage and analysis of customer information.

There are three aspects of CRM which can each be implemented in isolation from each other:

- Operational CRM- automation or support of customer processes that include a company's sales or service representative.
- Collaborative CRM- direct communication with customers that does not include a company's sales or service representative ("self service").
- Analytical CRM- analysis of customer data for a broad range of purposes.

PLC

Automation of many different processes, such as controlling machines or factory assembly lines, is done through the use of small computers called a programmable logic controller (PLC). This is actually a control device that consists of a programmable microprocessor, and is programmed using a specialized computer language. Before, a programmable logic controller would have been programmed in ladder logic, which is similar to a schematic of relay logic. A modern programmable logic controller is usually programmed in any one of several languages, ranging from ladder logic to Basic or C. Typically, the program is written in a development environment on a personal computer (PC), and then is downloaded onto the programmable logic controller directly through a cable connection. The program is stored in the programmable logic controller in non-volatile memory.

13.4 Office Automation

What is Office Automation(OA)?

Quite simply, Office Automation is the *customisation* of your current computer applications to make your everyday office tasks more easy, faster and consistent across your organisation.[1]

These applications may be your *word processing*, *spreadsheets*, e-mail, or presentation.

For word processing this could apply to simple items such as letterheads, faxes, *memos*, or quite complex including[2] customer documentation, technical documentation, annual reports or customising contracts.

For spreadsheets this could apply to monthly budgets, management accounts, *sales reps* expenses, or engineering calculations.

For E-mail this could apply to simple organisational memos, logging of system faults, or extensive customer complaint management.

For presentation this could apply to customer presentations, or reporting of the days events.

Take your current system of letters, faxes and memos. Using a simple screen interface we can provide you entire organisation with fast and consistent letters, faxes and memos. This system can be integrated into your word processor so it is available at the press of a button.

Your output will be consistent across your organisation! We can even direct different pages or documents to different printers.

Why Use Office Automation?

Office Automation can improve both the speed at which[3] your staff can prepare documentation, and more importantly will ensure that it is completed in a consistent manner.

In other words you decide how you would like your documents to look, to be structured, what data should be contained, and not only will your documents look that way, but also they will be completed more quickly. This will also allow for new staff to be inducted more effectively and faster.

Development Cycle

For every Office Automation project we aim to assess the unique requirements of your organisation and provide you with a *tailored* solution to meet your needs. We do this by designing a suitable software development cycle, that will meet the size and complexity requirements of your project. This design will usually be contained in the proposal for your project. The development cycle will outline the project flow of the development.

Development Environments

We are currently offering Office Automation development using[4] Microsoft Excel, Microsoft Word, Microsoft Outlook, Microsoft PowerPoint, Microsoft Access (for more information Custom Databases), plus other VBA compliant applications.

For other software development services check out our custom databases, and unique computer programs.

Words and Expressions

customisation 同 customization
customization [ˈkʌstəˌmaiˈzeiʃən] n. [计] 用户化；专用化；定制
word processing [计] 字处理软件；字处理
spreadsheet [ˈspredʃiːt] n. [计] 电子制表软件；电子数据表，电子表格
memo [ˈmeməu] n. 备忘录
sales rep 即 sales representative 商品经销代理，营业代表
tailor [ˈteilə] vt. 定制，专门制作；裁制；（为某一特定目的而）剪裁，制作

Abbreviations

VBA (Visual Basic for Applications) 适合于应用程序（开发）的 VB
OA (Office Automation) 办公自动化

Notes

1. 本句中的 of 表示动作的对象。the customisation of 意为"对……进行定制"。across 意为"横过，贯穿"。
2. including 引导一个同位语，修饰其前的 complex。
3. which 引导的定语从句修饰其前的 speed。
4. using 是现在分词，引导的短语作定语，修饰 development。

Exercises

1. Translate the text of Lesson 13.1 into Chinese.
2. Topics for oral workshop.
- Do you think that the distance education is an interesting topic in the computer applications? Why?
- What is Electronic Business? Talk about the areas in which e-commerce will help us a lot.

- Do you think that the efficiency in your work is improved by using office automation?
3. Translate the following into English.

（a）有各种各样的教育应用程序可用。教育软件设计用来教一门或多门技能，诸如阅读、数学、拼写、外语、世界地理，或者帮助准备一些标准化的测试。

（b）电子商务是使用信息技术和电子通信网络，以电子的、无纸的形式交换商务信息和实施诸事务处理（transaction）。

（c）字处理软件允许使用计算机来创建、编辑、存储和打印文档。你能够容易地插入、删除和移动一些词、句子和段落——任何时候（ever）都不使用橡皮擦（eraser）。

（d）电子表格（spreadsheet）是什么？电子表格软件从会计排成栏的（columnar）工作表（worksheet）取名，它仿制（imitate）该工作表。电子表格是由行与列的相交点形成的单元格（cell）集合组成的工作表。每个单元格能存储一项信息：数、单词或短语、公式。

4. Listen to the video "Virtual ancient Rome" and write down the text.

Unit 14　Computer Applications Ⅲ

14.1　Geographic Information Systems(GIS):
　　　A New Way to Look at Business Data

Geographic information systems are one of the fastest growing business applications and later this decade may be as common as word processing software and spreadsheet applications. [1]

A GIS, as defined by the *National Science Foundation*, is a computerized database management system used for the capture, storage, retrieval, analysis and display of spatial (e.g. locationally defined) data.

A GIS consists of the following three parts:

—GIS software.

—Hardware. Hardware needed to run a GIS depends on three interrelated variables:(1) Scope: the number of uses, number of applications and number of users;(2)*Scale* of data: the more detailed the maps, the more powerful the hardware needed; and(3)Functionality: the number of functions or operations to be performed on the data and the complexity of the functions. [2]

—Databases, both internal and external.

A key in developing a GIS system is *geocoding*. Geocoding is the process of linking attribute data to maps. Street address geocoding is the foundation technology of business geographic. It is said that about 80% of business data has some type of geographic component. [3]

Geocoding is much trickier to do well than it may appear. [4] Looking up address in a directory is easy in concept but can fail due to weaknesses in software, the geocoding reference directory or the data addresses themselves.

GIS systems allow a series of maps to be overlaid onto one another. By viewing the combination of computerized maps, a retailer could immediately see where his sales are high or low and where his competitors are strong or weak.

GIS systems allow data to be accessed in a variety of ways. Most *full-function* GIS systems combine three basic types of capabilities:(1)presentation mapping;(2)using maps as an organizing tool; and(3)spatial analysis.

The areas GIS systems are helping businesses include:(1) real estate,(2) direct mail marketing,(3) insurance,(4) banking,(5) service providers,(6) manufacturing,(7) transportation and distribution, and(8)retailing. [5]

Analysts say the GIS market is being driven by several factors:

1. Cheaper, faster hardware. PC prices are dropping rapidly and desktop computers are becoming more powerful.

2. Improved GIS software. GIS software today is more friendly, allowing users who have no experience with GIS to quickly learn the systems.

3. Reduced software prices. Low-end GIS systems can be purchased for a few thousand dollars and simple desktop mapping systems with limited data analysis functionality are available for less than $ 500.

4. Greater variety of *demographic* data.

5. Wider GIS use by businesses.

Worldwide, there are estimated to be about 100 vendors offering a variety of GIS systems. Of those, about 60 are based in North America and about two dozen are headquartered in Europe. All analysts agree that the two GIS market share leaders are Intergraph Corp. of Alabama and Environmental Systems Research Institute, Inc. (ESRI), California.

There are two kinds of GIS applications on the market. "Open Systems" allow direct *import* of data from your spreadsheet or database programs; "Closed Systems" do not. In general, open systems offer more utility. But they are more challenging to use. The *onus* is on the user to prepare the attribute data for import into the GIS. [6] Closed systems are easier to use. The attribute data arrives on your doorstep *neatly* bundled.

The GIS design process involves four basic elements: Geographic data; attribute data, both internal and external; mapping software; and hardware. Regarding geographical data, the first question to ask is, "What geographic area am I interested in?" Attribute data must be compatible with the GIS so it can be imported into the GIS. Mapping software should support data *entry*, data analysis, data output and display, and data management. As for hardware, a GIS quickly will *outgrow* the minimum system requirements recommended by the GIS vendor. Go for more than you need at the present time.

The non spatial, internal and external attribute data is another major cost of developing a GIS. Buying external data such as commercial demographic databases and developing internal databases can account for as much as 80% of the total cost of a GIS systems.

In designing a GIS, a key concern is selecting the right attribute data for the job. [7] In the United States, many commercial databases are based on the U. S. *Census Bureau's* data. Leading vendors of commercial demographic databases use sophisticated segmentation techniques for both consumer and business data that add value to basic, raw demographic data. External data also is available at little or no cost from government agencies, *trade associations*, universities, nonprofit groups and other organizations.

GIS packages range from PC based products that do little more than display data, sometimes called desktop mapping systems, to advanced systems that do sophisticated data modeling and can display maps in *vivid* detail and 3D quality.

GIS business applications can be divided into three categories:

1. Operations: In these applications, GIS supports the operational activities and decisions of the business.

2. Tactical: This type of GIS application involves "semi-structured" decision making by middle managers.

3. Strategic: These types of GIS applications are designed for upper management, who generally make "unstructured decisions".

Words and Expressions

National Science Foundation （美国）国家科学基金会
scale [skeil] n. 规模
geocoding [dʒiːəuˈkəudiŋ] n. 地理信息编码(术)
full-function 全功能
demographic [ˌdeməˈɡræfik] adj. 人口统计(学)的；人口的 n. 人口统计学；人口统计数据
import [imˈpɔːt] n. 移入 vt. 移入；进口
onus [ˈəunəs] n. 责任，义务；负担
neatly [niːtli] adv. 整齐地；整洁地
entry [ˈentri] n. 录入，登录；条目，项目；词条；入口
outgrow [autˈɡrəu] vt. 生长速度超过，长得比……快
Census Bureau （美国）人口统计局
trade association 行业协会，贸易协会，同业公会
vivid [ˈvivid] adj. 生动的；鲜明的；清晰的；逼真的

Notes

1. later 是 late 的比较级。全句可译为：地理信息系统是发展最快的商务应用软件之一，这十年（指 20 世纪 90 年代）后期它可能像字处理软件和电子表格应用软件一样普及。
2. to be performed 是被动形式的不定式，它引导的不定式短语作定语，修饰前面的 functions or operations。它所在部分可译为"(3)功能度：要对数据执行的功能或操作的数目与各种功能的复杂性"。
3. 全句可译为：据说大约 80% 的商务数据有某种类型的地理成分。
4. 这是形容词的比较级，用来表示"比……更……得多"，其中副词 much 作状语，含意为"（……得）多"。全句可译为：地理信息编码看来简单，但要做好，却要复杂得多。
5. GIS 前省略了 that，引导的限制性定语从句修饰 areas。可译为：GIS 系统正有助于其商务开展的领域包括(1)房地产、(2)直销、(3)保险、(4)金融、(5)服务行业、(6)制造、(7)运输和分发，以及(8)零售。
6. to prepare 引导的不定式短语作定语，修饰 the user。全句可译为：准备移入 GIS 的属性数据的责任落在了用户肩上。
7. selecting 引导一个动名词短语作表语。全句可译为：在设计 GIS 时，主要关心的是为作业选择恰当的属性数据。

14.2 Introduction to GPS

Why GPS?
Trying to figure out where you are and where you're going is probably one of man's oldest

pastimes. [1] *Navigation* and positioning are crucial to so many activities and yet the process has always been quite cumbersome. Over the years all kinds of technologies have tried to simplify the task but every one has had some disadvantage. Finally, the U.S. Department of Defense decided that the military had to have a super precise form of worldwide positioning. And fortunately they had the kind of money ($ 12 billion!) it took to build something really good. [2] The result is the Global Positioning System, a system that's changed navigation forever.

What is GPS?

The Global Positioning System (GPS) is a worldwide radio-navigation system formed from a *constellation* of 24 satellites and their ground stations. [3] GPS uses these "man-made stars" as reference points to calculate positions accurate to *a matter of* meters. In fact, with advanced forms of GPS you can make measurements to better than a centimeter! In a sense it's like giving every square meter on the planet a unique address. [4]

GPS receivers have been *miniaturized* to just a few integrated circuits and so are becoming very economical. And that makes the technology accessible to virtually everyone.

These days GPS is finding its way into cars, boats, planes, construction equipment, movie making *gear*, farm machinery, even laptop computers.

Soon GPS will become almost as basic as the telephone. Indeed, at *Trimble*, we think it just may become a universal utility.

How GPS Works?

Here's how GPS works in five logical steps:

1. The basis of GPS is *"triangulation"* from satellites.

We're using the word "triangulation" very loosely here because it's a word[5] most people can understand, but *purists* would not call what GPS does "triangulation" because no angles are involved. It's really *"trilateration"*. Trilateration is a method of determining the relative positions of objects using the geometry of triangles.

2. To "triangulate", a GPS receiver measures distance using the travel time of radio signals.

3. To measure travel time, GPS needs very accurate timing which it achieves with some tricks.

4. Along with distance, you need to know exactly where the satellites are in space. [6] High orbits and careful monitoring are the secret.

5. Finally you must correct for any delays the signal experiences as it travels through the atmosphere. [7]

Why We Need Differential GPS?

Basic GPS is the most accurate radio-based navigation system ever developed. And for many applications it's plenty accurate. But it's human nature to want MORE! So some *crafty* engineers *came up with* "Differential GPS", a way to correct the various inaccuracies in the GPS system, pushing its accuracy even farther. [8]

Differential GPS or "DGPS" can yield measurements good to a couple of meters in moving applications and even better in stationary situations.

That improved accuracy has a profound effect on the importance of GPS as a resource. With it, GPS becomes more than just a system for navigating boats and planes around the world. It becomes a universal measurement system capable[9] of positioning things on a very precise scale.

Real-world Applications of GPS

These applications fall into five broad categories.

• Location — determining a basic position

The first and most obvious application of GPS is the simple determination of a "position" or location. GPS is the first positioning system to offer[10] highly precise location data for any point on the planet, in any weather.

• Navigation — getting from one location to another

GPS was originally designed to provide navigation information for ships and planes. So it's no surprise that while this technology is appropriate for navigating on water, it's also very useful in the air and on the land.

• Tracking — monitoring the movement of people and things

If navigation is the process of getting something from one location to another, then tracking is the process of monitoring it as it moves along.

Commerce relies on fleets of vehicles to deliver goods and services either across a crowded city or through nationwide corridors. So, effective fleet management has direct bottom-line implications, such as telling a customer when a package will arrive, spacing buses for the best scheduled service, directing the nearest ambulance to an accident, or helping tankers avoid hazards.

GPS used in conjunction with communication links and computers can provide the backbone for systems tailored to applications in agriculture, mass transit, urban delivery, public safety, and vessel and vehicle tracking. [11]

• Mapping — creating maps of the world

It's a big world *out there*, and using GPS to survey and map it precisely saves time and money in this most stringent of all applications. Today, Trimble GPS makes it possible for a single *surveyor* to accomplish in a day what used to take weeks with an entire team. [12] And they can do their work with a higher level of accuracy than ever before.

Mapping is the art and science of using GPS to locate items, then create maps and models of everything in the world. GPS is mapping the world.

• Timing — bringing precise timing to the world

Although GPS is well-known for navigation, tracking, and mapping, it's also used to *disseminate* precise time, time intervals, and frequency. Knowing that a group of timed events is perfectly synchronized is often very important. GPS makes the job of "synchronizing our watches" easy and reliable. Astronomers, power companies, computer networks, communications systems, banks, and radio and television stations can benefit from this precise timing.

Words and Expressions

pastime [ˈpɑːstaim] *n.* 消遣；娱乐
navigation [ˌnævɪˈgeɪʃən] *n.* 导航；航海，航空；领航；航行
constellation [ˌkɔnstəˈleɪʃən] *n.* 〔天〕星群，星座；灿烂的一群
a matter of 大约，大概
miniaturize [ˈmɪnɪətʃəraɪz] *vt.* 使小型化
these days 现在，目前
gear [gɪə] *n.* 装备；工具；用具；传动装置；齿轮 *vt.* 装备；用齿轮连接
Trimble Trimble 导航公司，注册商标
triangulation [traɪˌæŋgjuˈleɪʃən] *n.* 三角测量；三角剖分
purist [ˈpjʊərɪst] *n.* 纯化论者；纯粹派艺术家；语言纯正癖者
trilateration [traɪˌlætəˈreɪʃən] *n.* 〔测〕三边测量（术）
crafty [ˈkrɑːftɪ] *adj.* 〔英方〕灵巧的，巧妙的；狡猾的，诡计多端的
come up with 提出，提供；想出；赶上
out there 向那边；〔口〕到战场
surveyor [səˈveɪə] *n.* 测量员，勘测员；检查员
disseminate [dɪˈsemɪneɪt] *vt.* 传播；散布 *vi.* 广为传播

Abbreviation

GPS (Global Positioning System) 全球定位系统

Notes

1. 本句中 trying 是动名词，引导的短语作主语。全句可译为：设法弄清楚你在何处和你正去何处，可能是人们最早的消遣之一。
2. 本句中 it 前省略了 that，它引导限制性定语从句，修饰前面的 money；it 是先行代词，代替不定式短语 to build …；really good 修饰其前的 something。
3. formed 是过去分词，引导的短语作定语，修饰其前的 system。下一句中的 accurate 是形容词，引导的短语修饰其前的 positions。
4. 本句中的 giving 是动名词，引导一个短语作介词 like 的宾语，the planet 指的是地球。全句可译为：在某种意义上，像是给地球上的每个平方米一个唯一的地址。
5. word 之后省略了关系代词 that，它引导一个限制性定语从句，修饰此 word。
6. 本句中 where 是关系副词，引导一个从句作 know 的宾语。
7. 本句中 delays 之后是修饰 delays 的定语从句，此处省略了 that。全句可译为：最后你必须对信号通过大气层时经历的任何延迟作校正。
8. 本句中 Differential GPS (DGPS) 意为"微分 GPS"，a way 引导的是同位语，解释之前的 DGPS；to correct 是不定式，引导一个短语作定语，修饰其前的 a way。现在分词 pushing 引导的短语作状语，表示伴随情况。

9. 形容词 capable 引导的短语作定语，修饰其前的 system，其后的动名词 positioning 引导的短语作 of 的宾语。
10. 不定式 to offer 引导一个短语作定语，修饰其前的 system。
11. 本句中的 used 是过去分词，引导的短语作定语，修饰其前的 GPS，tailored 也是过去分词，引导的短语作定语，修饰其前的 systems。全句可译为：与通信连接装置和计算机一起使用的 GPS 能为那样一些系统提供骨干网，这些系统是为农业、公共交通、市内递送、公共安全，以及船与车辆追踪等应用定制的系统。
12. 本句中的 it 是先行代词，代替 to accomplish 引导的不定式短语；what 是关系代词，作从句中的主语。全句可译为：今天，Trimble GPS 使得一个测量员有可能在一天内就完成以往整个组花几个星期完成的工作。

Terms

CAI

CAI(short for Computer Aided Instruction or Computer Assisted Instruction) refers to instruction or remediation presented on a computer. Many educational computer programs are available online and from computer stores and textbook companies. They enhance teacher instruction in several ways.

Computer programs are interactive and can illustrate a concept through attractive animation, sound, and demonstration. They allow students to progress at their own pace and work individually or problem solve in a group.

CAI improves instruction for students with disabilities.

CAT

CAT(short for Computer Aided Translation) is the broadest term used to describe computer technology applications that automate or assist the act of translating text from one spoken language to another. Today CAT tool technology is being used by both business users and professional translators. Business users conducting business internationally are now finding benefit in using CAT tools when communicating across languages particularly when a translator is not available and the user has limited multilingual skills. Professional translators are finding CAT tools highly effective in improving their translation productivity and quality for work that lends itself to the use of the different technologies.

There are three major categories of tools currently available to assist or automate the translation process. Specifically they are
- Terminology Managers,
- Machine Translation (MT) Tools, and
- Machine Assisted Human Translation (MAHT) or Translator Workbench Tools.

CAE

CAE stands for *computer-aided engineering*, computer systems that analyze engineering designs. Most CAD systems have a CAE component, but there are also independent CAE systems that can analyze designs produced by various CAD systems. CAE systems are able to simulate a design under a variety of conditions to see if it actually works.

CAPP

CAPP stands for Computer Aided Process Planning. Process planning can be defined as an act of preparing processing documentation for the manufacturing of a piece, part or an assembly. Depending on the production environment it can be
— Rough
— Detailed

When process planning is done using a computer, it is Computer Aided Process Planning.

14.3 Management Information System (MIS)

The management information system (MIS) concept has been defined in dozens of ways. Since one organization's model of an MIS is likely to differ from that of another, it's not surprising that their MIS definitions would also vary in scope and breadth.[1] For our purposes, an MIS can be defined as a network of computer-based data processing procedures developed in an organization and integrated as necessary with manual and other procedures for the purpose of providing timely and effective information to support decision making and other necessary management functions.[2]

Although MIS models differ, most of them recognize the concepts shown in Figure 14-1. In addition to what might be termed the horizontal management structure shown in Figure 14-1(a), an organization is also divided vertically into different specialities and functions which require separate *information flows* (see Figure 14-1(b)). Combining the horizontal managerial levels with the vertical specialities produces the complex organizational structure shown in Figure 14-1(c). Underlying this structure is a data base consisting, ideally, of internally and externally produced data relating to past, present, and predicted further *events*.[3]

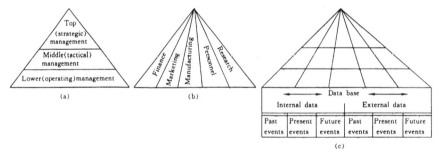

Figure 14-1 MIS design considerations

The *formidable* task of the MIS designer is to develop the information flow needed to support decision making (see Figure 14-2). Generally speaking, much of the information needed by managers who occupy different levels and who have different responsibilities is obtained from a collection of existing information systems (or subsystems).[4] These systems may be tied together very closely in an MIS. More often, however, they are more loosely *coupled*.

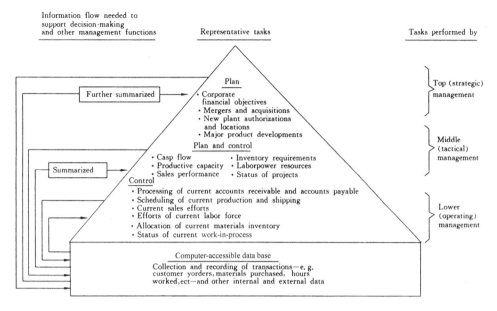

Figure 14-2 The task of MIS designers is to develop the information flow needed to support decision making

Words and Expressions

information flow   信息流
event [i'vent] *n.* 事件
formidable ['fɔːmidəbl] *adj.* 难对付的；艰难的
couple ['kʌpl] *vt.* 结合，连接；使耦合   *vi.* 结合

Abbreviation

MIS (Management Information System)   管理信息系统

Notes

1. from that of another 中的 that 代替 model of an MIS。全句可译为：因为一个组织的 MIS 模型多半与另一组织的 MIS 模型不同，毫不奇怪，它们的 MIS 定义也将随应用范围和广度的不同而变化。
2. 全句可译为：就我们的场合而言，MIS 可以定义为由一些基于计算机的数据处理过程所组成的网络，这些过程是在一个组织中开发的，并在必要时与手工的和其他的过程结合起来，以便为支持决策和其他必要的管理功能提供及时而有效的信息。
3. underlying 引导的动名词短语作主语。relating 引导的分词短语作定语，修饰 data。全句可译为：构成这个结构的基础是数据库，理想情况下它由内部和外部产生的，且与过去、现在和预期的将来事件相关的数据组成。
4. generally speaking 是插入语。全句可译为:一般来说,处于不同层次和负有不同责任的管理人员所需要的大部分信息是从一组现有信息系统(或子系统)获得的。

14.4 Enterprise Resource Planning

Enterprise Resource Planning (ERP) integrates internal and external management information across an entire organization, *embracing* finance/accounting, manufacturing, sales and service, etc. ERP systems automate this activity with an integrated software application. Its purpose is to facilitate the flow of information between all business functions inside the boundaries of the organization and manage the connections to outside *stakeholders*.

ERP systems can run on a variety of hardware and network configurations, typically employing a database to store data.

ERP systems typically include the following characteristics:
- An integrated system that operates in real time (or *next to* real time), without relying on periodic updates. [1]
- A common database, which supports all applications.
- A consistent look and feel throughout each module.
- Installation of the system without *elaborate* application/data integration by the Information Technology (IT) department.

By the mid 1990s[2] ERP systems addressed all core functions of an enterprise. Beyond corporations, governments and non-profit organizations also began to employ ERP systems.

ERP systems initially focused on automating *back office* functions that did not directly affect customers and the general public. *Front office* functions such as customer relationship management (CRM) dealt directly with customers, or e-systems such as e-commerce, e-government, e-telecom, and e-finance, or supplier relationship management (SRM) became integrated later, when the Internet simplified communicating with external parties. [3]

"ERP II" is *coined* in the early 2000s. It describes web based software that allows both employees and partners (such as suppliers and customers) real time access to the systems. "Enterprise application suite" is an alternate name of such systems.

Most systems are modular to permit automating some functions but not others. Some common modules, such as finance and accounting, are adopted by nearly all users; others such as human resource management are not. For example, a service company probably has no need for a manufacturing module. Other companies already have a system that they believe to be *adequate*. Generally speaking, the greater the number of modules selected, the greater the integration benefits, but also the greater the costs, risks and changes involved.

ERP systems connect to real time data and transaction data in a variety of ways. These systems are typically configured by systems integrators, who bring unique knowledge on process, equipment, and vendor solutions.

Direct integration — ERP systems connectivity (communications to *plant floor* equipment) is as part of their product offering. This requires the vendors to offer specific support for the plant floor equipment that their customers operate. ERP vendors must be expert in their own products, and connectivity to other vendor products, including competitors.

Database integration—ERP systems connect to plant floor data sources through *staging* tables in a database. Plant floor systems deposit the necessary information into the database. The ERP system reads the information in the table. The benefit of staging is that ERP vendors do not need to master the complexities of equipment integration. Connectivity becomes the responsibility of the systems integrator.

Configuring an ERP system is largely a matter of balancing the way the customer wants the system to work with the way it was designed to work.[4] ERP systems typically build many changeable parameters that modify system operation. For example, an organization can select the type of inventory accounting — FIFO or LIFO — to employ, whether to recognize revenue by geographical unit, product line, or distribution channel and whether to pay for shipping costs when a customer returns a purchase. When the system doesn't offer a particular feature, the customer can rewrite part of the code, or interface to an existing system. Both options add time and cost to the implementation process and can *dilute* system benefits. *Customization* inhibits seamless communication between suppliers and customers who use the same ERP system uncustomized.

Key differences between customization and configuration include:
- Customization is always optional, whereas the software must always be configured before use (e. g., setting up cost/profit center structures, organisational trees, purchase approval rules, etc.).
- The software was designed to handle various configurations, and *behaves* predictably any allowed configuration.
- The effect of configuration changes on system behavior and performance is predictable and is the responsibility of the ERP vendor. The effect of customization is less predictable, is the customer's responsibility and increases testing activities.
- Configuration changes survive upgrades to new software versions. Some customizations (e. g. code that uses pre defined "hooks" that are called before/after displaying data screens) *survive* upgrades, though they require retesting. Other customizations (e. g. those involving changes to fundamental data structures) are overwritten during upgrades and must be reimplemented.

Customization can be expensive and complicated, and can delay implementation. Nevertheless, customization offers the potential to obtain competitive advantage *vis a vis* companies using only standard features.[5]

The fundamental advantage of ERP is that integrating the *myriad* processes by which businesses operate saves time and expense. Decisions can be made more quickly and with fewer errors. Data becomes visible across the organization. Tasks that benefit from this integration include:

- Sales forecasting, which allows inventory optimization.
- Order tracking, from acceptance through fulfillment.
- Revenue tracking, from invoice through cash receipt.
- Matching purchase orders (what was ordered), inventory receipts (what arrived), and costing (what the vendor *invoiced*).

ERP systems centralize business data, bringing the following benefits:
- They eliminate the need to synchronize changes between multiple systems — *consolidation* of finance, marketing and sales, human resource, and manufacturing applications.
- They enable standard product naming/coding.
- They provide a comprehensive enterprise view (no "islands of information"). They make real time information available to management any where, anytime to make proper decisions.
- They protect sensitive data by consolidating multiple security systems into a single structure.

The disadvantages of ERP are as following:
- Customization is problematic.
- Re-engineering business processes to fit the ERP system may damage competitiveness and/or divert focus from other critical activities.
- ERP can cost more than less integrated and/or less comprehensive solutions.
- High switching costs increase vendor negotiating power vis a vis support, maintenance and upgrade expenses.
- Overcoming resistance to sharing sensitive information between departments can divert management attention.
- Integration of truly independent businesses can create unnecessary dependencies.
- Extensive training requirements take resources from daily operations.

Words and Expressions

embrace [im'breis] *vt.* 包括，包含
stakeholder ['steikhəuldə] *n.* 利益共享者；赌金持有人；赌金保管者
next to 几乎
elaborate [i'læbərit] *adj.* 精心制作的；详尽说明的
　　　　 [i'læbəreit] *vt.* 精心制作；详尽说明
back office 后台　front office 前台
coin [kɔin] *vt.* 创造；铸造；杜撰（新词、新语等）　*n.* 硬币
adequate ['ædikwit] *adj.* 适当的；充分的，足够的；可以胜任的
plant floor　工厂车间
staging ['steidʒiŋ] *n.* 中间集结；分段运输　staging table 暂存表，中间临时表

configure [kən'figə] vt. (尤指对计算机设备进行)配置；设定；使成形，使具形体
dilute [dai'lju:t] vt. 冲淡，稀释 adj. 稀释的，冲淡的
customization ['kʌstəˌmai'zeiʃən] n. 定制；[计]用户化；专用化
behave [bi'heiv] vi. 举止；表现；运转 vt. 使运转正常
survive [sə'vaiv] vt. 从……中逃生，幸免于；比……活得长 vi. 活下来；幸存
vis a vis 同……相比；面对面；对于
myriad ['miriəd] adj. 无数的 n. 无数，极大数量
invoice ['in'vɔis] vt. & vi. (把……)开发票；(把……)开清单 n. 发票；装货清单
consolidation [kənˌsɔli'deiʃən] n. 巩固；加强；强化；变坚固

Abbreviations

ERP (Enterprise Resource Planning) 企业资源计划
FIFO (First In First Out)   先进先出
LIFO (Last In First Out)   后进先出

Notes

1. 本句中，过去分词 integrated 作定语，含意为"集成式的"。全句可译为：一个实时(或几乎实时)运转的集成式系统，不依赖于周期的更新。
2. by the mid-1990s，意为"到 20 世纪 90 年代中期时(为止)"。
3. 本句中，最前的 such as 引导的短语修饰前面的 functions；dealt 是 deal 的过去分词形式，引导的短语作限制性定语从句，修饰其前的 CRM。全句可译为：诸如直接处理客户(信息)的客户关系管理(CRM)(系统)，e-商务、e-政务、e-通信和 e-金融这样的 e-系统，或供应商关系管理(SRM)(系统)这样一些前台功能，后来当因特网简化与外部各方的通信时变成了集成式的。
4. 本句中，configuring 引导的动名词短语作主语，balancing 引导的动名词短语作介词 of 的宾语，balance ... with ... ，含意为"使……与……平衡"。句中的两个 way 后面都省略了 that。全句可译为：配置一个 ERP 系统主要是使客户希望系统工作的方式与系统被设计成的工作方式达成平衡。
5. using 引导的分词短语修饰前面的 companies，含意为"使用……的"。全句可译为：然而，与仅使用一些标准特性的公司相比，定制提供了获得竞争优势的潜力。

Exercises

1. Translate the text of Lesson 14.1 into Chinese.
2. Topics for oral workshop.
• What is the essential difference between GIS and GPS?
• Do you use GPS in your live? What about it?
• Explain the MIS briefly. And describe its major applications.
3. Translate the following into English.

(a) 地理信息系统(GIS)是捕捉、存储、分析、管理和参照(with reference to)地理位置数据展示(present)数据的系统。简而言之(in the simplest terms)，GIS 是制图(cartography)、统计分析和数据库技术的融合(merge)。

(b) GPS 有一些应用超越了(beyond)导航和定位。GPS 可用于制图(cartography)、林业、矿产勘查、野生动物聚居地(habitation)管理、监控人和事物的运动，以及把精确定时带(bring)给世界。

(c) 管理信息系统是一个多学科方法(multi-disciplinary approach)业务管理的产品。它是(这样)一个产品，需要持续不断地(keep under a constant …)复审和修改以满足企业对信息的需求。

(d) 除了支持决策、合作与控制，信息系统也可以帮助管理人员和工作人员分析问题、把一些复杂的主题可视化，以及创造一些新产品。

4. Listen to the video "VPN" and write down the first paragraph.

Unit 15 Software Development

15.1 Overview of Software Engineering

Businesses around the world depend more and more on software in the very basics of their *operations*.[1] U.S. firms alone have 100 billion lines of program code in use today. This code cost $ 2 *trillion* to create and costs $ 30 billion a year to maintain. The typical Fortune 1,000 company maintains 35 million lines of code. Quality of software design and quality of business service are increasingly linked. We *take for granted* the everyday convenience we gain from reservation, telephone, *automated teller*, and credit card *authorization* applications.[2] We can take these conveniences for granted until they "crash" or have a "bug". Software engineers (SEs) developed those systems. The engineering skills they apply to developing applications go far beyond the writing of good programs.[3] The skills SEs need are to deploy and manage the data, software, hardware, and communications business *assets* of a corporation. These computer-related assets now account for almost half of all U.S. business investment.

Software engineers are skilled professionals who can make a real difference to business *profitability*.[4] The word professional is key here. Software development is *notoriously* difficult to manage; software projects are routinely over budget and behind schedule. Computer programmers are *legendary* for their lack of understanding of, or interest in, business.[5] SEs who are professionals are more likely to manage and deliver a quality project on time and within budget. One goal of this text is to challenge you to set high standards for personal excellence: to become a professional and to make a difference.

This Lesson introduces you to the book and the topics to be covered in more detail in later Lessons. The objectives of this Lesson are to: (1) review what you might already know, (2) give you a vocabulary for discussing applications, and (3) introduce the topics of this text. Use this Lesson to learn basic definitions and to begin building a mental picture of how different approaches to software engineering work.[6] You will learn the details in later Lessons.

Software engineering is the systematic development, operation, maintenance, and *retirement* of software. Software engineers (SEs) have a mental '*tool kit*' of techniques to use in developing applications. As students of information systems, you know *bits and pieces* of the tool kit. This text will show you how to use the tools together, and will add to what you already know. For instance, you should already know data flow diagrams (DFDs). DFDs are one of many tools, including new diagrams such as process hierarchies, process dependencies, and object diagrams. No one tool is ideal or complete. The SE knows how to select the tools, understanding their strengths and weaknesses. Most of all, an SE is not limited to a single tool he or she tries to force-fit to all situations.

Software engineering is important because it gives you a foundation on which to develop a career as an information systems development professional. At the end of the course, you will understand a variety of approaches to analyzing, designing, programming, testing, and maintaining information systems in organizations.[7] You will know the alternatives for developing applications, and you will know how and when to select from among them. You will be able to compare and contrast methodology differences and will know the major computer-aided software engineering (CASE) tools that support each methodology. Finally, you will have an appreciation of the roles of software engineers and how they work with project managers in application development. This conversation might be overheard in a manager's office:

Consultant Manager: "All right, Mary, tomorrow you start work on the rental processing application we are developing for ABC's Video Company. Mary, you are the project manager. Are you ready?"

Mary: "Yes, our first job is to find out more about the application. Then, Sam and I will decide our approach to development and the documentation that is needed. ABC's manager, Vic, is willing to provide us with whatever we need. Then, we will complete a feasibility analysis and…"

Mary is describing the first steps used by a modern software engineer in the development of a computer-based application. Software is the sequences of instructions in one or more programming languages that comprise a computer application to automate some business function.[8] Engineering is the use of tools and techniques in problem solving. Putting the two words together, software engineering is the systematic application of tools and techniques in the development of computer-based applications.

A software engineer is a person who applies a broad range of application development knowledge to the systematic development of application systems for organizations. Software engineers used to think of their job as *conscientious* development of well-structured computer programs. But, as the field evolved, systems analysis as a task appeared along with systems analysts, the people who perform that task.[9] Now, there is a *proliferation* of techniques, tools, and technologies to develop applications. Software engineers' jobs have evolved to now include evaluation, selection, and use of specific systematic approaches to the development, operation, maintenance, and retirement of software. Development begins with the decision to develop a software product and ends when the product is delivered. Operations is the daily processing that takes place. Maintenance encompasses the changes made to the logic of the system and programs to fix errors, provide for business changes, or make the software more efficient. Retirement is the replacement of the current application with some other method of providing the work, usually a new application.

Words and Expressions

operation [ˌɔpəˈreiʃən] *n.* 经营；运转；操作

trillion [ˈtriljən] *num.* 万亿　　*n.* 大量，无数

take sth. for granted 认为某事当然，认为……不成问题
automated teller 自动出纳机，自动柜员机
authorization [ˌɔːθəraiˈzeiʃən] n. 授权；委任；认可
asset [ˈæset] n. 资产；财产；有用的东西
profitability [ˌprɔfitəˈbiliti] n. 有利(可图)；有益性；有用性
notoriously [nəuˈtɔːriəsli] adv. 臭名昭著地，声名狼藉地
legendary [ˈledʒəndəri] adj. 传说中的，传奇中的
retirement [riˈtaiəmənt] n. 退役，引退
tool kit 工具箱，工具包
bits and pieces 零碎东西
conscientious [ˌkɔnʃiˈenʃəs] adj. 认真的；诚心诚意的；谨慎的
proliferation [prəˌlifəˈreiʃən] n. 激增；扩散；增殖，增生

Abbreviations

DFD (Data Flow Diagram) 数据流图
SE (Software Engineering) 软件工程
CASE (Computer-Aided Software Engineering) 计算机辅助软件工程

Notes

1. around 引导的介词短语作定语，修饰 businesses。全句可译为：世界各地的企业在基本操作上越来越依赖于软件。
2. 本句第 2 个 we 前省略了关系代词 that，它引导的限制性定语从句修饰 the everyday convenience。全句可译为：我们认为每天从预订服务、电话、自动出纳机和信用卡授权应用所获得的便利是天经地义的。
3. they 前省略了关系代词 that，它引导的定语从句修饰 the engineering skills。go far beyond 含意为"远远超过……"。全句可译为：他们用来开发应用程序的工程技能远不只是编写高质量的程序。
4. who 引导的限制性定语从句修饰 professionals。全句可译为：软件工程师是能使企业盈利能力发生真正变化的熟练专业人员。
5. legendary 含意为"传说中的、传奇中的"。全句可译为：计算机程序员因缺乏对企业的理解或兴趣，他们(所制作的软件)是不切实际的。
6. mental 含意为"智力的、脑力的"。全句可译为：通过这节课来了解基本定义并开始在脑海中构筑不同软件工程方法如何工作的画面。
7. analyzing 等是动名词，作介词 to 的宾语。全句可译为：在课程结束时，你将了解对机构中的信息系统进行分析、设计、程序设计、测试和维护的各种方法。
8. that 引导的限制性定语从句修饰 the sequences of instructions。全句可译为：软件是用一种或多种程序设计语言编写的指令序列，它们构成了使某个企业功能自动化的计算机应用(程序)。
9. the people 引导 systems analysts 的同位语。全句直译为：然而，随着活动范围的演变，

系统分析作为一个任务与执行该任务的系统分析员一起出现。宜译为"然而，随着活动范围的演变，系统分析员把系统分析作为一项任务来执行"。

15.2　Unified Modeling Language

Unified Modeling Language (UML) is a standardized general-purpose modeling-language in the field of object-oriented software engineering. The standard is managed, and was created by the Object Management Group.

UML includes a set of graphic notation techniques to create visual models of object-oriented *software-intensive systems*.

The Unified Modeling Language (UML) is used to specify, visualize, modify, construct and document the *artifacts* of an object-oriented software-intensive system under development. UML offers a standard way to visualize a system's architectural blueprints, including elements such as: activities, actors, business processes, database schemas, (logical) *components*, programming language statements, and reusable software components.

UML combines techniques from data modeling (entity relationship diagrams), business modeling (work flows), object modeling, and component modeling. It can be used with all processes, throughout the software development *life cycle*, and across different implementation technologies.[1] UML has *synthesized* the notations of the Booch method[2], the Object-modeling technique (OMT) and Object-oriented software engineering (OOSE) by *fusing* them into a single, common and widely usable modeling language. UML aims to be a standard modeling language which can model concurrent and distributed systems. UML is a *de facto* industry standard, and is evolving under the *auspices* of the Object Management Group (OMG).

UML models may be automatically transformed to other representations (e.g. Java) by means of QVT-like[3] transformation languages. UML is extensible, with two mechanisms for customization: *profiles* and *stereotypes*.

UML is not a development method by itself; however, it was designed to be compatible with the leading object-oriented software development methods of its time (for example OMT, Booch method, Objectory[4]). Since UML has evolved, some of these methods have been recast to take advantage of the new notations (for example OMT), and new methods have been created based on UML, such as IBM Rational Unified Process (RUP). Others include Abstraction Method and Dynamic Systems Development Method.

It is important to distinguish between the UML model and the set of diagrams of a system. A diagram is a partial graphic representation of a system's model. The model also contains documentation that drives the model elements and diagrams (such as written use cases).

UML diagrams represent two different views of a system model:
- Static (or structural) view: emphasizes the static structure of the system using objects, attributes, operations and relationships. The structural view includes class diagrams and composite structure diagrams.

- Dynamic (or behavioral) view: emphasizes the dynamic behavior of the system by showing collaborations among objects and changes to the internal states of objects. This view includes sequence diagrams, activity diagrams and state machine diagrams.

UML models can be exchanged among UML tools by using the XMI interchange format.

UML 2.2 has 14 types of diagrams divided into two categories. Seven diagram types represent structural information, and the other seven represent general types of behavior, including four that represent different aspects of interactions.

UML does not restrict UML element types to a certain diagram type. In general, every UML element may appear on almost all types of diagrams; this flexibility has been partially restricted in UML 2.0. UML profiles may define additional diagram types or extend existing diagrams with additional notations.

In keeping with the tradition of engineering drawings, a comment or note explaining usage, constraint, or intent is allowed in a UML diagram.[5]

Structure diagrams emphasize the things that must be present in the system being modeled.[6] Since structure diagrams represent the structure, they are used extensively in documenting the software architecture of software systems.

Behavior diagrams emphasize what must happen in the system being modeled. Since behavior diagrams illustrate the behavior of a system, they are used extensively to describe the functionality of software systems.

Interaction diagrams, a subset of behavior diagrams, emphasize the flow of control and data among the things in the system being modeled:

The Protocol State Machine is a *sub-variant* of the State Machine. It may be used to model network communication protocols.

Although UML is a widely recognized and used modeling standard, it is frequently criticized for the following: standards *bloat*, problems in learning and adopting, *linguistic incoherence*, capabilities of UML and implementation language mismatch, and *dysfunctional* interchange format.

Words and Expressions

software-intensive system　软件密集型系统
artifact ['ɑːtiˌfækt] *n.* 人工制品，制造物
component [kəm'pəunənt] *n.* 组件，构件
life cycle 生命周期
synthesize ['sinθiˌsaiz] *vt.* 综合；用合成法合成　*vi.* 综合；合成
fuse [fjuːz] *vt.* 熔合，熔(化)；融合　*n.* 熔丝，保险丝
de facto 事实上
auspice ['ɔːspis] *n.* [复] 赞助，主办；预兆，前兆
profile ['prəufail] *n.* 配置文件，概要文件；侧面；轮廓，外形；纵剖面(图)，纵断面(图)
stereotype ['stiəriətaip] *n.* 版型；铅版，铅版制版法；陈规，老套

in keeping with　与……一致；与……协调
sub-variant [ˌsʌb'vɛəriːənt,'vær-] n. 亚变种
bloat [bləut] vt. & vi. (使)膨胀
linguistic incoherence　语言不连贯，语言难理解，语言不清楚
dysfunctional [dis'fʌŋkʃənəl] adj. 机能失调的，功能失调的

Abbreviations

UML (Unified Modeling Language)　统一建模语言
OOSE (Object-Oriented Software Engineering)　面向对象软件工程
RUP (Rational Unified Process)　统一软件开发过程

Notes

1. 本句中，throughout 的意思是"遍及，贯穿"，across 的意思是"横过，穿过"。全句可译为：它能在整个软件开发生命周期和各种不同的实现技术中，用于所有的处理。
2. Booch method，Booch 法，这是用在软件工程中的一种技术，是广泛使用在面向对象的分析与设计中的对象建模语言和方法论。
3. QVT 指 Query/View/Transformation，即询问/视图/变换，在模型驱动的体系结构中，是由对象管理组(Object Management Group)定义的模型变换标准。QVT-like，通常译为"QVT 类的"。
4. Objectory 是一种面向对象的方法论。
5. 本句中，explaining 引导分词短语，修饰前面的 a comment or note。全句可译为：与传统的工程作图法一致，在 UML 图表中允许有解释用途、限制或意图的注解或注。
6. being modeled 修饰前面的 system，含意为"正被建模的"。全句可译为：结构图表强调正被建模的系统中必须存在的东西。

15.3 Integrated Computer Aided Software Engineering

Nowadays everything has to go faster. Because of the increasing speed of changing market-demands new products replace old ones much earlier than before, so the development of new products has to go faster. Thus the production lines have to be developed faster, too. A very important role in this development is software engineering because many production processes are 'computer aided', so software has to be designed for this production system. It seems very important to do the software engineering right and fast. In the past, software systems were build using traditional development techniques, which relied on hand-coding applications.[1] Software engineers had to design software without help of computers, by programming each step at one time. This way is much too costly and time-consuming. In order to speed up the process the bottlenecks in building software systems are to be found. This is hard to do because of the increasing role of computers in our society. Technology is developed further every day, so faster and bigger computers enter the scene. The software running[2] on these computers can be more extensive because they can handle more information in the same time, so there is an increasing amount of data to *go with* it. Finding the right data out of this increasing amount of information is getting harder and harder, so finding the bottleneck is harder to do. [3]

To speed up the software system building process, a new concept of designing software is introduced in the '70s, called Computer Aided Software Engineering (CASE). [4] This term is used for a new generation of tools that applies rigorous engineering principles to the development and analysis of software specifications. Simply, computers develop software for other computers in a fast way by using specific tools. When *implanted* in a Concurrent Engineering environment, this process is taking place while some parts of the developed[5] software are running already. It's a sort of on-line software engineering. There are a lot of problems with these kinds of systems because they are very complex, not easily maintainable and fragile. Some of the tools, which[6] were very *progressive* back then, are *obsolete* right now, so they have to be updated, which can be a problem because the software engineering process is fit around these tools.

The tools developed right now are evolutional products out of earlier tools. The first tools, developed in the mid '70s, were mainly introduced to automate the production and to maintenance structured diagrams. When this automation covers the complete life-cycle process, we speak of Integrated Computer Aided Software Engineering (I-CASE). When only one specific part of the life-cycle process is covered we speak of -just- Computer Aided Software Engineering (CASE).

Later on, integrated CASE (I-CASE) products were introduced. This was an important step because I-CASE products are capable of being used to generate entire applications from design specifications.

Recently, CASE tools have entered a third phase: the introduction of new methodologies based on capabilities of I-CASE tools. These new methodologies utilize *Rapid Prototyping* techniques to develop applications faster, at lower cost and higher quality. By using Rapid Prototyping a prototype can be made fast, so the developed system can be tested more often in between the development-phases because it doesn't cost much time to create a prototype. Mistakes can be detected and corrected earlier this way. [7] The earlier this can be done, the better because correcting these mistakes gets harder and more expensive when the system is developed further. So a lot of time and money can be saved using Rapid Prototyping.

As said above, a new set of tools is necessary. These tools should automate each phase of the life-cycle process and tie the application development more closely to the strategic operations of the business. A lot of different tools have been developed over the years and are being developed right now. There are so many tools that we could easily get confused. To survey all these CASE tools we divide them in the following categories:

1. *Information engineering-supporting products*. These are life-cycle processes, derived from the strategic plans of the enterprise and which provide a *repository* to create and

maintain enterprise models, data models and process models.[8]

2. *Structured diagramming-supporting products*. These are derived from several development methodologies like Gane-Sarson or Jackson. These products at least support data flow, control flow and entity flow, which are the three basic structured software engineering diagramming types.

3. *Structured development aids-providing products*. These products are providing aids for a structured development of the process. These products are very suitable to be used by the system analysts, because they are helped very much by a structured process, because those can be analysed faster and more accurately.

4. *Application-code-generating products*. These are products that generate application-code for a specific goal, set by the designer.[9] Most of the products in this area are using a COBOL-generator, which is a tool that generates programming code in a specific language out of specifications set by the system-designer.

The heart of a well-designed I-CASE system is a repository, which is used as a *knowledge base* to store information about the organization, its structure, enterprise model, functions, procedures, data models etc. The meaning represented[10] by diagrams and their detail windows is stored in the repository. The repository steadily accumulates information relating to the planning, analysis, design, construction and maintenance of systems. In other words: *The repository is the heart of a CASE system*.

Two types of mechanisms have been used in CASE software to store design information:

1. A *dictionary*, which contains names and descriptions of data items, processes, etc.

2. A *repository*, which contains this dictionary information and a complete coded representation of plans, models and designs, with tools for *cross-checking*, correlation analysis and *validation*.

Before implanting CASE and designing tools, a series of steps should be followed:

1. Conduct a technology-impact study to determine how the basic business of the organization should change to maximize the opportunities presented by rapid technological change

2. Evaluate how the organization should be *re-engineered* to take advantage of new technology

3. Establish a program for replacing the old systems with the most effective new technology

4. Commit to an overall integrated architecture

5. Select a development methodology

6. Select a CASE-tool

7. Establish a culture of reusability

8. Strive for an environment of open interconnectivity and software portability across the entire enterprise

9. Establish *inter-corporate* network links to most trading partners

10. Determine how to provide all knowledge to workers with a high level of computerized

knowledge and processing power

11. Determine the changes in management-structure required[11] to take full advantage of innovative systems, architectures, methodologies and tools

When an enterprise takes these actions ahead of its competition, it will gain a major competitive advantage. An enterprise should also be *up-to-date* because the rapid advances in computer technology allows the competition to *get ahead of* you. Some significant trends in the development of new system environments include:

1. Low-cost MIPS, the price of fast processors is decreasing and even faster everyday
2. Distributed computing environment, end-users are moving towards a *multilayered distributed computer architecture*
3. CASE and I-CASE tools, highly integrated, repository-driven, computer-aided systems engineering tools are making it possible to generate application code directly from graphical specifications
4. *Forward/reverse* engineering tools, re-engineering tools enable analysts to convert low-level data definition and unstructured process code into standardized data elements and structured code.
5. New development methodologies, more efficient development life-cycle processes are making it possible to develop applications more rapidly and in closer coordination with end users.
6. Growth of standards, standards are emerging that will govern the future evolution of hardware and software.[12]

We now have some insight in what I-CASE is, how it should be implanted and used and which tools are available.

Words and Expressions

go with　伴随；与……持同一看法；与……相配
implant [im'plɑ:nt] *vt.* 移植；植入，插入　*n.* 移植物，植入管
progressive [prə'gresiv] *adj.* 进步的，先进的；渐进的，累进的
obsolete ['ɔbsəli:t] *adj.* 已废弃的；老式的，过时的　*n.* 被废弃的事物
rapid prototyping　快速原型（法）
information engineering-supporting product　支持信息工程的产品
repository [ri'pɔzitəri] *n.* 贮藏所，仓库；陈列室；资源丰富地区
structured diagramming-supporting product　支持结构图的产品
structured development aids-providing product　提供结构化开发辅助的产品
application-code-generating product　应用程序代码生成的产品
knowledge base　知识库
cross-checking　交叉检验
validation [ˌvæli'deiʃən] *n.* 确认，证实；生效，合法化
re-engineer [ˌri:endʒi'niə] *vt.* 再工程；重组；重新设计，重新建造

inter-corporate　公司间
up-to-date　直到最近的，现代的，新式的，时新的
get ahead (of)　超过，胜过
multilayered distributed computer architecture　多层分布式计算机体系结构
forward/reverse　正向/逆向，正向/反向

Abbreviations

CE (Concurrent Engineering)　并行工程
I-CASE (Integrate Computer Aided Software Engineering)　集成式计算机辅助软件工程
MIPS (Million Instructions Per Second)　每秒百万条指令

Notes

1. 本句中，using 引导的分词短语作状语，表示方式。关系代词 which 引导的非限制性定语从句修饰前面的整个主句。hand-coding 意为"手工编码的"。
2. running 引导的分词短语作定语，修饰其前的 software。
3. 两处动名词 finding 引导的短语作主语。全句可译为：从这日益增加的信息量中找到合适的数据变得越来越困难，因此更难找到瓶颈了。
4. 本句中，动名词短语 designing software 是 concept 的同位语。called 引导的分词短语作非限制性定语，修饰其前的 concept，其作用接近同位语。全句可译为：为加速软件系统构建过程，在(20 世纪)70 年代引进了设计软件的新概念，称为计算机辅助软件工程(CASE)。
5. developed 是过去分词，作定语，修饰 software，意为"所开发的软件"。
6. 此关系代词 which 引导的非限制性定语从句修饰其前的 some。后面的 which 引导的非限制性定语从句修饰其前的句子"so they …"。
7. this way 作状语，表示方式。全句可译为：这样错误就能较早地被发现并校正。
8. 本句中，derived 引导的分词短语作非限制性定语，which 引导非限制性定语从句，两者都修饰 these。前面的 information engineering(IE) 意为"信息工程"。IE 是一种设计和开发信息系统的软件工程实现方法。
9. 本句中，that 引导的限制性定语从句修饰其前的 products；set 是过去分词，引导的分词短语修饰其前的 goal。下一句中的 set 情况类似。
10. represented 引导的分词短语作定语，修饰其前的 meaning。下一句中的 relating 引导的分词短语作定语，修饰其前的 information。
11. required 引导的分词短语作定语，修饰其前的 changes。全句可译为：确定为充分利用创新的系统、体系结构、方法论及工具而在管理结构中所需的改变。
12. 关系代词 that 引导的限制性定语从句修饰其前的 standards。全句可译为：标准的增长，正出现一些标准，这些标准将支配硬件和软件的未来发展。

15.4　Agile Software Development Methods

Probably the most noticeable change to software process thinking in the last few years has been the appearance of the word '*agile*'. We talk of agile software methods, of how to introduce agility into a development team, or of how to resist the impending storm of agilists determined to change well-established practices. [1]

From Nothing, to Monumental, to Agile

Most software development is a chaotic activity, often characterized by the phrase "code and fix". The software is written without much of an underlying plan, and the design of the system is *cobbled* together from many short term decisions. This actually works pretty well as the system is small, but as the system grows it becomes increasingly difficult to add new features to the system. Furthermore bugs become increasingly prevalent and increasingly difficult to fix. A typical sign of such a system is a long test phase after the system is "feature complete". Such a long test phase *plays havoc with* schedules as testing and debugging is impossible to schedule. [2]

The original movement to try to change this introduced the notion of *methodology*. These methodologies impose a *disciplined* process upon software development with the aim of making software development more predictable and more efficient. They do this by developing a detailed process with a strong emphasis on planning inspired by other engineering disciplines — which is why I like to refer to them as engineering methodologies (another widely used term for them is *plan-driven* methodologies).

Engineering methodologies have been around for a long time. They've not been noticeable for being terribly successful. [3] They are even less noted for being popular. The most frequent criticism of these methodologies is that they are *bureaucratic*. There's so much *stuff* to do to follow the methodology that the whole pace of development slows down. [4]

Agile methodologies developed as a reaction to these methodologies. For many people the appeal of these agile methodologies is their reaction to the bureaucracy of the engineering methodologies. These new methods attempt a useful compromise between no process and too much process, providing just enough process to gain a reasonable *payoff*.

The result of all of this is that agile methods have some significant changes in emphasis from engineering methods. The most immediate difference is that they are less document-oriented, usually emphasizing a smaller amount of documentation for a given task. *In many ways* they are rather code-oriented: following a route that says that the key part of documentation is source code.

However I don't think this is the key point about agile methods. Lack of documentation is a symptom of two much deeper differences:

• Agile methods are adaptive rather than predictive. Engineering methods tend to try to plan out a large part of the software process in great detail for a long span of time, this works well until things change. So their nature is to resist change. The agile methods, however, welcome change. They try to be processes that adapt and *thrive* on change, even *to the point of* changing themselves.

• Agile methods are people-oriented rather than process-oriented. The goal of engineering methods is to define a process that will work well whoever happens to be using it. Agile methods assert that no process will ever *make up* the skill of the development team, so the role of a process is to support the development team in their work.

Flavors of Agile Development

The term 'agile' refers to a philosophy of software development. Under this broad umbrella sits many more specific approaches such as extreme programming (XP), Scrum, Lean Development, etc. Each of these more particular approaches has its own ideas, communities and leaders. Each community is a distinct group of its own but to be correctly called agile it should follow the same broad principles.[5] Each community also borrows from ideas and techniques from each other. Many practitioners move between different communities spreading different ideas around—*all in all* it's a complicated but vibrant *ecosystem*.

XP (extreme programming)

XP begins with five values (Communication, Feedback, Simplicity, Courage, and Respect). It then elaborates these into fourteen principles and again into twenty-four practices.

XP's approach here, often described under the heading of Test Driven Development (TDD) has been influential even in places that haven't adopted much else of XP.

Scrum

Scrum also developed in the 80's and 90's primarily with OO development circles as a highly iterative development methodology.

Scrum concentrates on the management aspects of software development, dividing development into thirty day iterations (called 'sprints') and applying closer monitoring and control with daily scrum meetings.[6] It places much less emphasis on engineering practices and many people combine its project management approach with extreme programming's engineering practices. (XP's management practices aren't really very different.)

Context Driven Testing

As it turns out, several people in the testing community have been questioning much of mainstream testing thinking for quite a while. This has led to a group known as *context-driven testing*. The best description of this is the book Lessons Learned in Software Testing. This community is also very active on the web, take a look at sites hosted by Brian Marick (one of the authors of the agile manifesto), Brett Pettichord, James Bach, and Cem Kaner.

Lean Development

I remember a few years ago giving a talk about agile methods at the Software Development conference and talking to an eager woman about parallels between the agile ideas and lean movement in manufacturing. The lean movement in manufacturing was pioneered by Taiichi Ohno at Toyota and is often known as the Toyota Production System. Lean production was an inspiration to many of the early agilists. In general I'm very wary of these kinds of reasoning by analogy, indeed the engineering separation between design and construction got us into this mess *in the first place*. But there are interesting ideas from the lean direction.

(Rational) Unified Process

Another well-known process to have come out of the object-oriented community is the

Rational Unified Process (sometimes just referred to as the Unified Process). The original idea was that like the UML unified modeling languages the UP could unify software processes. Since RUP appeared about the same time as the agile methods, there's a lot of discussion about whether the two are compatible.

Should you go agile?

A lot of people claim that agile methods can't be used on large projects. We (ThoughtWorks) have had good success with agile projects with around 100 people and multiple continents. Despite this I would suggest picking something smaller to start with. Large projects are *inherently* more difficult anyway, so it's better to start learning on a project of a more manageable size.

Perhaps the most important thing you can do is find someone more experienced in agile methods to help you learn. Whenever anyone does anything new they inevitably make mistakes. Find someone who has already made lots of mistakes so you can avoid making those yourself.

One of the open questions about agile methods is where the boundary conditions lie. One of the problems with any new technique is that you aren't really aware of where the boundary conditions until you cross over them and fail. Agile methods are still too young to see enough action to get a sense of where the boundaries are. This is further compounded by the fact that it's so hard to decide what success and failure mean in software development, as well as too many varying factors to easily *pin down* the source of problems.

So where should you not use an agile method? I think it primarily comes down the people. If the people involved aren't interested in the kind of intense collaboration that agile working requires, then it's going to be a big struggle to get them to work with it.[7] In particular I think that this means you should never try to impose agile working on a team that doesn't want to try it.

There's been lots of experience with agile methods over the last ten years. At ThoughtWorks we always use an agile approach if our clients are willing, which most of the time they are.[8] I (and we) continue to be big fans of this way of working.

Words and Expressions

agile ['ædʒail] *adj.* 敏捷的，轻快的，灵活的
cobble ['kɔbl] *vt.* 修；粗粗地补　*n.* 圆石，鹅卵石
play havoc with　使陷入大混乱；对……造成严重破坏
methodology [meθə'dɔlədʒi] *n.* 方法学，方法论
discipline ['disiplin] *vt.* 使有纪律；训练　*n.* 纪律；训练；学科
plan-driven　计划驱动的
bureaucratic [ˌbjuərəu'krætik] *adj.* 官僚主义的，官僚政治的
stuff [stʌf] *n.* 资料；材料，原料；素质；要素　*vt.* 填，塞；把……装满
payoff ['peiɔf] *n.* 报偿；报应；报酬；发工资；盈利；决定因素
　　　　　adj. 决定的

293

in many ways　在许多方面
thrive [θraiv] *vi.* 茁壮成长；兴旺，繁荣；旺盛
to the point of　达到……的程度
make up　弥补，补偿；配制；编制（书、表等）；组成；和解
all in all　总的说来；头等重要的（东西）；一切的一切
ecosystem [i:kə'sistəm] *n.* 生态系统
context-driven testing　上下文驱动测试法
in the first place　首先，第一点；原先，本来
inherently [in'hiərəntli] *adv.* 内在地；固有地，生来地
pin down　使受约束；阻止，牵制；压住

Abbreviations

XP（eXtreme Programming）　极限程序设计
TDD（Test Driven Development）　测试驱动的开发

Notes

1. impending storm 含意为"即将来临的风暴"。本句中 determined 是过去分词，引导的短语作定语，修饰其前的 agilists。全句可译为：我们谈及敏捷软件方法，谈及如何把敏捷性引进开发组，或者说如何抵御即将来临的、决心改变已确立的习惯做法的敏捷主义风暴。

2. 本句中的 as 含意为"因为"。全句可译为：因为测试和调试是不可能排定时间表的，这样一个长测试阶段会使时间表陷入大混乱。

3. 本句可译为：它们没有因非常成功而引人注目。下一句可类似地译为：它们即使很流行也很少被人注意。

4. 本句为 so … that 型。全句可译为：遵循该方法论，有如此多的资料（要）去准备，以至于整个开发步伐减慢。

5. 本句是由连接词 but 连接的并列句，可译为：每个团体都是独特的群体，但要称得上是敏捷的，它便应该遵循一些同样广泛（适用）的原则。

6. 本句中的 dividing 和 applying 是现在分词，引导的短语作状语。全句可译为：Scrum 专注于软件开发的管理方面，将开发划分为若干个 30 天迭代（称为"冲刺"），并通过每天的 Scrum 会议实施更密切的监视和控制。

7. 本句中的 involved 是过去分词，含意为"所涉及的"，修饰前面的 people。that 是关系代词，引导限制性定语从句，修饰其前的 kind。全句可译为：如果所涉及的人员对敏捷工作方式需要的此类密切合作不感兴趣，那么，让他们使用它来工作需要付出极大努力。

8. 本句中的 which 是关系代词，引导的从句修饰前面的子句，可译为"他们大部分时间是那样的"。

15.5 Middleware

Middleware is computer software that connects software or applications. The software consists of a set of services that allows multiple processes running on one or more machines to interact. Middleware is sometimes called *plumbing* because it connects two applications and passes data between them. Middleware allows data contained in one database to be accessed through another. This definition would fit enterprise application integration (EAI) and data integration software.

This technology evolved to provide for interoperability in support of the move to coherent distributed architectures, which are used most often to support and simplify complex, distributed applications.[1] It includes web servers, application servers, and similar tools that support application development and delivery. Middleware is especially integral to modern information technology based on XML, SOAP, Web services, and service-oriented architecture (SOA).[2]

Middleware sits "in the middle" between application software that may be working on different operating systems. It is similar to the middle layer of a three-tier single system architecture, except that it is *stretched* across multiple systems or applications. Examples include EAI software, telecommunications software, *transaction* monitors, and *messaging-and-queuing* software.

The distinction between operating system and middleware functionality is, to some extent, arbitrary. While core kernel functionality can only be provided by the operating system itself,[3] some functionality previously provided by separately sold middleware is now integrated in operating systems. A typical example is the TCP/IP stack for telecommunications, nowadays included in virtually every operating system.

Object-Web defines middleware as: "The software layer that lies between the operating system and applications on each side of a distributed computing system in a network."

Use of Middleware

Middleware services provide a more functional set of application programming interfaces to allow an application to:
- Locate transparently across the network, thus providing interaction with another service or application
- Be independent from network services
- Be reliable and always available

when compared to the operating system and network services.[4]

Middleware offers some unique technological advantages for business and industry. For example, traditional database systems are usually deployed in closed environments where users access the system only via a restricted network or intranet (e.g., an enterprise's internal network). With the *phenomenal* growth of the World Wide Web, users can access virtually any database for which they have proper access rights from anywhere in the world. Middleware addresses the problem of varying levels of interoperability among different database structures. Middleware facilitates transparent access to *legacy* database management

295

systems (DBMSs) or applications via a web server without regard to database-specific characteristics.[5] Businesses frequently use middleware applications to link information from departmental databases, such as payroll, sales, and accounting, or databases *housed* in multiple geographic locations.

In the highly competitive *healthcare* community, laboratories make extensive use of middleware applications for data mining, laboratory information system (LIS) backup, and to combine systems during hospital mergers.[6] Middleware helps bridge the gap between separate LISs in a newly formed healthcare network following a hospital *buyout*.

Wireless networking *developers* can use middleware to meet the challenges associated with wireless sensor network (WSN), or WSN technologies. Implementing a middleware application allows WSN developers to integrate operating systems and hardware with the wide variety of various applications that are currently available.[7] Middleware can help software developers avoid having to write to application programming interfaces (API) for every control program, by serving as an independent programming interface for their applications.

Finally, e-commerce uses middleware to assist in handling rapid and secure transactions over many different types of computer environments. In short, middleware has become a critical element across a broad range of industries, thanks to its ability to bring together resources across *dissimilar* networks or computing platforms.

Types of Middleware

Hurwitz's classification system organizes the many types of middleware that are currently available. These classifications are based on *scalability* and recoverability:
- Remote Procedure Call—Client makes calls to procedures running on remote systems. Can be asynchronous or synchronous.
- Message Oriented Middleware—messages sent to the client are collected and stored until they are acted upon, while the client continues with other processing.
- Object Request Broker—this type of middleware makes it possible for applications on different machines to send objects and request services in an object-oriented system.
- SQL-oriented Data Access—middleware between applications and database servers.
- Embedded Middleware—communication services and integration interface software/firmware that operates between embedded applications and the real time operating system.

Other sources include these additional classifications:
- Transaction processing monitors—provides tools and an environment to develop and deploy distributed applications.
- Application servers—software installed on a computer to facilitate the serving (running) of other applications.
- Enterprise Service Bus—an abstraction layer on top of an Enterprise Messaging System.

Words and Expressions

plumbing [ˈplʌmiŋ] *vt.* 铺设管道；探测（plumb 的现在分词） *n.* 管道；管道工程
stretch [stretʃ] *vt.* 展开；伸展
transaction [trænˈzækʃən] *n.* 事务；处理；会报；交易
messaging-and-queuing 消息和队列
phenomenal [fiˈnɔminl] *adj.* 非凡的，出众的；显著的；现象的
legacy [ˈlegəsi] *n.* 传代物；遗产，遗赠物
house [haus] *vt.* 收藏；安置；给……房子住 *vi.* 住 *n.* 住宅；房
healthcare [ˈhelθkɛə] *n.* 医疗保健，卫生保健
buyout [ˈbaiˌaut] *n.* 全部买下
developer [diˈveləpə] *n.* 开发者
dissimilar [diˈsimilə] *adj.* 不同的，相异的
scalability [skeiləˈbiliti] *n.* 可伸缩性

Abbreviations

RPC (Remote Procedure Call) 远程过程调用
ORB (Object Request Broker) 对象请求代理
WSN (Wireless Sensor Network) 无线传感器网络

Notes

1. in support of ... 含意为"支持，支援"；coherent ... 含意为"黏在一起的，一致的，连贯的"；the move to ... 直译为"向……移动"。全句可译为：这个技术发展为提供可互操作性以支持分布式体系结构，这些体系结构经常用来支持和简化复杂的分布式应用。
2. integral 含意为"整体的，构成整体所必需的"。全句可译为：中间件尤其是为基于 XML、SOAP、Web 服务和面向服务体系结构（SOA）的现代信息技术所必需的。
3. 前半句可译为：虽然 OS 的核心内核功能只能由 OS 自己来提供。
4. independent from 含意为"独立于，与……无关"。全句可译为：与 OS 和网络服务比较，中间件服务提供一组更实用的应用程序设计接口，允许一个应用程序：
   - 在网上透明地定位（另一个应用程序），从而提供与另一个服务或应用程序的交互；
   - 独立于网络服务；
   - 可靠并且总是可用的。

   注："在网上透明地定位"是指，用户不需知道中间件是怎么找到所需程序的。
5. without regard to 意为"不考虑"，database-specific characteristics 意为"数据库专有的一些特性"。legacy database management systems or applications 是指"用已过时的软件编写的 DBMS 或应用程序"。

   下一句中前后两个 databases 是对等的，即各部门的数据库、安置在多个地理位置的数据库。

6. 句中使用"middleware applications for … , and to combine … "结构。全句可译为：在高度竞争的医疗保健界，实验室在数据挖掘、实验室信息系统支持，以及在医院合并期间联合诸系统时广泛地使用中间件应用程序。

下一句中 bridge the gap 可译为"连接，填补空白"。全句可译为：在一个医院全部买下后，最近形成的医疗保健网络中，中间件有助于连接各个实验室信息系统。

7. "to integrate operating systems and hardware with … "是"integrate … with … "形式，意思是"把 OS 和硬件与种类繁多、各种各样当前可用的应用程序集成。"

下一句中 by serving as … 意为"通过将中间件作为其各种应用程序的一个独立程序设计接口"。

Terms

AOP

In computing, aspect-oriented programming (AOP) is a programming paradigm which isolates secondary or supporting functions from the main program's business logic. It aims to increase modularity by allowing the separation of cross-cutting concerns, forming a basis for aspect-oriented software development.

AOP includes programming methods and tools that support the modularization of concerns at the level of the source code, while "aspect-oriented software development" refers to a whole engineering discipline.

Groupware

The term "groupware" refers to specialized software applications that enable group members to share and sync information and also communicate with each other more easily.

Groupware can allow both geographically dispersed team members and a company's on-site workers to collaborate with each other through the use of computer networking technologies (i. e., via the Internet or over an internal network/intranet). As such, groupware is especially important for remote workers and professionals on the go, since they can collaborate with other team members virtually.

Systems development life cycle

Systems development life cycle (SDLC or sometimes just SLC) is defined by the U. S. Department of Justice (DoJ) as a software development process, although it is also a distinct process independent of software or other Information Technology considerations. It is used by a systems analyst to develop an information system, including requirements, validation, training, and user ownership through investigation, analysis, design, implementation, and maintenance. SDLC is also known as information systems development or application development. An SDLC should result in a high quality system that meets or exceeds customer expectations, within time and cost estimates, works effectively and efficiently in the current and planned information technology infrastructure, and is cheap to maintain and cost-effective to enhance. SDLC is a systematic approach to problem solving and is composed of several

phases, each comprising multiple steps:
- The software concept—dentifies and defines a need for the new system.
- A requirements analysis—analyzes the information needs of the end users.
- The architectural design — creates a blueprint for the design with the necessary specifications for the hardware, software, people and data resources.
- Coding and debugging—creates and programs the final system.
- System testing—evaluates the system's actual functionality in relation to expected or intended functionality.

PDLC

Creating application programs is referred to as program development. The steps involved with program development are called the *program development life cycle* (PDLC). These steps begin with *problem analysis*, where the system specifications are reviewed by the system analyst and programmer to understand what the proposed system — and corresponding new program—must do. In the next step—*program design*—the program specifications from step one are refined and expanded into a complete set of design specifications. Two common approaches to program design are *structured programming* and *object-oriented programming*.

Once analysts have finished the program design for an application, the next stage is to code the program. Coding, which is the job of programmers, is the process of writing a program from scratch from a set of design specifications.

Debugging, part of the fourth step in the program development life cycle, is the process of making sure that a program is free of errors, or "bugs".

Program maintenance, the final step in the program development life cycle, is the process of updating software so that it continues to be useful. Program maintenance is costly.

Outsourcing

Outsourcing refers to getting things done from out-house instead of getting them done in-house. History of outsourcing is as old as the history of mankind. Since the individual started to form groups, small communities, and societies, the outsourcing began. The decision to outsource is often made in the interest of lowering firm costs, redirecting or conserving energy directed at the competencies of a particular business, or to make more efficient use of worldwide labor, capital, technology and resources. Though often used interchangeably, outsourcing differs from offshoring in that outsourcing is relative to the restructuring of the firm while offshoring is relative to the nation, though the two are not mutually exclusive, especially under conditions of globalization. Fundamentally and historically, outsourcing is a term relative to the organization of labor within and between societies.

MVC

MVC (short for Model View Control) is an architectural pattern used in software engineering. In complex computer applications that present a large amount of data to the user, a developer often wishes to separate data (model) and user interface (view) concerns, so that changes to the user interface will not affect data handling, and that the data can be reorganized without changing the user interface. The model view controller solves this problem by decoupling data access and business logic from data presentation and user interaction, by introducing an intermediate component: the controller.

Exercises

1. Translate the text of Lesson 15.1 into Chinese.
2. Topics for oral workshop.
- Talk about the need of software engineering: its origin and roles.
- Which type of language is the Unified Modeling Language? And its usage?
- Describe your understanding of Middleware briefly. How to adopt it in developing applications?

3. Translate the following into English.

(a) 软件开发过程有时称为软件开发生命周期(SDLC),因为它描述一个软件产品的生命,从其概念到其实现、交付、使用和维护。

(b) UML 提供一种标准方法(way)来书写系统的蓝图,包括概念上的一些东西,诸如业务过程(business process)和系统功能,以及程序设计语言语句、数据库模式和可重用软件组件这样一些具体东西。UML 代表最佳工程实践的汇集,这些工程实践在对一些大而复杂的系统的建模中已证明是成功的。

(c) 对象建模法是一种标识系统环境内部对象与那些对象之间关系的技术。面向对象的系统开发方法(approach)是基于若干个概念的,诸如对象、属性、行为、封装、类、继承、多态、持久性(persistence)等。

(d) 中间件是连接软件组件或企业应用程序的软件。中间件使得软件开发人员更容易执行通信和输入/输出,以便他们可以专注于特定目的的应用程序。通常,它支持复杂的、分布式的业务软件应用程序。

4. Listen to the video "VPN" and write down the second paragraph.

Unit 16　Network Security

16.1　What Do I Need to Know about Viruses?

A computer *virus* is a program that *maliciously* causes unwanted behavior on a computer. Viruses replicate themselves on computer systems by incorporating themselves into other programs shared among computer systems[1]. Viruses are not "accidents"; they are created by *rogue* programmers who distribute them and watch them spread.

The term "computer virus" was *coined* because of its similar attributes to a biological virus. Computer viruses have three defining qualities in common:

— the ability to replicate themselves.
— the requirement of a "carrier" or "host".
— the virus damages or causes unexpected behavior on *infected* computer systems.

The intentions of a computer virus program range from being harmless and humorous to being *overtly* damaging. Some viruses cause the computer screen to merely display silly animations or messages; others delete the contents of an entire hard disk! In any case, the effects of a virus are not intended by the computer user.

Computer viruses do exist and are passed around, both intentionally and unintentionally, but it is not very common for a computer system to become accidentally infected. Your computer can catch a virus from an infected program downloaded from the Internet, from a bulletin board system, or handed to you on a disk by a friend. Essentially viruses are passed through programs. You can get a virus by accessing an infected USB drive or hard disk or by running an infected program that you downloaded from the Internet via FTP, Web sites, or e-mail attachments.

There are six categories of viruses: *parasitic*, *bootstrap sector*, *multi-partite*, link, *companion*, and data file. Parasitic viruses infect files or programs. Bootstrap sector viruses replace either the programs that store information about a disk's contents or the programs that start the computer. Multi-partite viruses combine parasitic and bootstrap sector viruses. A companion virus creates a new program with the same name as a *legitimate* program. Link viruses modify the way the operating system finds a program. Data-file viruses infect programs that can open, manipulate, and close data files.

Nevertheless, there are precautions you can take to prevent an unwanted viral infection. *Anti-virus ammunition* includes understanding how viruses work, *habitually* using *virus scanners*, and implementing system protection methods.

You cannot get a virus from reading documents including Web pages, e-mail messages, and Usenet newsgroup *postings*. In fact, there have been some rumors and virus *"hoaxes"* that caused some people to panic. One such hoax was the so-called "Good Times" virus, which *allegedly could be carried and transmitted by e-mail*; supposedly, just reading a message with the words "Good Times" in the subject line would erase a hard disk or even destroy a computer's processor. [2] But there is no "Good Times" virus, and viruses just don't work this way.

Of the viruses that are real, each one is unique; this means you should use a reliable, up-to-date virus scanner. [3] A virus scanner is a program that is designed to check an entire computer system for known viruses or suspicious activity. There are many free virus scanners available, such as Disinfectant for the Macintosh or McAfee VirusScan for DOS/Windows.

Finally, be sure to learn about hardware write-protect methods and implement a reliable backup system to use in case of an emergency. A viral infection can be *contained* by immediately isolating computers on networks, halting the exchange of files, and using only write-protected disks. [4] Some antivirus software attempts to remove detected viruses, but more reliable results are obtained by turning off the computer, restarting it from a write-protected USB drive, erasing viruses, deleting infected files, and replacing files from backup disks. Usually regular backups of legitimate software and data can be used to restore a computer.

Words and Expressions

virus ['vaiərəs] *n.* 病毒　viral ['vair(ə)l] *adj.* 病毒的
maliciously [mə'liʃəsli] *adv.* 恶意地；预谋地
rogue [rəug] *n.* 流氓，无赖；淘气
coin [kɔin] *vt.* 创造；铸造；杜撰（新词、新语等）　*n.* 硬币
infect [in'fekt] *vt.* 感染
overtly [əu'və:tli] *adv.* 公开地；明显地；公然
parasitic [,pærə'sitik] *adj.* 寄生的
bootstrap sector　引导扇区
multi-partite ['mʌlti-'pɑ:tait] *adj.* 复合型的
companion [kəm'pæniən] *n.* 伙伴；成对
legitimate [li'dʒitimit] *adj.* 合法的；合理的；正统的　*vt.* 使合法
anti-virus　防病毒，抗病毒
ammunition [,æmju'niʃn] *n.* 手段；军火；武器
habitually [hə'bitʃuəli] *adv.* 习惯地
virus scanner　病毒扫描程序
posting [pəustiŋ] *n.* 帖子，张贴的新闻　*vt.* 贴出（布告等）；（用布告）宣布（post 的分词形式）
hoax [həuks] *vt.* 欺骗；戏弄　*n.* 骗局；恶作剧
allegedly [ə'ledʒidli] *adv.* 被说成是，据说
contain [kən'tein] *vt.* 抑制，遏制；包含

Notes

1. 本句中 shared among computer systems 是分词短语作定语，修饰 programs。
2. 本句中 supposedly 意为"想象上，按照推测，假定"；subject line "主题行"。本分句译为：只要读一则主题行中有"Good Times"这些词的消息就会删去硬盘内容，甚至破坏计算机的处理器。
3. 本句中 of 表示部分和全体的关系，of the viruses 即"在病毒中"。全句译为：真正的病毒，每一种都不同。这意味着你应使用可靠的、最新的病毒扫描程序。
4. 这里 contain 意为"遏制，抑制"。本句译为：病毒感染可以通过立即把计算机与网络断开、停止文件交换和只用写保护的磁盘来抑制。

注：本文介绍有关病毒的一般知识。你可以从 Infection Connection 网站访问病毒扫描程序资源。

16.2 Modern *Cryptography* — Data Encryption

Encryption can be used to protect data in transit as well as data in storage. Some vendors provide hardware encryption devices that can be used to encrypt and *decrypt* data. There are also software encryption packages which are available either commercially or as free software.

Encryption can be defined as the process of taking information that exists in some readable form (*plaintext*) and converting it into a form (*ciphertext*) so that it cannot be understood by others.

If the receiver of the encrypted data wants to read the original data, the receiver must convert it back to the original through a process called decryption. Decryption is the inverse of the encryption process. In order to perform the decryption, the receiver must be in possession of a special piece of data called the *key*.

The two main competing cryptography schemes are known as the *secret-key* (symmetric) system and the *public-key* (asymmetric) system. The secret-key system uses a single, wholly secret sequence both to encrypt and to decrypt messages. The public-key system uses a pair of mathematically related sequences, one each for encryption and decryption.[1]

Because secret-key system use only one key, care must be taken that the key is communicated only to *authorized parties*. This means that it must be sent either in person or through a set of protocols. In contrast, users of public-key systems can distribute their public keys to allow others to encrypt messages that can then be read only with the private key. Therefore, they do not require the complicated key-exchange protocols needed by secret-key system. Public-key systems also allow *electronic signatures* and data-integrity checks.

One disadvantage of public-key system is that they are slower than secret-key systems. Some applications try to combine public-key and secret-key cryptography to achieve security and performance. Many encryption packages that are basically secret-key systems use public-key encryption to communication the secret *session keys*. The secret-key system then encrypt the data.

In both public-key and secret-key systems there is a problem of keeping the secret/private keys safe. Some systems have tried encoding keys on *smart cards*, but these must be kept physically secure and each computer must be equipped with a card reader.

Secret-key Encryption

One of the most popular secret-key encryption schemes is IBM's Data Encryption System (DES), which became the U.S. federal standard in 1997. The standard form uses a 56-bit key to encrypt 64-bit data blocks.

The following is a notation for relating plaintext, ciphertext, and keys. We will use $C=E_k(P)$ to mean that the encryption of the plaintext P using key k gives the ciphertext C. Similarly, $P=D_k(C)$ reperesents of decryption of C to get the plaintext again. It then follows that
$$D_k(E_k(P))=P$$
DES has been studied by many of the world's leading cryptographers, but no weaknesses have been uncovered. To *crack* a DES-encrypted message a hacker or commercial spy would need to try 2^{55} possible keys. This type of search would need days of computer time on the world's fastest supercomputers. Even then, the message may not be cracked if the plaintext is not easily understood. [2]

Developers using DES can improve security by changing the keys frequently, using temporary session keys, or using *triple-encryption* DES. With triple DES, each 64-bit block is encrypted under three different DES keys. Recent research has confirmed that triple-DES is indeed more secure than single-DES. The User Data *Masking* Encryption Facility is an *export-grade* algorithm substituted for DES in several IBM products, such as the Distributed Computing Environment (DCE). [3]

Public-key Encryption

The key distribution problem has always been the weak link in the secret-key systems. Since the encryption key and decryption key are the same (or easily derived from one another) and the key has to be distributed to all users of the system, it seemed as if there was an inherent built-in problem: keys had to be protected from theft, but they also had to be distributed, so they could not just be locked up in a bank *vault*.

In public-key *cryptosystem*, the encryption and decryption keys were different, and plaintext encrypted with the public key can only be deciphered with the private key from the same pair. Conversely, plaintext encrypted with the private key can be decrypted only with the public key (it is used in electronic signatures). [4] The notations for these are as follows:
$$C=E_k(P), \qquad P=D_{k_1}(C)=D_{k_1}(E_k(P)) \quad \text{or}$$
$$C=D_{k_1}(P), \qquad P=E_k(C)=E_k(D_{k_1}(P))$$
here k is public key and k_1 is private key (or secret key). Users can make their public keys freely available or place them at a key distribution center for others to access. However, the private key must be kept safe. In public-key systems there is no need to find a safe channel for

communicating a shared secret key.

The most-widely adopted public-key cryptosystem is RSA. It is known by the initials of the three discoverers (Rivest, Shamir, Adleman) at M. I. T. The method stem from prime number theory. The public and private keys are based on a pair of very-large prime numbers that are not known by anyone intercepting the message. Because, at the bit-length of typical keys, very few numbers are primes, public-key systems must use very large keys to prevent *exhaustive searches* that try to guess the prime numbers used or to factor the keys.[5] Commonly the keys used in state-of-the-art public-key systems are 512 or 1024 bits.

Words and Expressions

cryptography [krip'tɔgrəfi] *n.* 密码学
encryption [in'kripʃən] *n.* 加密
decrypt [di:'kript] *vt.* 解密；译码；解码
decipher [di'saifə] *vt.* 破译(密码)；解读　*n.* 密电译文
plaintext [plein'tekst] *n.* 明文；纯文本
ciphertext ['saifətekst] *n.* 密文
key [ki:] *n.* 钥匙；关键；[计]密钥
secret-key　密钥，秘密密钥
public-key　公钥，公开密钥
authorized party　授权方；特许方
electronic signature　电子签名
session key　会话密钥
smart card　智能卡
crack [kræk] *vi.* 裂开，断裂；解开；破译　*n.* 裂缝，破裂声
triple-encryption　三重加密
mask [mɑ:sk] *vt.* 掩蔽；屏蔽；掩饰，伪装；戴面具　*vi.* 掩饰；化装；
　　　　　　n. 屏蔽；掩码；(假)面具；口罩；防护面具；伪装；遮盖物；面模；假面舞会
export-grade　出口级
vault [vɔ:lt] *n.* 地下室
cryptosystem ['kriptəuˌsistəm] *n.* 密钥系统，密钥体系
exhaustive search　穷尽查找，穷举搜索

Notes

1. 本句中 one each for ... 可译为"一个用于……，另一个用于……"。全句译为：公开密钥系统采用一对数学上相关的序列，一个用于加密，另一个用于解密。
2. 本句中 even then 指"即使花了时间算出来"，因此本句译为：如果未加密的明文是不容易理解的，即使算出，报文也可能解不开。
3. facility "设施，功能，装备"。本句译为：用户数据屏蔽加密设施是一种在IBM的一些产品中，如分布式计算环境(DCE)，代替DES的出口级算法。

4. 本句中 encrypted with 引导的短语修饰 plaintext。本句译为：相反，用私钥加密的明文只能用公钥解密。
5. at the bit-length of typical keys 这里指"以典型密钥的位长来说"；factor 意为"因子分解"。全句译为：以典型密钥的位长而言，因为素数很少，所以公开密钥系统必须使用非常大的密钥，以防止用穷举搜索法猜出所用的素数或者对密钥进行因子分解。

下一句中 state-of-the-art 译为"最新的"。

注：本短文介绍数据加密的基本思想，包括秘密密钥系统和公开密钥系统。

16.3 Firewalls and Proxies

When you connect your LAN to the Internet, you are enabling your users to reach and communicate with the outside world. At the same time, however, you are enabling the outside world to reach and interact with your LAN. Firewalls are just a modern *adaptation* of that old *medieval* security *standby*: digging a deep *moat* around your castle.[1] This design forced everyone entering or leaving the castle to pass over a single *drawbridge*, where they could be inspected by the I/O police. With networks, the same trick is possible: a company can have many LANs connected in arbitrary ways, but all traffic to or from the company is forced through an electronic drawbridge (firewall).

Basically, a firewall is a standalone process or a set of integrated processes that runs on a router or server to control the flow of networked application traffic passing through it. Typically, firewalls are placed on the entry point to a public network such as the Internet. They could be considered *traffic cops*. The firewall's role is to ensure that all communication between an organization's network and the Internet conform to the organization's security policies. Primarily these systems are TCP/IP based and, depending on the implementation, can enforce security roadblocks as well as provide administrators with answers to the following questions:

- Who's been using my network?
- What were they doing on my network?
- When were they using my network?
- Where were they going on my network?
- Who failed to enter my network?

In general, there are three types of firewall implementations, some of which can be used together to create a more secure environment. These implementations are: *packet filtering*, application proxies, and circuit-level or *generic-application proxies*.

Packet filtering is often achieved in the router itself. Application proxies, on the other hand, usually run on standalone servers. Proxy services take a different approach than packet filters, using a (possibly) modified client program that connects to a special intermediate host that actually connects to the desired service.[2]

Packet Filtering

Consider your network data a neat little package that you have to deliver somewhere.

This data could be part of an e-mail, file transfer, etc. With packet filtering, you have access to deliver the package yourself. [3] The packet filter acts like a traffic cop; it analyzes where you are going and what you are bringing with you. However, the packet filter does not open the data package, and you still get to drive it to the destination if allowed. [4]

Most commercial routers have some kind of built-in packet filtering capability. However, some routers that are controlled by ISPs may not offer administrators the ability to control the configuration of the router. In those cases, administrators may opt to use a standalone packet filter behind the router.

Either way, an administrator needs to understand how to identify data packages in terms the packet filer can understand. Since all Internet traffic is based on IP (Internet Protocol), each application or "package" can be identified through a specific TCP (Transmission Control Protocol) or UDP (User Datagram Protocol) port. These ports are registered and defined in RFC (Request for Comment) 1700 which can be found on the Internet. For example, port 23 is for Telnet. A company could block incoming packets for all IP addresses combined with port 23. In this way, no one outside the company could log in via Telnet.

Application Proxy

To understand the application proxy, consider this scenario where you needed to deliver your neat little package of network data. With application-level proxies, the scenario is similar, but now you need to rely on someone else to deliver the package for you. Hence the term proxy illustrates this new scenario. The same rules apply as they do for packet filtering, except that you don't get to deliver your package past the gate. [5] Someone will do it for you, but that agent needs to look inside the package first to confirm its contents. If the agent has permission to deliver the contents of the package for you, he will.

Most commercial routers do not have proxy capabilities today, although I believe that proxy technology will be integrated with router code in the future. Until then, you need to rely on a standalone system that can support application-level proxy services.

Since an application proxy needs to communicate on behalf of the sender, it needs to understand the specific language or protocols associated with a particular application. Take as an example the widely used HTTP (HyperText Transfer Protocol) proxy. If you are using a browser on your network, it is highly likely that your IS group has an HTTP proxy configured to allow you to access the Web via a central server. That single machine understands HTTP conversations and can speak on behalf of the requesting client. This is application-level proxying.

Of curse, security and encryption also come into play, since the proxy must be able to open the "package" to look at it or decode its contents. [6] These are important issues obviously, but to *do* them *justice* would require another article.

***Circuit-Level* or Generic-Application Proxy**

As with application-level proxies, you need to rely on someone to deliver your package for you. The difference is that if these circuit-level proxies have access to deliver the package to your requested destination, they will. They don't need to know what is inside.

Circuit-level proxies (specifically SOCKS) work outside of the application layers of the protocol. These servers allow clients to pass through this centralized service and connect to whatever TCP port the clients specify. SOCKS servers also have the ability to *authenticate* the source address of connection requests and can block unauthorized clients from connecting out onto the Internet.[7] Most TCP-based applications can become "*SOCKSified*." You do this by recompiling and linking them with a SOCKS client library. DLL-based TCP stacks have the additional benefit of being able to provide applications with SOCKS client capabilities through the use of *shims*, eliminating the need to recompile.

Words and Expressions

adaptation [ədæp'teiʃən] *n.* 改编，改制；适应
medieval [medi'i:vəl] *adj.* 中世纪的；古老的
standby ['stændbai] *n.* 可依靠的东西；备用的东西
moat [məut] *n.* 护城河
drawbridge ['drɔ:bridʒ] *n.* 吊桥
traffic cops   交通警
packet filtering   包过滤
generic-application proxies   通用应用程序代理
do ... justice   公平处理，说句公道话
circuit-level   电路级
SOCKSified   SOCKS 化
shim [ʃim] *n.* 薄垫片   *vt.* 用垫片填
authenticate [ɔ:'θentikeit] *v.* 鉴别；认证；证实

Abbreviation

RFC (Request For Comments)   请求评论；请求注解

Notes

1. 本句译为：防火墙只是古代中世纪防御方法——在城堡的周围挖一条很深的护城河——的一种现代应用。
2. 说明：代理服务的模式为

3. 本句中 have access to 意为"可以接近，有权访问"。本句译为：使用包过滤，你自己来传送此数据包。

说明：这是相对于应用程序代理而言的，那种方式要通过中间人来传递数据包。

4. 本句中 to drive it to 意为"把它送到"。后半句译为：如果允许，你仍要把它送到目的地。

5. 本句中 they 代表"same rules"；the gate 指应用程序代理。本句译为：包过滤适用的规则也适用于应用程序代理，除了一点不同，即你不能越过应用程序代理来递交你的包。

6. come into play 这里译成"随之而来，开始起作用"；decode its content 译为"对其内容进行译码"。下一句中的 do them justice 指"说清这些问题"。全句译为：显然这些问题很重要，但说清这些问题需要另一篇文章。

7. 本句中 block ... from ... 译为"阻断……做……"。后半句译为：能阻断未经批准的客户机外接到 Internet。

注：本节介绍三类防火墙实现方案——包过滤、代理、通用应用程序代理——的工作原理。本文属于内容和翻译并重的选文。

Terms

Hacker（黑客）

Hacker mostly refers to computer criminals, due to the mass media usage of the word since the 1980s. This includes what hacker slang calls "script kiddies", people breaking into computers using programs written by others, with very little knowledge about the way they work.

Cookie

A **cookie** is a piece of text stored on a user's computer by their web browser. A cookie can be used for authentication, the identifier for a server-based session, storing site preferences, shopping cart contents, or anything else that can be accomplished through storing text data.

A cookie consists of one or more name-value pairs containing bits of information, which may be encrypted for information privacy and data security purposes. The cookie is sent as a field in the header of the HTTP response by a web server to a web browser and then sent back unchanged by the browser each time it accesses that server.

AES

The Advanced Encryption Standard (AES) is a specification for the encryption of electronic data established by the U.S. National Institute of Standards and Technology (NIST) in 2001.

AES is based on the Rijndael cipher developed by two Belgian cryptographers, Joan Daemen and Vincent Rijmen, who submitted a proposal to NIST during the AES selection process. Rijndael is a family of ciphers with different key and block sizes.

For AES, NIST selected three members of the Rijndael family, each with a block size of 128 bits, but three different key lengths: 128, 192 and 256 bits.

AES has been adopted by the U.S. government and is now used worldwide. It supersedes the Data Encryption Standard (DES), which was published in 1977. The algorithm described by AES is a symmetric-key algorithm, meaning the same key is used for both encrypting and decrypting the data.

The name Rijndael (Dutch pronunciation: ['reindɑːl]) is a play on the names of the two inventors (Joan Daemen and Vincent Rijmen).

WPA (Wi-Fi Protected Access, Wi-Fi 网络安全存取)

Wi-Fi Protected Access (WPA) and Wi-Fi Protected Access II (WPA2) are two security protocols and security certification programs developed by the Wi-Fi Alliance to secure wireless computer networks. The Alliance defined these in response to serious weaknesses researchers had found in the previous system, WEP (Wired Equivalent Privacy).

The Wi-Fi Alliance intended WPA as an intermediate measure in anticipation of the availability of the more secure and complex WPA2. WPA2 is a common shorthand for the full IEEE 802.11i standard.

The WPA protocol implements much of the IEEE 802.11i standard. Specifically, the Temporal Key Integrity Protocol (TKIP) was adopted for WPA. WEP used a 40-bit or 104-bit encryption key that must be manually entered on wireless access points and devices and does not change. TKIP employs a per-packet key, meaning that it dynamically generates a new 128-bit key for each packet and thus prevents the types of attacks that compromised WEP.

WPA also includes a message integrity check. This is designed to prevent an attacker from capturing, altering and/or resending data packets. This replaces the cyclic redundancy check (CRC) that was used by the WEP standard. CRC's main flaw was that it did not provide a sufficiently strong data integrity guarantee for the packets it handled. WPA uses a message integrity check algorithm called Michael to verify the integrity of the packets. Michael is much stronger than a CRC, but not as strong as the algorithm used in WPA2.

WPA2

WPA2 has replaced WPA. WPA2 implements the mandatory elements of IEEE 802.11i. In particular, it includes mandatory support for CCMP, an AES-based encryption mode with strong security. Certification began in September, 2004; from March 13, 2006, WPA2 certification is mandatory for all new devices to bear the Wi-Fi trademark

PSK

Pre-shared key mode (PSK, also known as Personal mode) is designed for home and small office networks that don't require the complexity of an 802.1X authentication server. Each wireless network device encrypts the network traffic using a 256 bit key. This key may be entered either as a string of 64 hexadecimal digits, or as a passphrase of 8 to 63 printable ASCII characters. If ASCII characters are used, the 256 bit key is calculated by applying the PBKDF2 key derivation function to the passphrase, using the SSID as the salt and 4096 iterations of HMAC-SHA1.

Blockchain(区块链)

A blockchain, originally block chain, is a growing list of records, called *blocks*, which are linked using cryptography. Each block contains a cryptographic hash of the previous block, a timestamp, and transaction data (generally represented as a Merkle tree root hash).

By design, a blockchain is resistant to modification of the data. It is "an open, distributed ledger that can record transactions between two parties efficiently and in a verifiable and permanent way". For use as a distributed ledger, a blockchain is typically managed by a peer-to-peer network collectively adhering to a protocol for inter-node communication and validating new blocks. Once recorded, the data in any given block cannot be altered retroactively without alteration of all subsequent blocks, which requires consensus of the network majority. Although blockchain records are not unalterable, blockchains may be considered secure by design and exemplify a distributed computing system with high Byzantine fault tolerance. Decentralized consensus has therefore been claimed with a blockchain.

Blockchain was invented by a person using the name Satoshi Nakamoto in 2008 to serve as the public transaction ledger of the cryptocurrency bitcoin. The identity of Satoshi Nakamoto is unknown. The invention of the blockchain for bitcoin made it the first digital currency to solve the double-spending problem without the need of a trusted authority or central server. The bitcoin design has inspired other applications, and blockchains which are readable by the public are widely used by cryptocurrencies. Blockchain is considered a type of payment rail. Private blockchains have been proposed for business use.

Exercises

1. Translate the text of Lesson 16.1 into Chinese.
2. Topics for oral workshop.
- What is the computer virus?
- Talk about your experience of catching computer viruses.
- How is a computer infected with viruses?
- How to protect data during transmission? Define the term data encryption and explain how it works.
3. Translate the following into English.

安全是一个包含多个方面的(all-encompassing)术语,它描述了为保护信息不受未经授权的访问的所有概念、方法(techniques)和技术。对信息安全有几个要求。

机密性:通常用加密来隐藏数据以阻止未经授权的查看和存取(访问)。

鉴别(Authenticity):知道你正在与之通信的人或系统是谁或你认为它是什么的能力。

访问控制:一旦个人或系统已通过身份验证,访问控制就确定了他们访问数据和使用系统的能力。

数据完整性(integrity):提供信息系统或数据是真的保证。

可用性:确保信息以安全的方式提供给用户。

为了防止未经授权的访问,必须使用某种类型的识别过程——从密码到物理访问对象(访问卡等),再到验证某种类型的个人特征(如指纹)的生物识别设备。

一些最安全的访问控制系统既能识别身份,又能认证身份。识别身份涉及该人的姓名或其他身份特征是否被列为授权用户;认证身份是指确定这个人是否确实是他或她所声称的那个人。

4. Listen to the video "VPN" and write down the third paragraph.

Unit 17　Computer Systems

17.1　Embedded Systems

What is an Embedded System?

Embedded systems are electronic devices that incorporate microprocessors within their implementations. The main purposes of the microprocessor are to simplify system design and provide flexibility. Having a microprocessor in the device means that removing bugs, making modifications, or adding new features are only matters of rewriting the software that controls the device. Unlike PCs, however, embedded systems may not have a disk drive and so the software is often stored in a read-only memory (ROM) chip; this means that modifying the software requires either replacing or "reprogramming" the ROM.

As Table 17-1 indicates, embedded systems are found in a wide range of application areas. Originally they were used only for expensive industrial-control applications, but as technology brought down the cost of dedicated processors, they began to appear in *moderately* expensive applications such as automobiles, communications and office equipment, and televisions. Today's embedded systems are so inexpensive that they are used in almost every electronic product in our life.

Table 17-1　Examples of Embedded Systems

Application Area	Examples
Aerospace	Navigation systems, automatic landing systems, flight *altitude* controls, engine controls, space exploration (e.g., the *Mars Pathfinder*)
Automotive	*Fuel injection* control, passenger environmental controls, *antilock* braking systems, *air bag* controls, *GPS mapping*
Children's Toys	Nintendo's "Game Boy", Mattel's "My Interactive Pooh", Tiger Electronic's "Furby"
Communications	Satellites, network routers, switches, hubs
Computer Peripherals	Printers, scanners, keyboards, displays, modems, hard disk drives CD-ROM drives
Home	Dishwashers, microwave ovens, VCRs, televisions, stereos, fire/security alarm systems, lawn *sprinkler* controls, *thermostats*, cameras, *clock radios*, *answering machines*
Industrial	Elevator controls, *surveillance systems*, robots
Instrumentation	Data collection, *oscilloscopes*, signal generators, signal analyzers, power supplies
Medical	*Imaging systems* (e.g., XRAY, MRI, and ultrasound), patient monitors, *heart pacers*
Office Automation	FAX machines, *copiers*, telephones, *cash registers*
Personal	Personal Digital Assistants (PDAs), *pagers*, cell phones, wrist watches, video games, portable MP3 players, GPS

In many cases we're not even aware that a computer is present and so don't realize just how *pervasive* they have become. For example, although the typical family may own only one or two personal computers, the number of embedded computers found within their home and cars and among their personal belongings is much greater. [1]

What is often surprising is that embedded processors account for virtually 100% of worldwide microprocessor production![2] For every microprocessor produced for use in a desktop computer, more than 100 are produced for use in embedded systems. This shouldn't be surprising considering that the number of embedded microprocessors found in the average *middle-class household* in North America in 1999 was estimated to be between 40 and 50.

What's Unique about the Design Goals for Embedded Software?

The objective of this book is to help you learn how to design and implement software for embedded systems. Although you already have some experience writing desktop application programs in a high-level language, writing embedded application programs presents some new challenges regarding reliability, performance, and cost.

Reliability expectations will place greater responsibility on programmers to eliminate bugs and to design our software to tolerate errors and unexpected situations. Many embedded systems have to run 24 hours a day, seven days a week, 365 days a year. You can't just "reboot" when something goes wrong! For this reason, good coding practices and thorough testing *take on* a new level of importance in the *realm* of embedded processors.

Performance goals will force us to learn and apply new techniques such as multi tasking and scheduling. The need to communicate directly with *sensors*, *actuators*, *keypads*, displays, etc., will require programmers to have a better understanding of how alternative methods for performing input and output provide opportunities to trade speed, complexity, and cost. [3] Although we'll usually program in a high-level language for better productivity, use of these alternatives will occasionally require that we drop to the level of the computer and program directly in assembly language.

Computers use *fixed precision* and *2's complement* representation to store and process numeric values. This results in *subtle* differences from how humans handle numbers, and it is often an unexpected source of problems. [4] Our software will have to live with limitations regarding the range and resolution of numbers. But as programmers, we need to be sure that we have a comprehensive understanding of the consequence of exceeding these limitations and of how to handle such situations. [5]

Unlike processors embedded in large, expensive systems, consumer products are designed to be *mass* produced with a minimal production cost. To be competitive, they must achieve fast time to market and ideally require no modification once in production. [6]

What Does "Real-time" Mean?

Real-time systems process events. Events occurring on system inputs cause other events to occur as system outputs. Examples of input events include such things as the detection of a telephone ring signal, the application of force to the *brake pedal* of a car, or the opening of a microwave oven's door. The output events produced in response to these input events might be to take the phone line "*off-hook*", to apply *hydraulic* pressure to the automobile's brake system, and to turn off the microwave oven.[7]

One of the primary design goals of real-time systems is minimizing response time. A *soft* real-time system is one that is designed to compute the response as fast as possible but doesn't have an explicit deadline. If a deadline is *imposed*, the system is known as a hard real-time system. Keeping response times of hard real-time systems within the given deadlines is always important, or the entire system may fail to operate properly. For example, an antilock braking system must detect and respond to loss of *traction* within a few milliseconds; a delay of one or two seconds would be *intolerable* and potentially *deadly*.

Words and Expressions

moderately ['modəritli] *adv.* 适度地
pervasive [pəː'veisiv] *adj.* 普遍深入的；渗透性的；充满的
middle-class household 中产家庭
aerospace ['ɛərəuspeis] *n.* 航空宇宙；航空航天科学；航空航天工程；航空航天工业
altitude ['æltitjuːd] *n.* （尤指海拔）高度，高处
Mars [mɑːz] Pathfinder 火星探路者号
fuel injection 燃油喷射控制
antilock [ænti'lɔk] *adj.* （制动系统）防抱死的 *n.* （制动系统）防抱死,防锁死
air bag 安全气囊
GPS mapping GPS 绘图，GPS 制图
Nintendo's "Game Boy" 某公司的产品名
sprinkler ['spriŋklə] *n.* 洒水车，洒水装置
thermostat ['θəːməstæt] *n.* 恒温器，自动调温器，温度调节装置
clock radio 有定时自动开关功能的收音机；收音机闹钟
answering machine 电话应答机
surveillance [səː'veiləns] system 监视系统，监督系统
oscilloscope [ɔ'siləskəup] *n.* （物）示波器
imaging system 成像系统
heart pacer 心脏起搏器
copier ['kɔpiə] *n.* 复印机
cash register （美）收银机，现金出纳机
pager ['peidʒə(r)] *n.* 呼机，寻呼机
take on 披上；呈现；具有；雇用；承担
realm [relm] *n.* 领域
actuator ['æktjueitə] *n.* 传动装置；传动器；促动器；激励者

sensor ['sensə] *n.* 传感器
keypad ['ki:pæd] *n.* [计] 键区；小键盘
fixed precision　　固定精度
2's complement ['kɔmplimənt]　　2 的补码
subtle ['sʌtl] *adj.* 难以捉摸的，微妙的，微细的
mass [mæs] *n.* 大量；群众　*adj.* 大规模的；群众的
brake pedal　　刹车踏板
off-hook　　脱钩
hydraulic [hai'drɔ:lik] *adj.* 水力的，水压的
impose [im'pəuz] *vt.* 征税；强加
traction ['trækʃən] *n.* 牵引
intolerable [in'tɔlərəbl] *adj.* 无法忍受的，难耐的
deadly ['dedli] *adj.* 致命的；势不两立的

Abbreviation

MRI（Magnetic Resonance Imaging）　核磁共振成像

Notes

1. 本句中 found within their home ... 修饰前面的 computers。全句译为：例如，虽然一般的家庭可能只有一两台个人计算机，但是在他们的家里、汽车里以及个人物品中可找到的嵌入式计算机的数量要多得多。

2. 本句中 account for "占，说明"。What is often surprising = The thing that is often surprising。全句译为：常令人惊奇的是，嵌入式处理器实际上占了世界范围微处理器产量的 100%。下一句可译为：每生产一台用于台式机的微处理器，就会生产 100 台以上用于嵌入式系统的微处理器。

3. 本句中 trade 意为"交易，用……进行交换"。全句译为：要求与传感器、传动装置、键区、显示器等设备直接通信，这就要求程序员更好地理解执行输入和输出的替代方法如何提供机会以换取速度、复杂性和成本。下一句的后半句中，drop to ... and program ... 是并列结构，意思是，使用这些可选择的方法将偶尔要求我们降到机器一级，并直接用汇编语言编程序。

4. 要理解本句和下一句的意思必须知道，计算机用固定精度来存储数（如用 32 位），所以与人处理数有一些差别。本句中 unexpected 意为"想不到的，意外的，未预料到的"。全句译为：这就导致了与人处理数的一些微妙差别，并且它经常是一些意想不到的问题的根源。下一句中 live with 意为"承认，忍受，与……住在一道"；resolution "分辨，分辨率"，这里指数之间的分辨，也是由固定精度决定的。例如，若精度为两位，则 0.99 的下一个数为 1.00，而不是 0.990000…01。全句译为：我们的软件必须承认数的范围和分辨的限制。

5. 本句中 understanding of the consequence ... and of how to handle ... 是并列结构。全句译为：但是作为程序员，我们需要确信自己对超出这些限制的后果及如何处理这些情况有全面的理解。

6. 全句可译为：为了竞争，他们必须做到快速在市场上销售，并且理想化地要求，一旦生产就不需修改。
7. 全句可译为：响应这些输入事件所产生的输出事件可能是，把电话线挂断，把液压加到汽车的刹车系统，以及关微波炉。

注：虽然我们用了许多嵌入式系统，如 TV、电风扇、微波炉、手机、电梯，但许多人并未意识到。本文简单介绍嵌入式系统的概念、应用及其特点。

17.2 Distributed Systems

The motivation for constructing and using *distributed systems stems from* a desire to share resources *transparently*. The term "resource" is rather abstract one, but now it best characterizes the range of things that can usefully be shared in a networked computer system. It extends from hardware components such as computers, disks and printers to software-defined entities such as files, databases and data objects of all kinds. It includes the stream of video frames that emerges from a digital video camera and the audio connection that a mobile phone call represents.[1]

We define a distributed system as one in which hardware or software components located at networked computers communicate and coordinate their actions only by passing messages.[2] This simple definition covers the entire range of systems in which networked computers can usefully be deployed.

Computers that are connected by a network may be spatially separated by any distance. They may be on separate continents, in the same building or the same room. Our definition of distributed systems has the following characteristics:

Transparency: Differences between the various computers and the ways in which they communicate are hidden from users. The same holds for the internal organization of the distributed system.

Consistency: Users and applications can interact with a distributed system in a consistent and uniform way, regardless of where and when interaction takes place.

Extensibility (or scalability): Distributed systems should be relatively easy to expand or *scale*. This characteristic is a direct consequence of having independent computers, but at the same time, hiding how these computers actually take part in the system as a whole.

Independent failures: A distributed system will normally be continuously available, although perhaps certain parts may be temporarily out of order. Users and applications should not notice that parts are being replaced or fixed, or that new parts are added to serve more users or applications.

To support heterogeneous computers and networks while offering a single-system view, distributed systems are often organized by means of a layer of software that is logically placed between a higher-level layer consisting of users and applications, and a layer underneath consisting of operating systems, as shown in Figure 17-1.[3] Accordingly, such a distributed system is sometimes called middleware.

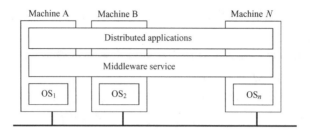

Figure 17-1   A distributed system organized as middleware

We give three examples of distributed systems:
- The Internet;
- An intranet, which is a portion of the Internet managed by an organization;
- Mobile and *ubiquitous* computing.

The Internet and Intranet (leave out)

Mobile and ubiquitous computing

Technological advances in device *miniaturization* and wireless networking have led increasingly to the integration of small and portable computing devices into distributed systems. These devices include:
- Laptop computers.
- Handheld devices, including personal digital assistants (PDAs), mobile phones, pagers, *video cameras* and digital cameras.
- *Wearable* devices, such as smart watches with functionality similar to a PDA.
- Devices embedded in appliances such as washing machines, *hi-fi* systems, cars and refrigerators.

The portability of many of these devices, together with their ability to connect conveniently to networks in different places, makes mobile computing possible. Mobile computing (also called *nomadic* computing) is the performance of computing tasks while the user is on the move, or visiting places other than their usual environment. In mobile computing, users who are away from their 'home' intranet (the intranet at work, or their residence) are still provided with access to resources via the devices they carry with them. They can continue to access the Internet; they can continue to access resources in their home intranet; and there is increasing *provision* for users to utilize resources such as printers that are conveniently nearby as they move around. The latter is also known as *location-aware* or *context-aware* computing.

Ubiquitous computing is the *harnessing* of many small, cheap computational devices that are present in users' physical environments, including the home, office and even natural *settings*.[4] The term 'ubiquitous' is intended to suggest that small computing devices will eventually become so pervasive in everyday objects that they are *scarcely* noticed. That is, their computational behavior will be transparently and *intimately* tied up with their physical function.

The presence of computers everywhere only becomes useful when they can communicate with one another. For example, it would be convenient for users to control their washing machine and their hi-fi system from a 'universal remote control' device in the home. Equally, the washing machine could *page* the user via a *smart badge* or watch when the washing is done.

Ubiquitous and mobile computing overlap, since the mobile user can in principle benefit from computers that are everywhere. But they are distinct, in general. Ubiquitous computing could benefit users while they remain in a single environment such as the home or a hospital. Similarly, mobile computing has advantages even if it involves only conventional, discrete computers and devices such as laptops and printers.

Figure 17-2 shows a user who is visiting a host organization. The figure shows the user's home intranet and the host intranet at the site that the user is visiting. Both intranets are connected to the rest of the Internet.

The user has access to three forms of wireless connection. Their laptop has a means of connecting to the host's wireless LAN. This network provides coverage of a few hundreds of meters (a floor of a building, say). It connects to the rest of the host intranet via a gateway. The user also has a mobile (cellular) telephone, which is connected to the Internet. The phone gives access to pages of simple information, which it presents on its small display. Finally, the user carries a digital camera, which can communicate over a personal area wireless network (with range up to about 10 m) with a device such as a printer.

With a suitable system infrastructure, the user can perform some simple tasks in the host site using the devices they carry. While journeying to the host site, the user can fetch the latest stock prices from a web server using the mobile phone. During the meeting with their hosts, the user can show them a recent photograph by sending it from the digital camera directly to a suitably enabled printer in the meeting room. This requires only the wireless link between the camera and printer. And they can in principle send a document from their laptop to the same printer, utilizing the wireless LAN and wired Ethernet links to the printer.

Mobile and ubiquitous computing are a lively area of research.

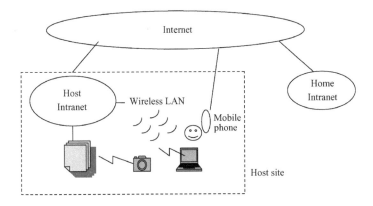

Figure 17-2 Portable and handheld devices in a distributed system

Words and Expressions

distributed system   分布式系统

stem from 起源于，来自
transparently [træns'pεərəntli] adv. 透明地；显然地；易觉察地
transparency [træns'pεərənsi] n. 透明性，透明，透明度
scale [skeil] vt. 扩缩；伸缩；按比例缩小或扩大；按比例缩减或增加
ubiquitous [juː'bikwitəs] adj. 普适的；普遍存在的，无所不在的，到处存在的；普及的
miniaturization [ˌminətʃərai'zeiʃn] n. 小型化
video camera 摄影机，摄像机
wearable ['wεərəbl] adj. 可穿戴(式)的，可(佩)带的，可穿用的
hi-fi ['hai'fai] n. 高度保真的音响装置 ＝high fidelity
     adj. 音响高度保真性的
nomadic [nəu'mædik] adj. 游牧的
provision [prə'viʒən] n. 供应；(一批)供应品；预备，准备
context-aware 上下文感知；情景感知；觉察上下文；环境敏感
location-aware 位置感知；位置敏感；位置相关；位置知晓
harness ['hɑːnis] vt. 利用，治理；给(马等)上挽具 n. 马具；挽具
setting ['setiŋ] n. 环境；安置；安装
scarcely ['skεəsli] adv. 几乎不，简直没有
intimately ['intimitli] adv. 密切地
page [peidʒ] vt. 呼叫，播叫
smart badge [bædʒ] 智能徽章

Notes

1. 全句译为：它包括从数字摄像机来的视频帧流和移动电话通话所代表的音频连接。
2. 本句中 located at … 修饰前面的 components。networked computers 可译为"连网的计算机"。全句译为：我们定义分布式系统为这样的系统，在这个系统中，连网的计算机硬件或软件组件只有通过传送消息进行通信并协调它们的活动。
3. 本句中 consisting of … 修饰前面的 layer。全句译为：为了支持异构的(不同的)计算机和网络，而同时提供单一系统视图，分布式系统经常用一层软件来组织，该层软件逻辑上位于由用户和应用程序组成的高层(软件)和由 OS 组成的底层(软件)之间，如图 17-1 所示。
4. 全句可译为：普适(或无处不在的)计算是利用许多小的、便宜的计算设备进行的计算，这些计算设备是在用户的物理环境(包括家里、办公室甚至自然环境)中存在的。下一句中 suggest 译为"表明、暗示、意思是"。全句译为：术语"无处不在"意在暗示，小的计算设备将最终遍布在各种日常物品中，以至于它们几乎不会被注意到。

  注：分布式系统是松耦合的多机系统。现在绝大多数应用系统都是建立在多机上的分布式应用系统。本文介绍分布式系统的基本概念、特性，以及随着无线通信技术的发展而发展的移动计算和普适(或无所不在)计算。

  大家熟悉的 C/S 系统是共享服务器资源的分布式系统；P2P 系统是共享分布的、对等式资源的分布式系统；网格系统是共享一领域内各组织的各种资源的分布式系统；而 CORBA 和 Web 服务则用于共享不同平台上的程序(或称服务)，它们是开发分布式应用系统的规范。

17.3 Cloud Computing and Cloud Storage

17.3.1 Cloud Computing

Cloud computing is computing in which large groups of remote servers are networked to allow centralized data storage and online access to computer services or resources. Clouds can be classified as public, private or hybrid.

Cloud computing relies on sharing of resources to achieve *coherence* and *economies of scale*, similar to a *utility* (like the electricity grid) over a network. At the foundation of cloud computing is the broader concept of converged infrastructure and shared services.[1]

Cloud computing, or in simpler shorthand just "the cloud", also focuses on maximizing the effectiveness of the shared resources. Cloud resources are usually not only shared by multiple users but are also dynamically reallocated *per demand*. This can work for allocating resources to users. For example, a cloud computer facility that serves European users during European business hours with a specific application (e.g., E-mail)[2] may reallocate the same resources to serve North American users during North America's business hours with a different application (e.g., a web server). This approach should maximize the use of computing power, thus reducing environmental damage as well since less power, air conditioning, *rack space*, etc. are required for a variety of functions. With cloud computing, multiple users can access a single server to retrieve and update their data without purchasing licenses for different applications.

The name cloud computing was *inspired* by the cloud symbol that's often used to represent the Internet in flow charts and diagrams.

Cloud computing providers offer their services according to three fundamental models: Infrastructure-as-a-Service (IaaS), Platform-as-a-Service (PaaS) and Software-as-a-Service (SaaS), see Figure 17-3.

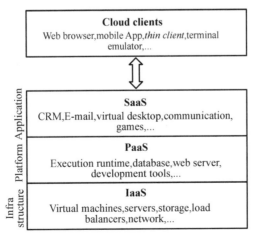

Figure 17-3 Service models

Infrastructure-as-a-Service like Amazon Web Services provides virtual server instances with unique IP addresses and blocks of storage on demand. Customers use the provider's application program interface (API) to start, stop, access and configure their virtual servers and storage. In the enterprise, cloud computing allows a company to pay for only as much capacity as is needed, and bring more online as soon as required. Because this pay-for-what-you-use model resembles the way electricity, fuel and water are consumed; it's sometimes referred to as utility computing.

Platform-as-a-service in the cloud is defined as a set of software and product development tools *hosted* on the provider's infrastructure. Developers create applications on the provider's platform over the Internet. PaaS providers may use APIs, website portals or gateway software installed on the customer's computer[3]. Google Apps are examples of PaaS. Some providers will not allow software created by their customers to be moved off the provider's platform.

In the **software-as-a-service** cloud model, the vendor supplies the hardware infrastructure, the software product and interacts with the user through a front-end portal. SaaS is a very broad market. Services can be anything from Web-based email to inventory control and database processing. Because the service provider hosts both the application and the data, the end user is free to use the service from anywhere.

A cloud service has three distinct characteristics that differentiate it from traditional hosting. It is sold on demand, typically by the minute or the hour; it is *elastic*—a user can have as much or as little of a service as they want at any given time; and the service is fully managed by the provider (the consumer needs nothing but a personal computer and Internet access). Significant innovations in *virtualization* and distributed computing, as well as improved access to high-speed Internet and a weak economy, have led to a growth in cloud computing. [4] Cloud vendors are experiencing growth rates of 50% per *annum*.

The goal of cloud computing is to provide easy, *scalable* access to computing resources and IT services.

17.3.2 Cloud Storage

Cloud storage is a model of data storage where the digital data is stored in logical pools, the physical storage *spans* multiple servers (and often locations), and the physical environment is typically owned and managed by a hosting company. These cloud storage providers are responsible for keeping the data available and accessible, and the physical environment protected and running. [5] People and organizations buy or lease storage capacity from the providers to store user, organization, or application data.

Cloud storage services may be accessed through a *co-located* cloud computer service, a web service application programming interface (API) or by applications that utilize the API, such as cloud desktop storage, a cloud storage gateway or Web-based content management systems. [6]

Architecture

Cloud storage is based on highly virtualized infrastructure and is like broader cloud computing in terms of accessible interfaces, *near-instant* elasticity and scalability, *multi-tenancy*, and *metered* resources. [7] Cloud storage services can be utilized from an *off-premises* service or deployed *on-premises*.

Cloud storage typically refers to a hosted object storage service, but the term has broadened to include other types of data storage that are now available as a service, like block storage.

Cloud storage is:

- Made up of many distributed resources, but still acts as one-often referred to as federated storage clouds, see Figure 17-4.
- Highly *fault tolerant* through *redundancy* and distribution of data
- Highly *durable* through the creation of *versioned* copies
- Typically eventually consistent with regard to data replicas[8]

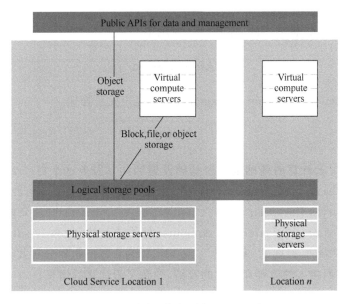

Figure 17-4　High level cloud storage architecture

Words and Expressions

coherence [kəu'hiərəns] *n.* 聚合性；一致性
economy of scale　规模经济
utility [ju:'tiləti] *n.* 实用程序；功用；公用事业；公用事业设备
per demand　按需
rack space　储存架区
inspire [in'spaiə] *vt.* 赋予灵感；鼓舞；激励
host [həust] *vt.* 托管；做主人，主持　　*n.* 主机，宿主机

thin client　瘦客户机；瘦客户端；精简型计算机
elastic [i'læstik] adj. 有弹性的，灵活的
virtualization ['vəːtʃuəˌlaizeiʃən] n. 虚拟化
annum ['ænəm] n. [拉]年，岁
scalable ['skeiləb(ə)l] adj. 可伸缩的，可缩放的
span [spæn] vt. 跨越；横跨　n. 跨度；跨径；一段时间
co-locate [kəu'ləukeit] v. (使)驻扎在同一地点；使位于一处
near-instant　近乎即时，在瞬间
multi-tenancy　多租用
meter ['miːtə] vt. 用仪表测量　vi. 用表计量　n. 计量器；计量仪
off-premises　远程；场所外；外部
on-premises　在场所内；内部部署；本地
fault tolerant　容错
redundancy [ri'dændənsi] n. 冗余
durable ['durəbl] adj. 耐用的，持久的
version ['vəːʃən] n. 版本　vt. 制作……新版本

Abbreviation

CRM (Customer Relationship Management)　客户关系管理

Notes

1. 全句可译为：云计算的基础是汇聚基础设施和共享服务这个更广义的概念。下一句的 maximizing the effectiveness of the shared resources 可译为：使共享资源的效用最大化。
2. 这里 that serves European users ... application (e.g., E-mail)从句修饰前面的 facility。
3. 这里 installed on the customer's computer 修饰前面的 gateway software。
4. weak 意为"差的，微弱的，疲软的"。全句可译为：虚拟化和分布式计算的重大创新，以及高速因特网接入的改善和疲软的经济，导致了云计算的增长。
5. 本句中 keeping 和 running 是并行结构，即 providers are responsible for keeping ... and for running。句中 protected 为过去分词，修饰前面的 physical environment，表示"物理环境被保护"。
6. 本句中 through 和 by 是并行结构，即 be accessed through 和 be accessed by。全句可译为：云存储服务可以通过位于一处的云计算机服务、Web 服务应用程序接口(API)来访问，或者通过利用 API 的应用程序(如云桌面存储、云存储网关或基于 Web 的内容管理系统)来访问。
7. 本句中 in terms of "按照，根据，从……来看"。multi-tenancy 的含义是"多租户"，意指"一个软件实例服务多个租户"，这是相对于"一个软件实例服务一个租户"而言的。全句可译为：云存储基于高度虚拟化的基础设施，并且从可访问接口、近乎即时的灵活性和可缩放性、多租户，和计量资源这几方面来说，它像更广义的云计算。下一句可译为：可以从场所外服务或部署的场所内使用云存储服务。

8. eventually consistent 意为"最终一致的",是分布式存储的一致性模型中的一种。因为在分布式系统中,一个存储页(或块)可以有多个 replicas "复制品,副本",因此在对多副本的存储块进行写操作时就存在这些副本的一致性问题。这句话可译为:关于多个数据副本,通常最终一致的。

注:云计算属于分布式计算,云存储也属于分布式存储。云计算的提供者把他们管理的基础设施、平台或软件作为服务提供给用户。用户实际上是租户,按租用数量付钱使用这些资源。云存储是广义上的云计算,本篇给出了云计算的三种服务模式,以及云存储的高层体系结构。

Terms

Smartphone(智能手机)

A smartphone (or smart phone) is a mobile phone with more advanced computing capability and connectivity than basic feature phones. Smartphones typically include the features of a phone with those of another popular consumer device, such as a personal digital assistant, a media player, a digital camera, and/or a GPS navigation unit. Later smartphones include all of those plus the features of a touchscreen computer, including web browsing, Wi-Fi, 3rd-party apps, motion sensor, mobile payment and 3G.

In 2007, Apple Inc. introduced the iPhone, one of the first mobile phones to use a multi-touch interface. The iPhone was notable for its use of a large touchscreen for direct finger input as its main means of interaction, instead of a stylus, keyboard, or keypad typical for smartphones at the time.

Wearable computer(可穿戴式计算机)

Wearable computers, also known as body-borne computers or wearables are miniature electronic devices that are worn by the bearer under, with or on top of clothing. This class of wearable technology has been developed for general or special purpose information technologies and media development. Wearable computers are especially useful for applications that require more complex computational support than just hardware coded logics.

If one is asked to give a simple, yet modern, example for wearable technology, that will be the Nike+ system which allows you to track your time, distance, pace and calories via a sensor in the shoe. Another example can be Google Glass, which combine innovative displays with some novel gestural movements for interaction.

Many issues are common to the wearables as with mobile computing, ambient intelligence and ubiquitous computing research communities, including power management and heat dissipation, software architectures, wireless and personal area networks.

Web service

A web service provides a service interface enabling clients to interact with servers in a more general way than web browsers do. Clients access the operations in the interface of a web service by means of requests and replies formatted in XML and usually transmitted over HTTP. Web services can be accessed in a more ad hoc manner than CORBA-based services, enabling them to be more easily used in Internet-wide applications.

Like CORBA and Java, the interfaces of web services can be described in an IDL (in Web services, it is called WSDL). But for web services, additional information including the encoding and communication protocols in use and the service location need to be described.

SOA (Service Oriented Architecture)

SOA is an evolution of distributed computing based on the request/reply design paradigm for synchronous and asynchronous applications. An application's business logic or individual functions are modularized and presented as services for consumer/client applications. What's key to these services is their loosely coupled nature; i.e., the service interface is independent of the implementation. Application developers or system integrators can build applications by composing one or more services without knowing the services' underlying implementations. For example, a service can be implemented either in .Net or J2EE, and the application consuming the service can be on a different platform or language.

P2P system (Peer-to-Peer system)

Peer-to-peer systems represent a paradigm for the construction of distributed systems and applications in which data and computational resources are contributed by many hosts on the Internet, all of which participate in the provision of a uniform service. Their emergence is a consequence of the very rapid growth of the Internet, embracing many millions of computers and similar numbers of users requiring access to shared resources.

Peer-to-peer applications have been used to provide file sharing, web caching, information distribution and other services exploiting the resources of tens of thousands of machines across the Internet. They are at their most effective when used to store very large collections of immutable data. Their design diminishes their effectiveness for applications that store and update mutable data objects.

A key problem for peer-to-peer systems is the placement of data objects across many hosts and subsequent provision for access to them in a manner that balances the workload and ensures availability without adding undue overheads.

Grid

Grid is an infrastructure that gathers a large diversity of distributed physical resources and software, such as
- supercomputers and parallel machines
- clusters of PCs
- massive storage systems
- databases and data sources

- sensors
- special devices
- software

Grid facilitates controlled sharing of resources across organizational boundaries.

Software agent

A software agent is an autonomous process capable of reacting to, and initiating changes in its environment, possibly in collaboration with users and other agents.

Virtualization(虚拟化)

Virtualization is the creation of a virtual (rather than actual) version of something, such as an operating system, a server, a storage device or network resources.

You probably know a little about virtualization if you have ever divided your hard drive into different partitions. A partition is the logical division of a hard disk drive to create, in effect, two separate hard drives.

Operating system virtualization is the use of software (called hypervisor) to allow a piece of hardware to run multiple operating system images at the same time. The hypervisor is actually controlling the host processor and resources and allocates what is needed to each operating system, making sure that the guest operating systems (called virtual machines) cannot disrupt each other.

There are three areas of IT where virtualization is making headroads, network virtualization, storage virtualization and server virtualization:

- Network virtualization is a method of combining the available resources in a network by splitting up the available bandwidth into channels, each of which is independent from the others, and each of which can be assigned (or reassigned) to a particular server or device in real time. The idea is that virtualization disguises the true complexity of the network by separating it into manageable parts, much like your partitioned hard drive makes it easier to manage your files.
- Storage virtualization is the pooling of physical storage from multiple network storage devices into what appears to be a single storage device that is managed from a central console. Storage virtualization is commonly used in storage area networks (SANs).
- Server virtualization is the *masking* of server resources (including the number and identity of individual physical servers, processors, and operating systems) *from* server users. The intention is to spare (用不着) the user from having to understand and manage complicated details of server resources while increasing resource sharing and utilization and maintaining the capacity to expand later.

Quantum computer (量子计算机)

A quantum computer is a device for computation that makes direct use of quantum mechanical phenomena, such as superposition(叠加、叠置)and entanglement(交织、纠缠)to perform operations on data. The basic principle behind quantum computation is that quantum properties can be used to represent data and perform operations on these data.

Quantum computing

Quantum computing utilizes nuclear spins to store and process information whereas classical computers operate using solid state electronics, notably the transistor. Quantum computing is not inherently any faster than classical computing. The difference is that quantum computing allows for parallel processing.

To explain, if you asked a classical computer to perform two calculations, it would do them in sequence, returning one answer after the other. A quantum computer, when asked to do the same thing, would return both answers at once. While it performed the actual computation faster, it takes an equal amount of time in the end, because you have to figure out which answer goes to which question with the quantum computer.

Certain algorithms have been developed for quantum computers (which can capitalize on purely quantum mechanical behavior such as convolution) which allow for specialized functions to be sped up. The two most common examples are directory lookups and number factoring. Because of the latter, quantum computers hold importance in the field of cryptography.

Recently IBM created a quantum computer which factored 15 into 5 and 3. The technology is still in its infancy, but it is steadily moving forward.

Supercomputer

A supercomputer is a computer with a high level of performance compared to a general-purpose computer. The performance of a supercomputer is commonly measured in floating-point operations per second (FLOPS) instead of million instructions per second (MIPS). Since 2017, there are supercomputers which can perform up to nearly a hundred quadrillion FLOPS. Since November 2017, all of the world's fastest 500 supercomputers run Linux-based operating systems. Additional research is being conducted in China, the United States, the European Union, Taiwan and Japan to build even faster, more powerful and more technologically superior exascale supercomputers (exascale: exa-scale; exa=2^{60}).

Supercomputers play an important role in the field of computational science, and are used for a wide range of computationally intensive tasks in various fields, including quantum mechanics, weather forecasting, climate research, oil and gas exploration, molecular modeling (computing the structures and properties of chemical compounds, biological macromolecules, polymers, and crystals), and physical simulations (such as simulations of the early moments of the universe, airplane and spacecraft aerodynamics, the detonation of nuclear weapons, and nuclear fusion). Throughout their history, they have been essential in the field of cryptanalysis.

Supercomputers were introduced in the 1960s, and for several decades the fastest were made by Seymour Cray at Control Data Corporation (CDC), Cray Research and subsequent companies bearing his name or monogram. The first such machines were highly tuned conventional designs that ran faster than their more general-purpose contemporaries. Through the 1960s, they began to add increasing amounts of parallelism with one to four processors being typical. From the 1970s, vector processors operating on large arrays of data came to dominate. A notable example is the highly successful Cray-1 of 1976. Vector computers remained the dominant design into the 1990s. From then until today, massively parallel supercomputers with tens of thousands of off-the-shelf processors became the norm.

The US has long been the leader in the supercomputer field, first through Cray's almost uninterrupted dominance of the field. Japan made major strides in the field in the 1980s and 90s, but since then China has become increasingly active in the field.

DNA computing

DNA computing is a branch of computing which uses DNA, biochemistry, and molecular biology hardware, instead of the traditional silicon-based computer technologies. Research and development in this area concerns theory, experiments, and applications of DNA computing. The term "molectronics" has sometimes been used, but this term has already been used for an earlier technology, a then-unsuccessful rival of the first integrated circuits; this term has also been used more generally, for molecular-scale electronic technology.

17.4 Quantum Computing

Quantum computing is the use of *quantum-mechanical* phenomena such as *superposition* and *entanglement* to perform computation. A quantum computer is used to perform such computation, which can be implemented theoretically or physically.

The field of quantum computing is actually a sub-field of quantum information science, which includes quantum cryptography and quantum communication. Quantum Computing was started in the early 1980s when Richard Feynman and Yuri Manin expressed the idea that a quantum computer had the potential to simulate things that a classical computer could not. In 1994, Peter Shor published an algorithm (which is a quantum algorithm) that is able to efficiently solve some problems that are used in *asymmetric* cryptography that are considered hard for classical computers.

There are two main approaches to physically implementing a quantum computer currently, analog and digital. Analog approaches are further divided into quantum simulation, quantum *annealing*, and *adiabatic* quantum computation[1]. Digital quantum computers use quantum logic gates to do computation. Both approaches use *quantum bits* or qubits.

Qubits are fundamental to quantum computing and are somewhat analogous to bits in a classical computer. Qubits can be in a 1 or 0 quantum state.[2] But they can also be in a superposition of the 1 and 0 states. However, when qubits are measured the result is always either a 0 or a 1; the *probabilities* of the two outcomes depends on the quantum state they were in.[3]

A small number of qubits quantum computers have been developed by a number of companies, including IBM, Intel, and Google. D-Wave Systems has been developing its own version of a quantum computer that uses annealing.[4] Today's physical quantum computers are very noisy and quantum error correction is a *burgeoning* field of research. Unfortunately existing hardware is so noisy that *fault-tolerant* quantum computing [is] still a rather distant dream. *As of* April 2019 neither large scalable quantum hardware has been demonstrated nor have commercially useful algorithms for today's small noisy quantum computers been published. There is an increasing amount of investment in quantum computing by governments, established companies, and *start-ups*. Both applications of near-term intermediate-scale device and the demonstration of quantum *supremacy* are actively pursued in academic and industrial research.[5] Quantum computers are not intended to replace classical computers, they are expected to be a different tool we will use to solve complex problems that are beyond the capabilities of a classical computer.

The *Bloch sphere* is a representation of a qubit, the fundamental building block of quantum computers. A binary bit can be at either of the two poles of the sphere, but a qubit can exist at any point on the sphere, see Figure 17-5.

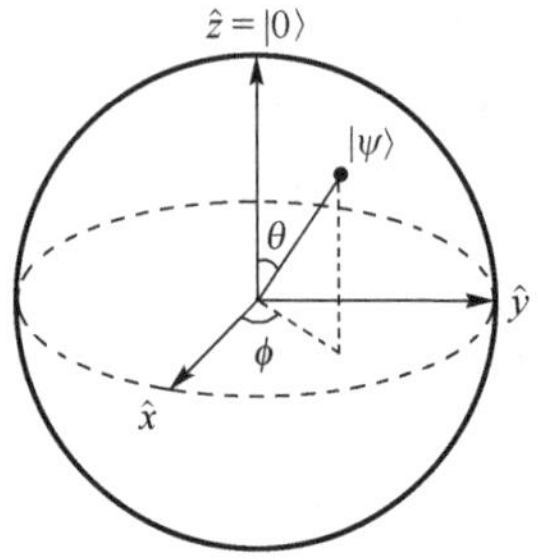

Figure 17-5 Block sphere

Basics

A classical computer has a memory made up of bits, where each bit is represented by either a one or a zero. A quantum computer, on the other hand, maintains a sequence of qubits, which can represent a one, a zero, or any quantum superposition of those two qubit states;[6] a pair of qubits can be in any quantum superposition of 4 states, and three qubits in any superposition of 8 states. In general, a quantum computer with n qubits can be in any superposition of up to 2^n different states. (This compares to a normal computer that can only be in one of these 2^n states at any one time).

A quantum computer operates on its qubits using quantum gates and measurement (which also alters the observed state). An algorithm is composed of a fixed sequence of quantum logic gates and a problem is encoded by setting the initial values of the qubits, similar to how a classical computer works. The calculation usually ends with a measurement, *collapsing* the system of qubits into one of the 2^n *eigenstates*, where each qubit is zero or one, decomposing into a classical state.[7] The outcome can therefore be at most n classical bits of information if the algorithm did not end with a measurement; the result is an unobserved quantum state. (Such unobserved states may be sent to other computers as part of distributed quantum algorithms.)

Quantum algorithms are often probabilistic, in that they provide the correct solution only with a certain known probability. Note that the term non-deterministic computing must not be used in that case to mean probabilistic (computing), because the term non-deterministic has a different meaning in computer science.

Principles of Operation

A quantum computer with a given number of qubits is fundamentally different from a classical computer composed of the same number of classical bits. For example, representing the state of an n-qubit system on a classical computer requires the storage of 2^n *complex coefficients*, while to characterize the state of a classical n-bit system it is sufficient to provide the values of the n bits. Although this fact may seem to indicate that qubits can hold exponentially more information than their classical counterparts, care must be taken not to overlook the fact that the qubits are only in a probabilistic superposition of all of their states. This means that when the final state of the qubits is measured, they will only be found in one of the possible *configurations* they were in before the measurement. It is generally incorrect to think of a system of qubits as being in one particular state before the measurement. The qubits are in a superposition of states before any measurement is made, which directly affects the possible outcomes of the computation.

An example of an implementation of qubits of a quantum computer could start with the use of particles with two *spin* states: "down" and "up" (typically written $|\downarrow\rangle$ and $|\uparrow\rangle$ or $|0\rangle$ and $|1\rangle$)[8], see Figure 17-6. This is true because any such system can be mapped onto an effective spin-1/2 system[9].

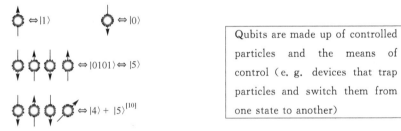

Figure 17-6   Classically allowed states[11]

Potential and Obstacles

Potential uses of quantum computing include cryptography, quantum search, quantum simulation, solving linear equations, Quantum supremacy etc.

There are a number of technical challenges in building a large-scale quantum computer, and thus far quantum computers have yet to solve a problem faster than a classical computer. David DiVincenzo, of IBM, listed the following requirements for a practical quantum computer:

- scalable physically to increase the number of qubits;
- qubits that can be initialized to arbitrary values;
- universal gate set;
- qubits that can be read easily.

A recent review by Mikhail Dyakonov in IEEE Spectrum argues that practical quantum computers are not likely to be implemented. He says: "There is a tremendous gap between the *rudimentary* but very hard experiments that have been carried out with a few qubits and the extremely developed quantum-computing theory, which relies on manipulating thousands to millions of qubits to calculate anything useful. That gap is not likely to be closed anytime soon".

Words and Expressions

quantum computing　量子计算
quantum-mechanical　量子力学的
superposition [ˌsjuːpəpəˈziʃən] n. 叠加；重叠；叠合；重合
entanglement [inˈtæŋglmənt] n. 纠缠；牵连
asymmetric [ˌeisiˈmetrik] adj. 不对称的，非对称的
anneal [əˈniːl] n. 退火；锻炼；磨炼（意志）　vt. 退火；韧炼，锻炼
adiabatic [ˌædiəˈbætik] adj. 绝热的；隔热的
quantum bit (qubit)　量子比特，量子位
probability [ˌprɔbəˈbiləti] n. 概率；可能性；或然性；几率；或然率；很可能发生的事
burgeon [ˈbəːdʒən] vi. 迅速发展；激增；发芽；生出蓓蕾　n. 嫩芽，蓓蕾
fault-tolerant [ˈfɔːlt ˈtɔlərənt]　容错的
as of　到……时为止；在……时；从……时起
start-up [ˈstɑːt ʌp]　刚成立的公司；初创企业（尤指互联网公司）；启动
supremacy [suːˈpreməsi] n. 优势；至高无上；最大权力；最高权威；最高地位
Bloch sphere　布洛赫球面
collapse [kəˈlæps] v. 塌缩；倒塌，崩溃；折叠；（劳累后）倒下/坐下
　　　　　　　　　n. 垮台；倒塌；崩溃
eigenstate [ˈaigənˌsteit] n. 本征态
complex coefficient　复系数
configuration [kənˌfigəˈreiʃən] n. 配置；（位）态；构造，结构；形状，外形
spin [spin] v. (使)旋转；疾驰；纺（纱）；眩晕　n. 旋转；自旋；眩晕
rudimentary [ˌruːdiˈmentəri] adj. 基本的，初步的

Notes

1. 本句中 quantum annealing ——量子退火（QA）是一种元启发式方法，通过使用量子涨落的过程，在给定的一组候选解（候选态）上找到给定目标函数的全局最小值。量子退火主要用于解决搜索空间是离散的（组合优化问题）且具有多个局部最小值的问题。句中 adiabatic quantum computation ——绝热量子计算（AQC）是量子计算的一种形式，它依靠绝热定理进行计算，与量子退火密切相关，可以看作量子退火的一个子类。
2. 本句可译为：诸量子位可以是（处于）1 或 0 量子态。下一句可译为：但它们也可以是（处于）1 和 0 两个（状）态的叠加。
3. 要得知量子计算的结果，必须进行测量运算，见注释 7。测量时把叠加态变成一个确定态，即结果总是 0 或 1。这两个结果出现的概率取决于测量时它们所处的量子态。
4. D-Wave Systems, Inc. 是一家量子计算公司，总部设在加拿大不列颠哥伦比亚省的本

拿比。annealing 是指 quantum annealing。

5. 本句可译为：近期的中等规模设备（指量子计算设备）的应用和量子优势的演示都在学术和工业研究中积极进行。

6. 本句可译为：另一方面，量子计算机维持一个量子位序列，它可以表示 1、0 或这两个量子位态的任何量子叠加。量子叠加是量子力学的一个基本原理。它指出，就像经典物理学中的波一样，任何两个（或更多）量子态都可以加在一起（"叠加"），结果是另一个有效的量子态；相反，每个量子态都可以表示为两个或更多其他不同态的叠加（和）。

7. 在量子计算中有三类量子运算，第一类是酉变换（也译为"幺正变换"，unitary transformation），其作用是将一个量子态改变为另一个量子态，是可逆运算；第二类是张量积运算（tensor product），它用于由多个量子系统构成复合量子系统；第三类是测量（measurement）运算，它是沟通经典世界与量子世界的桥梁，要得知量子计算的结果，必须对诸量子位态进行测量运算。由于被测的微观系统（都是微观粒子结构，如光子、离子阱、超导环等）过于脆弱，因此测量会影响和干扰被测系统，致使被测系统发生塌缩。这种量子塌缩把叠加态变成一个确定态。

    本句中 collapsing 和 decomposing 是现在分词，引导的短语作状语，表示结果。整个短语可译为：把量子比特系统塌缩成 2^n 个本征态之一，其中每个量子比特是 0 或 1，从而分解成经典状态。

8. 在量子力学和粒子物理学中，自旋是基本粒子、复合粒子（强子）和原子核所携带的角动量的固有形式。全句可译为：量子计算机的量子比特实现的一个例子可以从使用具有两个自旋态——"向下"和"向上"——的粒子开始。

9. 本句可译为：这是真的，因为任何这样的系统都可以映射到有效的自旋-1/2 系统中。

10. 该行所示为 4 个量子位态的叠加（4 和 5）。

11. 在量子力学中，自旋是所有基本粒子的内在特性，自旋数描述粒子在一个完整自旋中有多少个对称面；1/2 自旋（即自旋为 1/2）意味着粒子必须完全旋转两次（旋转 720°），才能具有与开始时相同的位态。一个自旋为 0 的粒子像一个圆点：从任何方向看都一样。自旋为 2 的粒子像个双头的箭头：只要转过半圈（180°），看起来便是一样的了，即具有与开始时相同的量子位态了。

注：量子计算用量子力学原理进行运算。至今量子计算机的研发仍处于初期阶段，但它具有对大规模数据进行并行处理的能力，因此将能解决那些超出现今经典计算机处理能力的复杂问题。本选文用于对量子计算的大致了解。

Exercises

1. Translate the text of Lesson 17.2 into Chinese.
2. Topics for oral workshop.
   - What is an embedded system?
   - Talk about wide applications of embedded system.
   - What is distributed systems?
   - Talk about the distributed application system you use in your daily life.
3. Translate the following into English.

分布式系统是一组独立的计算机，这组计算机在系统用户看来好像是(appear ... as)一台单一计算机。几乎所有大的软件系统都是分布式的，例如，企业范围的业务系统必须支持多个用户运行跨越在不同地方的公共应用程序。

一个分布式系统包含各种各样的应用程序、它们下面的支撑软件、它们运行的硬件，以及连接分布式硬件的通信连接。最大最知名的分布式系统是组成 WWW 的计算机、软件和服务的集合，它很普遍(pervasive 渗透)并与大多数其他已存在的分布式系统共存和连接。最常用的分布式系统是联网的客户-服务器系统。分布式系统共有下面描述的一般特性。

- 多节点
- 消息传递或通信
- 资源共享
- 分散控制
- 并发(并行)性
- 容错(Fault tolerance)
- 异构型(Heterogeneity)
- 开放性

分布式系统有许多固有的优点，超过了集中式系统。某些应用也是固有分布式的。一般，分布式系统

- 产生更高性能；
- 允许增量式增大；
- 允许一个用户把一个程序并行地运行在许多不同的机器上；
- 提供更高的可靠性。

4. Listen to the video "VPN" and write down the fourth paragraph.

参 考 资 料

[1] 计算机世界. 计算机专业时文选读, 1998.3.2.

[2] Microsoft Bookshelf Internet Dictionary, 1998.

[3] The Visua C++ Handbook, 2nd Edition, 1995.

[4] Andrew S. Tanenbaum. Distributed Operating Systems. Prentice Hall, Inc., 1997.

[5] Andrew S. Tanenbaum. Computer Networks: Third Edition. Prentice Hall, Inc. 1997.

[6] Encarta 98 Desk Encyclopedia 1996-97 Microsoft Corporation.

[7] Peter Rob & Carlos Coronel. Database Systems Design, Implementation, and Management. Wadsworth Publishing Company, 1993.

[8] 易水. 计算机新技术时文精选. 北京：宇航出版社, 1997.

[9] Ralf Steinmetz & Klara Nahrstedt. Multimedia—Computing, Communications & Applications. Prentice Hall, Inc., 1997.

[10] Dan W. Patterson. Introduction to Artificial Intelligence and Expert Systems. Prentice Hall, Inc., 1990.

[11] William Ford & William Topp. Data Structures With C++. Prentice Hall, Inc., 1997.

[12] Winfried Karl Grassmann & Jean-Paul Tremblay. Logic and Discrete Mathematics—A Computer Science Perspective. Prentice Hall, Inc., 1996.

[13] 刘兆毓. 计算机英语. 北京：清华大学出版社, 1997.

[14] Sue A. Conger. The New Software Engineering. Wadsworth Publishing Company, 1994.

[15] John P. Reilly. Rapid PROTOTYPING Moving to BusinessE—Centrice Development. International Thomson Computer Press, 1996.

[16] Pericles Loucopoulos & Roberto Zicari. Conceptual Modeling, Databases, and CASE—An Integrated View of Information Systems Development. John Wiley & Sons, Inc., 1992.

[17] Computer & Control Abstracts, 1997, Number 10.

[18] 秦卫平, 陈伟. 计算机专业英语. 重庆：重庆大学出版社, 1997.

[19] 司爱侠, 张强. 计算机英语教程. 北京：电子工业出版社, 1996.

[20] Douglas E. Comer. Internetworking with TCP/IP. Prentice Hall, Inc., 1995.

[21] Douglas E. Comer. Computer Networks and Internet. Prentice Hall, Inc., 1999.

[22] Sandeep Singhal, Thomas, etc. The Wireless Application Protocol: Writing Application for the Mobile Internet. 2001.

[23] Getting to 3G, IEEE Internet Computing, January/February 2001.

[24] Joseph R. Kiniry. Wavelength Division Multiplexing. IEEE Internet Computing, March/April 1998.

[25] Howard Frazier, Howard Johnson. Gigabit Ethernet: From 100 to 1000 Mbps. IEEE Internet Computing, January/February, 1999.

[26] Andre Bergholz. Extending Your Markup: An XML Tutorial. IEEE Internet Computing, July/August, 2000.

[27] Rick Lehrbaum. Using Linux in Embedded and Real-time Systems, February. 2000.

[28] Business Administration: Introduction to computers. THAMES, 2001.

[29] Felisa Verdejo and Gordon Davies (ed.). The Virtual Campus Trends for higher education and training.

Chapman & Hall, 1998.

[30] Peter Cunningham and Friedrich Fröschl. Electronic Business Revolution Opportunities and challenges in the 21st Century. Spring-Verlag, Berlin, 1999.

[31] John Vince and Rae Earnshaw. Virtual worlds on the Internet. Computer Society, 1998.

[32] William Stallings. Computer Organization and Architecture—Designing for Performance(5th Edition). Prentice Hall, Inc. 2001.

[33] Aviel D. Rubin. Wireless Networking Security. Communications of the ACM, May 2003 Volume 46, Number 5.

[34] Russ Housley and William Arbaugh. Security Problem in 802.11-based Networks. Communications of the ACM, May 2003 Volume 46, Number 5.

[35] Daniel W. Lewis. Fundamentals of embedded software: Where C and assembly meet. Pearson Education, Inc., Prentice Hall, 2002.

[36] Andrew S. Tanenbaum, Maarten van Steen. Distributed systems—principles and paradigms. Prentice Hall, 2002.

[37] George Coulouris et al. Distributed systems—concepts and design(4th Edition). Pearson Education Limited, 2005.

[38] 计算机世界. 计算机专业时文选读. 2005.5.30, 2005.8.8, 2006.8.21.

[39] Ian Foster, Carl Kesselman. The Grid: Blueprint for new computing infrastructure(2nd Edition). Elsevier Inc. 2004.

[40] Charles S. Parker et al. Understanding computers: Today and tomorrow. Thomson Learning, Inc. 2003.

[41] Pang-Ning Tan, Michael Steinbach, Vipin Kumar. Introduction to Data Mining. Pearson Education, Inc., 2006.

[42] David M. Bourg, Glenn Seemann. AI for Game Developers. O'Reilly Media, Inc., 2004.

[43] Scott W. Ambler. Agile Database Techniques—Effective Strategies for the Agile Software Developer. Wiley Publishing, Inc., 2003.

[44] 刘艺, 王春生. 计算机英语(第3版). 北京: 机械工业出版社, 2009.

[45] 姜同强, 苗天顺. 计算机英语(第2版). 北京: 清华大学出版社, 2009.

[46] Adam Drozdek. Data structure and algorithms in Java(2nd Edition). Thomson Learning, Inc., 2005.

[47] Wikipedia, www.wiki.com.

[48] CCW 资料. 当医疗遇见 3D 打印. 2013.7.

[49] 资料, 中国先进战机突飞猛进 得益于制造技术领先世界, 2013.

[50] 张春红, 等. 物联网技术与应用. 北京: 人民邮电出版社, 2011.

[51] Donald Hearn, M. Pauline Baker. Computer graphics-with OpenGL(4th Edition). Pearson, 2012.2.

[52] 蔡士杰, 等译. 计算机图形学(第4版). 北京: 电子工业出版社, 2014.

[53] Azharz, INTRODUCTION TO ANDROID PROGRAMMING.

[54] From AZHARZ'S INSTRUCTABLES circuits, 2018.12.20.

[55] Deep learning, From Wikipedia, the free encyclopedia, 2018.12.25.

[56] Sophia (robot), From Wikipedia, the free encyclopedia, 2019.1.4.

[57] Sarah Knapton, AlphaGo Zero: Google DeepMind supercomputer learns 3,000 years of human knowledge in 40 days, The Telegraph, News 2017.10.18.

[58] Vangie Beal. Big Data Analytics, From Webopedia: Online Tech Dictionary for Students. Educators and IT Professionals, 2018.10.24.

[59] Jeff Vance. Big Data Analytics Overview. Datamation＞Aplications，2018.12.27.

[60] Augmented reality，From Wikipedia，the free encyclopedia，2016.1.7.

[61] A New English-Chinese Dictionary 新英汉词典 增补本，上海译文出版社，2003年2月第55次印刷.

[62] 徐家福，宋方敏. 量子计算机程序设计. 北京：科学出版社，2013.

[63] 吴楠，宋方敏. 量子计算和量子计算机. 计算机科技与探索，2007年第1期，2007.

[64] 高扬，卫峥. 白话深度学习与TensorFlow. 北京：机械工业出版社，2017.